HEALTH IMPACTS OF ULTRA-PROCESSED FOODS

HEALTH IMPACTS OF ULTRA-PROCESSED FOODS

Dr. Noopur Gautam

Delve
PUBLISHING

www.delvepublishing.com

Health Impacts of Ultra-Processed Foods

Dr. Noopur Gautam

Delve Publishing
4164 Lakeshore Road
Burlington, ON L7L 1A4
Canada
www.delvepublishing.com

Email: orders@arclereducation.com

ISBN: 978-1-77956-441-2 (Hardcover)

ABOUT THE AUTHOR

Dr Noopur Gautam is having twenty three years of experience in food technology sector and presently working as Associate Professor in Institute of Food Technology, Bundelkhand University, Jhansi UP. She has done her graduation degree from Kurukshetra University, and post graduation degree from Guru Jambeshwar University, Hisar. Her doctorate is from Bundelkhand University in Food Science and Technology. Under her supervision many students are doing their PhDs. She has also worked in Multinational Companies for five years and is a trained auditor for HACCP and ISO 9000 with wide experience in Quality Management System. She is in the panel of expert referee in Elsevier with an outstanding contribution in reviewing research papers and articles. She has authored three books and number of research papers. There is a amalgam of research and application in her approach which reflects in her work.

TABLE OF CONTENTS

LIST OF FIGURES

LIST OF TABLES

PREFACE

The concept of ultra-processed foods (UPFs) has increasingly become recognized as a key descriptor for unhealthy foods within dietary patterns, both in academic nutrition literature and in leading health reports. This growing recognition highlights how UPFs, which are foods significantly altered from their original form through industrial processes and containing numerous additives, can have detrimental effects on overall dietary quality. Their role in exacerbating health issues and increasing the risk of various diet-related diseases, disorders, and conditions is becoming more apparent. Despite this growing awareness, there remains a significant lack of focus on UPFs in public health strategies aimed at improving population health outcomes. To bridge this gap, a thorough examination of the current evidence linking UPF consumption to adverse health outcomes is crucial.

Dietary risk factors play a pivotal role in contributing to the global burden of disease (GBD). In 2017, dietary factors were responsible for approximately 11 million deaths from non-communicable diseases (NCDs), accounting for about 22% of all adult deaths. Additionally, these factors contributed to 15% of disability-adjusted life years (DALYs) lost, which measures the overall burden of disease in terms of years of healthy life lost due to premature mortality or disability. The leading contributors to diet-related deaths include major health conditions such as cardiovascular disease (CVD), cancer, and type 2 diabetes. For non-fatal chronic conditions that impact quality of life and contribute to DALYs, asthma, musculoskeletal disorders, and mental health issues are particularly significant.

Dietary risk factors are diverse and include a variety of nutrients, specific types of foods, and overall dietary patterns. Nutritional risks are associated with high intakes of sodium, saturated fats, trans fats, and added sugars. On the other hand, inadequate consumption of beneficial foods such as whole grains, fruits, vegetables, nuts, seeds, and fish is also linked to health risks. High consumption of harmful foods like red meat, processed meats, potato chips, and sugar-sweetened beverages (SSBs) further exacerbates these risks.

Dietary patterns that are linked to adverse health outcomes include low adherence to healthier dietary guidelines such as the Healthy Eating Index or the Alternative Healthy Eating Index, the Mediterranean Dietary Pattern, or the Dietary Approaches to Stop Hypertension (DASH) diet. These patterns are associated with better health outcomes and reduced risks of chronic diseases. In contrast, high adherence to the Western dietary pattern, which is typically characterized by high consumption of processed foods, sugary drinks, and unhealthy fats, is associated with increased health risks.

The textbook *Health Impacts of Ultra-Processed Foods* is a comprehensive examination of the various facets of ultra-processed foods and their effects on human health. Beginning with an introduction to ultra-processed foods, the book sets the stage by defining what constitutes ultra-processed foods, their prevalence in modern diets, and the driving forces behind their widespread consumption.

Following this foundational understanding, the second chapter examines the nutritional composition and labeling of these foods, providing insights into their typical ingredients, nutrient profiles, and the often misleading nature of food labels that can obscure the unhealthy aspects of these products.

The third chapter focuses on the metabolic effects and the challenges they present for weight management, exploring how ultra-processed foods contribute to obesity and related metabolic disorders through mechanisms like high calorie density, low satiety, and metabolic dysregulation.

Chapter four examines the cardiovascular health implications, detailing how the consumption of these foods is linked to increased risks of hypertension, heart disease, and stroke due to their high content of trans fats, sodium, and added sugars.

Mental health and cognitive function are addressed in the fifth chapter, which discusses the growing body of evidence linking ultra-processed food consumption to mental health issues such as depression, anxiety, and cognitive decline, highlighting the potential neuroinflammatory pathways involved.

In the sixth chapter, the focus shifts to cancer risk, reviewing studies that associate ultra-processed foods with higher incidences of various cancers due to factors like carcinogenic contaminants, nutrient deficiencies, and pro-inflammatory effects.

The seventh chapter explores the effects of these foods on children and adolescents, a particularly vulnerable population, addressing concerns about childhood obesity, developmental issues, and the establishment of unhealthy eating patterns early in life.

Digestive health and gastrointestinal issues are the focus of the eighth chapter, examining how ultra-processed foods can disrupt gut microbiota, contribute to digestive disorders, and impair gastrointestinal health.

The ninth chapter examines the impacts of food additives, which are ubiquitous in ultra-processed foods, on human health, discussing the potential adverse effects of artificial sweeteners, preservatives, colorings, and flavorings.

Finally, the tenth chapter considers the socioeconomic and cultural factors that influence the consumption of ultra-processed foods, exploring how economic constraints, marketing practices, and cultural shifts towards convenience foods shape dietary habits and health outcomes across different populations.

–Author

Chapter 1

Introduction to Ultra-Processed Foods

Contents

INTRODUCTION

Foods classified as ultra-processed—a term that has gained significant traction in recent decades—have undergone extensive industrial processing and are packed with artificial ingredients and additives. These foods are usually made from ingredients that are extracted from whole foods, like oils, sugars, and starches. These ingredients are then mixed with flavor enhancers, colorings, and preservatives to create products that are convenient and frequently very appetizing. Ready-to-eat meals, sugary drinks, sugary cereals, and snacks like potato chips are a few examples.

Ultra-processed food consumption is directly related to contemporary lifestyles, which are marked by a growing desire for quick fixes and convenient options. In a time when hectic schedules and a fast-paced lifestyle are the norm, ultra-processed foods present a convenient and speedy meal option. There is a price for this convenience, though. Studies are showing that eating these foods frequently has negative effects on one's health. Ultra-processed foods, which are high in unhealthy fats, added sugars, and sodium, are linked to several health problems, such as type 2 diabetes, obesity, and cardiovascular diseases. These foods frequently lack fiber and important nutrients that are found in larger, less processed foods.

The influence of ultra-processed foods on eating habits and general nutrition is one of the main problems with them. They frequently replace nutrient-dense foods like fruits, vegetables, and whole grains, which lowers nutrient intake. Because of their high sugar and fat content, these foods have an addictive quality that can contribute to the maintenance of unhealthy eating habits. To make matters worse, consuming ultra-processed foods regularly has been associated with increased inflammation and negative effects on gut health.

The consumption of whole, minimally processed foods is encouraged, as is dietary education, to counteract the detrimental effects of ultra-processed foods on health. While promoting healthier eating environments, policymakers, and health advocates are also looking into ways to lessen the accessibility and allure of highly processed foods. In our increasingly industrialized world, knowing the nature and effects of ultra-processed foods is essential to making informed dietary choices and promoting a healthier relationship with food (Monteiro et al., 2010; Ares et al., 2016; Monteiro et al., 2018; Monteiro et al., 2019).

1.1. DEFINITION AND CLASSIFICATION

An ultra-processed food (UPF) is an industrially formulated edible substance derived from natural food or synthesized from other organic compounds. The resulting products are designed to be highly profitable, convenient, and hyperpalatable, often through food additives such as preservatives, colorings, and flavorings. UPFs have often undergone processes such as molding/extruding, hydrogenation or frying.

Ultra-processed foods first became ubiquitous in the 1980s, though the term ultra-processed food gained prominence from a 2009 paper by Brazilian researchers as part of the Nova classification system. As of 2024, research into the effects of UPFs is rapidly evolving.

Consuming ultra-processed foods may increase your chances of developing a number of diseases, such as obesity, type 2 diabetes, cardiovascular disease, asthma, certain cancers, and all-cause mortality, according to epidemiological data. Eating a lot of highly processed food may also contribute to food addiction. Over 34 papers that offer evidence of these effects on public health were found in a 2024 meta-analysis that was published in The BMJ; however, the precise mechanism underlying the effects remained unclear. A considerable amount of developed countries' diets consist of ultra-processed foods, ranging from 10% in Romania to over 50% in the US and the UK.

Some writers have attacked the Nova classification system for placing too much emphasis on food type rather than quantity consumed, as well as the idea of ultra-processed foods for being ill-defined. Other writers, primarily in the nutrition field, have criticized the lack of a specific mechanism being linked to the health effects, emphasizing that the evidence currently available does not specify how eating a highly processed diet affects bodily systems.

1.1.1. Definitions

Food processing has raised concerns at least since the Industrial Revolution. Ultra-processed food is relatively new; in his seminal work The Omnivore's Dilemma, Michael Pollan described highly processed industrial food as "edible food-like substances." In a 2009 commentary, Carlos Augusto Monteiro mentioned Pollan as an inspiration for the term "ultra-processed food."

Monteiro's team developed the Nova classification for grouping unprocessed and processed foods beginning in 2010, whose definition of ultra-processing became the most widely accepted, refined through successive publications. The identification of ultra-processed foods, as well as the category itself, is a subject of debate among nutrition and public health scientists, and other definitions have been proposed.

A survey of systems for classifying levels of food processing in 2021 identified four "defining themes":

1. Extent of change (from the natural state).
2. Nature of change (properties, ingredients added).

3. Place of processing (where/by whom).
4. Purpose of processing (why, essential/cosmetic).

1.1.1.1. Monteiro's 2009 Commentary

Ultra-processed foods are essentially confections made with group 2 ingredients, which are substances taken out of whole foods and usually combined with clever additives to make them palatable, addictive, and edible. They may be shaped, labeled, and marketed to seem wholesome and "fresh," but they bear no real resemblance to group 1 foods, or minimally processed foods. Ultra-processed foods are generally not consumed with or as part of minimally processed foods, dishes, and meals, in contrast to the ingredients in group 2. Conversely, they are intended to be either ready-to-eat (sometimes with the addition of liquid, like milk) or ready-to-heat, and they are frequently eaten either by themselves or in combination (like bread and burgers or savory snacks and soft drinks).

This definition is socially as much as ingredient-based, which makes understanding ultra-processed foods extremely intuitive, even for consumers who are not well-versed in food science. The definition "lacks precision," according to a letter in response to Monteiro's 2009 commentary, because it does not include the quantifiable definitions found in conventional food science. As a result, experts cannot agree on whether the definition can serve as a reliable foundation for scientific control. Hyperpalatable food has a quantitative definition that researchers have created, but ultra-processed food does not.

1.1.2. Nova Classification

The ultra-processed foods were then grouped together and presented in the Nova food classification system by Monteiro's team. Rather than emphasizing food types or nutrients, the system concentrates on food processing. Foods are divided into four categories by Nova: minimally or fully processed foods, processed foods, processed ingredients, and ultra-processed foods. Rather than using a central advisory board, Nova is an open classification that gradually improves its definitions through scientific publications.

The food products produced industrially that contain a variety of ingredients (formulations) such as sugar, oils, fats, and salt (usually in higher amounts and combinations than in processed foods) as well as food ingredients that are either unusable or infrequently used in cooking (like protein isolates, hydrogenated oils, modified starches, and high-fructose corn syrup). Foods from Group 1 are either not present in the formulation or make up a very small percentage of its ingredients. The production of ultra-processed foods is made possible by industrial methods like extrusion, molding, and pre-frying; the use of additives, such as flavorings, colorants, emulsifiers, and non-sugar sweeteners, whose purpose is to make the finished product palatable or hyperpalatable; and sophisticated packaging, which is typically made of synthetic materials. The ingredients and processes used here are intended to produce tasteful, convenient, and highly profitable (long shelf life, low cost of ingredients, strong branding, ready-to-eat or drink) alternatives to all other Nova food groups as well as freshly made dishes and meals. The presence of food substances without culinary value (such as different types of sugars like fructose, high-fructose corn syrup, "fruit juice concentrates," invert sugar, maltodextrin, dextrose, and lactose), modified starches, modified oils like hydrogenated or interesterified oils, and protein sources like hydrolyzed proteins, soy protein isolate, gluten, casein, whey protein, and "mechanically separated meat") or cosmetic additives (flavors, flavor enhancers, colors, emulsifiers, sweeteners, thickeners, and anti-foaming, bulking, carbonating, foaming, gelling, and glazing agents) in the ingredient list can be used to identify ultra-processed foods. The Nova definition of ultra-processed food is not meant to be used for nutrient profiling, nor does it make any comments about the nutritional value of food.

1.1.2.1. International Agency for Research on Cancer (IARC)

The International Agency for Research on Cancer, an intergovernmental agency that forms part of the World Health Organization, classifies foods by their degree of processing. This system breaks food into "non-processed," "moderately processed," and "highly processed" food. The system does not take into account the nature or purpose of changes.

1.1.2.2. Siga Index

A French classification system for processed foods is called the Siga Index. It takes a holistic and reductionist stance and is based on the level of processing and nutritional value of foods. Every food product is given a score between 1 and 100 by the Siga Index; higher scores correspond to higher nutritional quality and lower processing. Ultra-processed foods (UPFs) are foods with a score of less than 40 on the Siga Index, meaning they are high in artificial ingredients, preservatives, and additives but low in nutritional value.

1.1.2.3. International Food Information Council (IFIC)

The International Food Information Council defines five levels of food processing: minimally processed, foods processed for preservation, mixtures of combined ingredients, ready-to-eat processed foods, and prepared foods/meals.

1.1.2.4. NUPENS

The Center for Epidemiological Research in Nutrition and Health at the University of São Paulo has proposed a variant on the Nova classification consisting of unprocessed, minimally, or moderately processed foods; processed foods; and ultra-processed foods.

1.1.3. What Is Ultra-Processed Food?

The term "ultra-processed foods" comes from the NOVA food classification system, which was developed by researchers at the University of São Paulo, Brazil.

The system places food into four categories based on how much they have been processed during their production:

1. **Unprocessed or Minimally Processed Foods**: This includes produce such as fruit, vegetables, milk, fish, pulses, eggs, nuts, and seeds that have no added ingredients and have been little altered from their natural state.
2. **Processed Ingredients**: This includes foods that are added to other foods rather than eaten by themselves, such as salt, sugar, and oils.
3. **Processed Foods**: These are foods that are made by combining foods from groups 1 and 2, which are altered in a way that home cooks could do themselves. They include foods such as jam, pickles, tinned fruit and vegetables, homemade breads and cheeses.
4. **Ultra-Processed Foods**: Ultra-processed foods typically have more than one ingredient that you never or rarely find in a kitchen. They also tend to include many additives and ingredients that are not typically used in home cooking, such as preservatives, emulsifiers, sweeteners, and artificial colors and flavors. These foods generally have a long shelf life.

The terms "processed" and "ultra-processed" are thrown around a lot and are usually associated with something negative, but what makes something processed or ultra-processed? Unless you're eating the broccoli you harvested directly from your garden, most of the food we eat daily has undergone some form of processing. But that's not necessarily a bad thing, as we'll discover.

Processing food means changing its natural state to prolong its shelf life, make it safe to store or eat, enhance its taste, or even increase its nutritional value. Methods like pasteurizing, canning, fermenting, freezing, and drying foods are forms of processing. When a food item is ultra-processed, it means that the producer uses industrial-scale methods and ingredients that you may not recognize and would not use in home cooking to produce the final product.

1.1.4. A List of Common Ultra-Processed Foods

Examples of ultra-processed foods include ice cream, ham, sausages, crisps, mass-produced bread, breakfast cereals, biscuits, carbonated drinks, fruit-flavored yogurts, instant soups, and some alcoholic drinks including whisky, gin, and rum.

1. **Unprocessed or Minimally Processed Foods**: Minimally processed or unprocessed food has not been altered or has no added ingredients. Washed and bagged spinach, pre-cut fresh fruit, or frozen vegetables are all minimally processed. They're made to be more convenient to consume, but their nutritional value hasn't been altered.
2. **Processed Culinary Ingredients**: Processed culinary ingredients are made from unprocessed foods through simple processing. This group includes oil, butter, sugar, salt, dried herbs, and spices. They are added to other foods, rather than eaten by themselves.
3. **Processed Foods**: Processed foods are partially altered by adding sugar, oil, fat, salt, and other culinary ingredients to minimally processed foods. Processed foods like cheese, homemade or artisanal bread, and tofu have been altered, but not in a way that's bad for our health.
4. **Ultra-Processed Foods**: Ultra-processed foods are entirely altered and have high levels of unhealthy fats, refined sugars, and salt. They also undergo industrial processes, like hydrogenation and molding, and contain additives like dyes, stabilizers, flavor enhancers, emulsifiers, and defoaming agents.

These foods are very calorie-dense and don't contain many, if any, valuable nutrients. Cookies, chips, and fast food are all ultra-processed. Ultra-processed foods are engineered to be convenient, extra tasty, and highly profitable for the companies that make them.

1.1.5. Why Are Ultra-Processed Foods Bad For Us?

When we consume ultra-processed foods, we make less room in our diets for more nutrient-dense foods because they frequently have high levels of sugar, salt, and saturated fat. Additionally, it has been proposed that these foods' additives may be to blame for any unfavorable health outcomes. The way that food is actually processed may also have an impact on how our bodies react to it. Research has demonstrated, for instance, that the body absorbs less fat when foods like nuts are consumed whole than when the nut is ground up and the oils are released. Diets heavy in ultra-processed foods may also have an impact on our gut health, according to a recent theory.

Further investigation is required to distinguish between these various components and determine which aspect of ultra-processed foods—if any—may be harmful to our health. At the moment, it's also unclear whether consuming a diet high in these foods indicates a lifestyle that is generally associated with worse health or whether the problem lies with any one of these components alone. However, reducing does seem reasonable considering the high levels of sugar, salt, and saturated fat found in the majority of these foods.

1.1.5.1. Are Cereal and Bread Considered Ultra-Processed Foods?

There are some less apparent examples of ultra-processed foods, but you might also think of chips, sweets, and vibrantly colored sugary drinks. Ultra-processed foods include commonplace items like mass-produced or packaged bread and breakfast cereals. This is due to the fact that additional ingredients like sweeteners, emulsifiers, and artificial flavors and colors are frequently added during production. The fact that items like these, which can be a component of a nutritious diet, are grouped with less nutrient-dense, highly processed foods is one of the criticisms leveled at the NOVA method of food classification. Ultra-processed, less-nutritious foods include premade meals, sausages, and nuggets, as well as baked goods, cakes, buns, biscuits, pastries, and chips.

Because they are highly processed, plant-based alternatives to meat and cheese may not be as healthful as they are claimed to be. Despite the issues associated with categorizing various food types as ultra-processed, the NOVA system remains a popular choice for researchers. Although there are benefits and drawbacks to its use, research has shown a connection between diets heavy in these foods and poorer health.

1.1.5.2. Should I Be Cutting Out Ultra-Processed Foods?

While it may seem sensible to return to consuming only minimally processed foods, most of us are unable to do so due to time and financial constraints. Even though there seems to be increasing evidence that ultra-processed foods are detrimental to our heart and circulation, it's still unclear whether we need to completely avoid them due to the type and quality of the research.

The Scientific Advisory Committee on Nutrition (SACN), which conducted a review of the data regarding processed foods and health in 2023, concluded that due to the limitations of the current research, care should still be taken when recommending a diet. Consider striking a balance in your diet rather than attempting to totally eliminate these foods. Aim to fit in time each week for cooking at home. Make sure there are minimally processed foods there as well. Eat fruit and vegetables with your meals and drink water rather than sugary drinks.

It's crucial to keep in mind that not all foods that are highly processed are created equal. When incorporating ultra-processed foods into your diet, opt for those that offer greater nutritional value, such as wholegrain bread and cereals or baked beans, rather than snacks, desserts, or pizza. Make it a habit to read food labels so you can quickly spot and reduce the amount of foods high in saturated fat, sugar, or salt.

1. Read more about how food labeling works.
2. Explore our collection of heart-healthy recipes.

1.1.5.3. How to Cut Down On Ultra-Processed Foods

Eating a lot of processed foods leaves less room for healthier foods like fruit and vegetables, fish, unsaturated oils, pulses, and nuts and seeds.

So, if you want to change your diet, try to make some swaps, such as:

1. Instead of flavored yogurts with added sugar or sweeteners, choose plain yogurt and add your own chopped fresh, frozen or dried fruit for sweetness.
2. Instead of buying sauces or ready meals, cook your favorites in larger amounts at home and freeze the extra portions to use another day.
3. Have porridge in the morning with fruit and nuts instead of sugary low-fiber breakfast cereals.
4. Eat fresh, baked or stewed fruit instead of shop-bought fruit pies or cakes.
5. Have some nuts instead of biscuits with your afternoon cuppa.

1.1.6. The Latest Research on Ultra-Processed Foods and Heart Health

Several studies have shown that eating higher amounts of ultra-processed foods is linked to a greater risk of:

1. high blood pressure.
2. cardiovascular disease.
3. heart attacks and strokes.

4. early death.

One study followed 10,000 Australian women for 15 years and was presented at the European Society of Cardiology Congress in August 2023. The likelihood of developing high blood pressure was found to be 39% higher in individuals whose diets contained the highest amounts of ultra-processed food than in those with the lowest. A second, more comprehensive analysis that was also given at the meeting examined 10 studies with a combined total of over 325,000 men and women. It revealed that the highest consumption of ultra-processed foods was associated with a 24% increased risk of major cardiac and circulatory events, such as angina, strokes, and heart attacks. A 6% increase in the risk of heart disease was associated with every 10% increase in daily consumption of ultra-processed foods. A 2019 study that tracked 19,899 Spanish university graduates also found a link between ultra-processed foods and an early death risk. Groups of participants were formed based on how much ultra-processed food they consumed. Less than two servings were consumed daily by the group consuming the fewest ultra-processed foods, and more than four servings were consumed daily by the group eating the most. After an average of 10.4 years, those in the highest intake group had a 62% higher chance of dying than those in the low consumption group.

Due to the observational nature of these three studies, they are limited to identifying correlations between variables. They are unable to establish a direct link between ultra-processed foods and harmful health outcomes like heart disease; alternative explanations may exist. When the researchers in the Australian and Spanish studies adjusted for other dietary factors like sugar, salt, and saturated fat, they discovered the same elevated risks. This implies that rather than the increased levels of fat, salt, and sugar, the processing itself might be hazardous.

It's crucial to keep in mind that observational studies like this one can only demonstrate an association, according to BHF Senior Dietitian Victoria Taylor. They are unable to reveal the cause of this. Because the researchers' classification of ultra-processed foods is so broad, there could be a variety of reasons why eating these foods is associated with a higher risk to our health, such as their nutritional value, food additives, or other aspects of an individual's lifestyle. It's critical that we fully comprehend this before considering any changes to guidance or policy.

A Mediterranean-style diet, which emphasizes an abundance of minimally or unprocessed foods like fruit, vegetables, fish, nuts, and seeds, beans, lentils, and whole grains, is already something we advise people to follow. This is beneficial for reducing the risk of heart and circulatory disease, along with regular exercise and quitting smoking.

1.1.7. Economics

The high amount of processing lends ultra-processed food to be subject to different economic constraints compared to natural food.

1.1.7.1. Profitability

Ultra-processed foods often use less expensive ingredients, allowing ultra-processed food to be priced lower. Furthermore, ultra-processed foods are more consistently available in stores. Global production networks of multinational food companies creating ultra-processed food are supported by high brand awareness, aggressive globalization tactics, and the purchasing of local companies selling similar products.

Companies selling ultra-processed food frequently target youth consumers and middle-income countries. Many of these companies use big data to choose which consumers to market to. Furthermore, the ultra-processed food industry uses indirect and direct lobbying in large countries to influence local food policy.

1.1.7.2. Affordability for Consumers

They often have an extended shelf life, an important consideration for lower-income consumers without reliable access to refrigeration. Among other reasons for the popularity of ultra-processed foods is the inexpensive cost of their main ingredients. The price of ultra-processed food fluctuated less than unprocessed food over a 12-year period.

1.1.8. History and Evolution of Ultra-Processed Foods

Ultra-processed food history and development are an intriguing journey reflecting larger shifts in industry, society, and technology. Ultra-processed foods have their roots in the early 1900s, a time of rapid technological advancements in food processing and a growing focus on convenience.

Food processing gained popularity in the early 1900s thanks to inventions like powdered goods and canned goods. The need for increased food safety, a longer shelf life, and simpler transportation motivated these early actions. More sophisticated processing techniques were made possible by the advancement of food preservation and stability techniques like pasteurization and refrigeration.

A major turning point occurred in the mid-20th century with the post-World War II economic boom. During this time, consumerism increased and people started living more convenient lives. The widespread use of home appliances like microwaves and refrigerators changed people's eating habits and increased demand for prepared meals and snacks. Businesses started concentrating on developing goods that were appealing to the increasing number of working women and busy families, in addition to being convenient.

This new consumer culture was reflected in the introduction of processed foods such as instant noodles and frozen dinners. Further developments in food processing technologies and a sharp rise in the range of ultra-processed products on the market occurred in the 1970s and 1980s. As artificial flavors, colors, and preservatives proliferated, producers were able to produce goods with longer shelf lives and more stable flavors. The creation of novel ingredients, such as hydrogenated oils and high-fructose corn syrup, was also essential in improving flavor and texture and cutting production costs.

The effects of ultra-processed foods on health were brought to light in the late 20th and early 21st centuries. Studies started to find connections between these foods' consumption and a number of health problems, such as diabetes, obesity, and cardiovascular disease. Many public health initiatives aimed at promoting healthier diet choices and reducing consumption were spurred by this mounting body of evidence. As a result, some food producers have begun to reformulate their goods to cut down on artificial additives, unhealthy fats, and added sugars.

Ultra-processed foods are still evolving today due to changes in consumer preferences and technological advancements. Transparency in food labeling is becoming more and more popular, and there is a greater emphasis on wholesome and sustainable substitutes. Notwithstanding these difficulties, the development of ultra-processed food history reveals a nuanced interaction among consumer demand, technological advancement, and health concerns that have shaped the current food environment (Monteiro et al., 2010; Ares et al., 2016; Monteiro et al., 2018; Monteiro et al., 2019).

1.2. HEALTH EFFECTS

There hasn't been a randomized controlled trial to date to look into the impact of ultra-processed foods on any health outcome other than weight gain, so the effect of these foods on health has mostly been studied using nutritional epidemiology. Although studies separating different types of ultra-processed food have found adverse effects primarily for only some sub-groups, such as soft drinks, these studies have shown an overall increased risk for disease.

Overall, there is a correlation between UPF consumption and many unfavorable health outcomes, such as poor mental, cardiometabolic, and overall health, as well as a shorter life expectancy. The high levels of fat, sugar, and salt in these foods may be one reason for their detrimental effects on health, though this isn't the case for all ultra-processed foods. Contaminants, specific food additives, and the high heat treatment of these foods are some other possible causes. However, as of right now, science cannot agree on anything.

1.2.1. Ultra-Processed Food Unhealthy

Everyone agrees that eating too much processed food is unhealthy. Extremely processed food diets have been associated with a higher risk of heart disease, weight gain, cancer, and even death. But why? Processing modifies the intricate nutrient structure of food, or what scientists refer to as the food matrix. Moreover, studies have demonstrated that modifications to the food matrix affect how our bodies react to food, which may increase our chance of developing certain medical disorders.

Let's examine the research on the effects of highly processed foods on our health. A seven-year study that examined the diets of over 44,000 French adults discovered a correlation between a high intake of ultra-processed foods and a higher risk of mortality. Similar results were obtained from a Spanish study involving nearly 20,000 adults. More ultra-processed food consumption was associated with an increased risk of heart disease, according to a large study conducted over 5 years involving over 100,000 French adults. A comparable analysis involving the same individuals discovered a correlation between a 10% rise in ultra-processed food intake and a 12% increased risk of cancer. Despite the importance of these studies, cause, and effect cannot be established due to their observational design.

1.2.2. Ultra-Processed Food Linked to 32 Harmful Effects on Health, Review Finds

Ultra-processed food (UPF) (Figure 1.1) is directly linked to 32 harmful effects on health, including a higher risk of heart disease, cancer, type 2 diabetes, adverse mental health and early death, according to the world's largest review of its kind.

The findings from the first comprehensive umbrella review of evidence come amid rapidly rising global consumption of UPF such as cereals, protein bars, fizzy drinks, ready meals and fast food. In the UK and US, more than half the average diet now consists of ultra-processed food. For some, especially people who are younger, poorer or from disadvantaged areas, a diet comprising as much as 80% UPF is typical.

Figure 1.1. *Ultra-processed foods undergo multiple industrial processes and often contain colors, emulsifiers, flavors, and other additives.*

Source: By AlejandroLinaresGarcia – Own work, CC BY-SA 4.0, https://commons.wikimedia.org/w/index.php?curid=7169744.

The review involved experts from many leading institutions, including Johns Hopkins Bloomberg School of Public Health in the US, the University of Sydney and Sorbonne University in France. Writing in the BMJ, they concluded: "Overall, direct associations were found between exposure to ultra-processed foods and 32 health parameters spanning mortality, cancer, and mental, respiratory, cardiovascular, gastrointestinal, and metabolic health outcomes."

1. Greater exposure to ultra-processed food was associated with a higher risk of adverse health outcomes, especially cardiometabolic, common mental disorders and mortality outcomes.
2. "These findings provide a rationale to develop and evaluate the effectiveness of using population-based and public-health measures to target and reduce dietary exposure to ultra-processed foods for improved human health."
3. Ultra-processed foods, including packaged baked goods and snacks, fizzy drinks, sugary cereals, and ready-to-eat or ready meals, undergo multiple industrial processes and often contain colors, emulsifiers, flavors, and other additives. These products also tend to be high in added sugar, fat, and/or salt but are low in vitamins and fiber.
4. Previous studies have linked UPF to poor health, but no comprehensive review provided a broad assessment of the evidence

in this area.

5. To bridge this gap, researchers carried out an umbrella review – a high-level evidence summary – of 45 distinct pooled meta-analyses from 14 review articles associating UPF with adverse health outcomes.
6. The review articles were all published in the past three years and involved 9.9 million people. None were funded by companies involved in the production of UPF.
7. Estimates of exposure to ultra-processed foods were obtained from a combination of food frequency questionnaires, 24-hour dietary recalls, and dietary history and were measured as higher versus lower consumption, additional servings per day, or a 10% increment.
8. The researchers graded the evidence as convincing, highly suggestive, suggestive, weak, or no evidence. They also assessed the quality of evidence as high, moderate, low, or very low.
9. Overall, the results show that higher exposure to UPF was consistently associated with an increased risk of 32 adverse health outcomes, The BMJ reported.
10. Convincing evidence showed that higher UPF intake was associated with about a 50% increased risk of cardiovascular disease-related death, a 48 to 53% higher risk of anxiety and common mental disorders, and a 12% greater risk of type 2 diabetes.
11. Highly suggestive evidence also indicated that higher PF intake was associated with a 21% greater risk of death from any cause, a 40 to 66% increased risk of heart disease related death, obesity, type 2 diabetes and sleep problems, and a 22% increased risk of depression.
12. There was also evidence for associations between UPF and asthma, gastrointestinal health, some cancers and cardiometabolic risk factors, such as high blood fats and low levels of "good" cholesterol, although the researchers cautioned the evidence for these links remains limited.
13. The researchers acknowledged several limitations to the umbrella review, including that they couldn't rule out the possibility that other unmeasured factors and variations in assessing UPF intake may have influenced their results.

Some experts not involved in the research also highlighted that much of the research included in the umbrella review was weak and also cautioned that the findings do not prove cause and effect.

However, Dr. Chris van Tulleken, an associate professor at University College London and one of the world's leading UPF experts, said the findings were "entirely consistent" with a now "enormous number of independent studies which clearly link a diet high in UPF to multiple damaging health outcomes including early death."

"We have a good understanding of the mechanisms by which these foods drive harm," he added. In part it is because of their poor nutritional profile – they are often high in saturated fat, salt, and free sugar. However, the way they are processed is also important – they're engineered and marketed in ways that drive excess consumption – for example, they are typically soft and energy-dense and aggressively marketed usually to disadvantaged communities.

1.2.3. Limiting Processed Foods Is Already Something We Know

Understanding how foods become ultra-processed can be a helpful step in remembering to minimize your intake of those foods. Diligent label reading can also help you choose products with fewer ingredients. Cooking at home goes a long way toward reducing consumption of ultra-processed foods. Restaurants (especially fast-food restaurants) are notorious for tinkering with their recipes to achieve a certain flavor, which could involve using less-than-nutrient-dense ingredients. However, in some cases, getting unprocessed, whole food isn't that simple, whether it's an issue of affordability, availability, or accessibility. Still, you can likely make small changes to tweak the amount of ultra-processed foods in your diet.

Chart to help you make some nutritious swaps:

Ultra-Processed	**Processed**	**Home Version**
Sweetened breakfast cereals	Plain bran cereal	Oat groats or steel-cut oats sweetened with honey
Soda	Artificially flavored sparkling water	Water with a splash of fruit juice or fruit slices
Flavored potato chips	Plain tortilla chips	DIY pita chips
White bread	Whole wheat bread with minimal ingredients	Homemade baked potato chips
Fried chicken	Deli rotisserie chicken	Roast chicken from scratch
Flavored candy bar with long ingredient list	Simple candy bar with short ingredient list	Dark chocolate squares

Frozen, blended coffee drink	Store-bought cold brew	Drip coffee
Mashed potato flakes	Frozen potatoes	Fresh, whole potatoes
Energy drink	Sweetened fruit juice	Fresh-squeezed orange juice
Flavored granola bars with added sugar and preservatives	Granola bars with minimal additives	DIY granola
Artificially flavored cheese crackers	Naturally flavored crackers	Apple and cheese slices

Thanks to years of exposure to diet culture, we have already read about which foods have been socially labeled as "bad" and "good." But it's really not that simple. Attaching positive or negative labels to foods hasn't necessarily changed what we consume.

Food is more than fuel and filler — it's a relationship. So, the next time you head to the grocery store, remember that not all "processed" foods are necessarily bad for you. Focus on consuming more fruits and vegetables, whether fresh, frozen, or canned.

Read labels and add foods to your shopping cart that have ingredient lists of things you recognize. Limit items that are high in added sugar and sodium. The American Heart Association recommends that men consume no more than about 9 teaspoons of added sugar, or 36 grams, per day and women aim for no more than 6 teaspoons, or 25 grams, per day.

And ultra-processed foods? When you already know they're not the best for you, it's going to be about quantity, not quality, and comfort over health. The best step is to a check in with your mind and gut before putting something in your shopping cart (Monteiro et al., 2010; Ares et al., 2016; Monteiro et al., 2018; Monteiro et al., 2019).

1.3. OVERALL DIET QUALITY MAY BE MORE IMPORTANT THAN HOW MUCH ULTRA-PROCESSED FOODS YOU EAT

According to research that was just published in the journal BMJ, eating certain ultra-processed foods is linked to an increased risk of death. In line with earlier research, the study's authors concluded that "our findings suggest that meat/poultry/seafood based ready-to-eat products and sugar-sweetened and artificially sweetened beverages are major factors contributing to the harmful influence of ultra-processed foods on mortality." For long-term health, the results support restricting the consumption of some ultra-processed food types. Based on our findings, restricting the overall intake of ultra-processed foods may not have a significant impact on early mortality; however, cutting back on the consumption of specific ultra-processed food subgroups, like processed meat, may be advantageous, they continued.

1.3.1. Details on the Study of Ultra-Processed Foods

Over 74,000 female nurses and 39,000 male healthcare workers had their health monitored for 30 years during the study. Participants in the study submitted data on their health and lifestyle choices every two years. The participants also filled out a food questionnaire every four years. They were then assigned a score for the overall quality of their diet. According to the study, individuals who consumed seven servings of ultra-processed food on a daily average were 4% more likely to die than their counterparts who consumed three servings on average. The highest ultra-processed food consumption was associated with an 8% increased risk of neurodegenerative deaths.

Dairy-based desserts were the most strongly associated with death from neurodegenerative causes among the ultra-processed food groups. There is mounting evidence that eating a lot of processed food increases the risk of dementia, lower cognitive function, and demyelination of the central nervous system—a condition that is a precursor to multiple sclerosis. A diet high in ultra-processed foods has been linked to blood-brain barrier impairment and neuroinflammation, which can result in neurodegeneration, according to studies, the study's authors wrote.

1.3.2. The Importance of Overall Diet Quality

The new study's researchers did note that the correlation between ultra-processed foods and death was not as strong after accounting for diet quality overall. They claimed that this shows dietary quality might have a bigger impact. The study, according to the researchers, shows that not all highly processed foods need to be outlawed. Rather, with an emphasis on overall diet quality, specific types of ultra-processed foods, such as ready-to-eat meals made of meat, poultry, and seafood, could be restricted.

Dana Hunnes, PhD, a senior dietitian supervisor at the RR-UCLA Medical Center in California and an unaffiliated third party to the study, told Medical News Today that "diet quality looks at the total overall quality of someone's diet." Does it contain ultra-processed and processed foods, or does it contain unprocessed foods like whole grains, legumes, nuts, seeds, and fruits and vegetables? In contrast to ultra-processed foods, which are unrecognizable and have had their nutritional value stripped (sometimes with it added back in through fortification), good-quality/high-quality diets consist primarily of whole foods as they grew from the ground and are low in calories and fat but high in nutrients (including fiber).

If 95% of your diet consists of whole, unprocessed foods with good nutritional quality, then you can occasionally eat ultra-processed foods," Hunnes continued. "Alternatively, you can consume 90% ultra-processed/processed foods that are unrecognizable, low in fiber, and of very low diet quality." There are times when ultra-processed foods make sense in a balanced diet. Being completely ascetic and eating only healthfully isn't feasible 100% of the time; we can't be that strict all the time.

1.3.3. How to Have a More Healthy Diet

The USA. S. A healthy diet that includes fruits, vegetables, whole grains, different types of proteins, fat-free or low-fat milk and milk products, and foods low in sodium, cholesterol, trans fat, and saturated fat is advised by the Dietary Guidelines for Americans (Trusted Source). Convenience foods can be made at home in place of prepackaged meals, the president of the Academy of Nutrition and Dietetics, who was not involved in the study.

Since the BMJ study is observational, it is not possible to draw definitive conclusions about cause and effect. According to the study's authors, some foods may be incorrectly categorized because the system for classifying ultra-processed foods doesn't fully take into account how complex food processing is. The authors of the study concluded that further research was necessary to enhance the categorization of ultra-processed foods and validate their results in diverse demographics (Monteiro et al., 2010; Ares et al., 2016; Monteiro et al., 2018; Monteiro et al., 2019).

1.4. ULTRA-PROCESSED FOODS, ESPECIALLY ARTIFICIAL SWEETENERS, MAY INCREASE DEPRESSION RISK

Depression is a common mental health condition that can be debilitating (Figure 1.2). Researchers are still seeking the best ways to treat it and the best ways to prevent its development. One area of interest is how diet plays a role.

Figure 1.2. *Ultra-processed foods have been linked to an increased risk of depression.*

Source: By Thayne Tuason – Own work, CC BY-SA 4.0, https://commons.wikimedia.org/w/index.php?curid=66493006.

1.4.1. Eating Ultra-Processed Foods Influences Depression Risk

Researchers of this study looked at the relationship between ultra-processed foods and depression. A dietitian who was not involved in the study, gave the following definition of ultra-processed foods:

- "Ultra-processed foods, by the NOVA definition, are foods that are made up of manufactured ingredients with the addition of salt, oil, or sugar to make [them] palatable and to help preserve [them]. They usually don't have any worthwhile nutritional benefits. Some examples include cold packaged snacks like chips or cookies, sodas, packaged pastries, many sweet breakfast cereals, candy, etc."
- "The highly processed nature of these foods often yields a cheaper product that is more shelf-stable and more palatable than a whole food item. This makes them easy to eat. They are also usually high in calories, fat, salt, and sugar which can lead to weight gain," she told Medical News Today.

They looked at food frequency questionnaires from participants. The researchers then looked at the amounts of ultra-processed foods that participants consumed based on the NOVA classification. This system helps group foods based on processing and helps identify ultra-processed foods.

The researchers accounted for certain possible or known risk factors for depression, including age, activity levels, alcohol intake, and smoking.

During the study, 2,122 participants developed depression when researchers defined depression by a strict definition, and 4,840 participants developed depression when researchers used a broader definition for depression.

1.4.2. Added Risk From Artificial Sweeteners

"Participants with high UPF intake were found to have greater BMI, higher smoking rates, and an increased prevalence of comorbidities such as diabetes, hypertension, and dyslipidemia. Moreover, they were less likely to engage in regular exercise."

1.4.3. Study Limitations

This research does have certain limitations. First, it only included women, meaning the results cannot be generalized. In addition, the majority of participants were non-Hispanic and white.

"The cohort under investigation consisted [almost] exclusively of non-Hispanic White women, aged 42 to 62 years, with no inclusion of men or individuals from diverse ethnic backgrounds. This homogeneity in the study population calls for caution when generalizing the findings to broader demographics, as the impact of dietary choices on mental health may vary across different gender and ethnic groups," Hulsey noted. "Future research should aim to encompass a more diverse range of participants to ensure the applicability of the findings to a wider population, thus providing a more comprehensive understanding of the relationship between diet and mental well-being," she added. The research also relies on certain participant self-reporting, which introduces the possibility of inaccurate data collection. The researchers also note that there's the possibility of misclassification of the outcome since no one conducted structured clinical interviews (Monteiro et al., 2010; Ares et al., 2016; Monteiro et al., 2018; Monteiro et al., 2019).

REFERENCES

1. Ares, G., Vidal, L., Allegue, G., Giménez, A., Bandeira, E., Moratorio, X., et al. (2016). Consumers' conceptualization of ultra-processed foods. *Appetite, 105*, 611–617.
2. Baker, P., Machado, P., Santos, T., Sievert, K., Backholer, K., Hadjikakou, M., et al. (2020). Ultra-processed foods and the nutrition transition: Global, regional and national trends, food systems transformations and political economy drivers. *Obesity Reviews, 21*(12), e13126.
3. Chajès, V., Biessy, C., Byrnes, G., Deharveng, G., Saadatian-Elahi, M., Jenab, M., et al. (2011). Ecological-level associations between highly processed food intakes and plasma phospholipid elaidic acid concentrations: Results from a cross-sectional study within the European prospective investigation into cancer and nutrition (EPIC). *Nutrition and Cancer, 63*(8), 1235–1250.
4. Chen, Z., Khandpur, N., Desjardins, C., Wang, L., Monteiro, C. A., Rossato, S. L., et al. (2023). Ultra-processed food consumption and risk of type 2 diabetes: Three large prospective U.S. cohort studies. *Diabetes Care, 46*(7), 1335–1344.
5. Fardet, A. (2018). Characterization of the degree of food processing in relation with its health potential and effects. In S. L. McCormick (Ed.), *Advances in Food and Nutrition Research* (Vol. 85, pp. 79–129). Elsevier.
6. Fardet, A., & Rock, E. (2019). Ultra-processed foods: A new holistic paradigm? *Trends in Food Science & Technology, 93*, 174–184.
7. Gibney, M. J. (2019). Ultra-processed foods: Definitions and policy issues. *Current Developments in Nutrition, 3*(2), nzy077.
8. Gibney, M. J., Forde, C. G., Mullally, D., & Gibney, E. R. (2017). Ultra-processed foods in human health: A critical appraisal. *The American Journal of Clinical Nutrition, 106*(3), 717–724.
9. Isaksen, I. M., & Dankel, S. N. (2023). Ultra-processed food consumption and cancer risk: A systematic review and meta-analysis. *Clinical Nutrition, 42*(6), 919–928.
10. Kliemann, N., Al Nahas, A., Vamos, E. P., Touvier, M., Kesse-Guyot, E., Gunter, M. J., et al. (2022). Ultra-processed foods and cancer risk: From global food systems to individual exposures and mechanisms. *British Journal of Cancer, 127*(1), 14–20.
11. LaFata, E. M., & Gearhardt, A. N. (2022). Ultra-processed food addiction: An epidemic? *Psychotherapy and Psychosomatics, 91*(6), 363–372.
12. Lane, M. M., Gamage, E., Du, S., Ashtree, D. N., McGuinness, A. J., Gauci, S., et al. (2024). Ultra-processed food exposure and adverse health

outcomes: Umbrella review of epidemiological meta-analyses. *BMJ, 384*, e077310.

13. Leite, F. H., Khandpur, N., Andrade, G. C., Anastasiou, K., Baker, P., Lawrence, M., et al. (2022). Ultra-processed foods should be central to global food systems dialogue and action on biodiversity. *BMJ Global Health, 7*(3), e008269.
14. Monteiro, C. A. (2009). Nutrition and health: The issue is not food, nor nutrients, so much as processing. *Public Health Nutrition, 12*(5), 729–731.
15. Monteiro, C. A., Cannon, G., Levy, R. B., Moubarac, J. C., Louzada, M. L., Rauber, F., et al. (2019). Ultra-processed foods: What they are and how to identify them. *Public Health Nutrition, 22*(5), 936–941.
16. Monteiro, C. A., Cannon, G., Moubarac, J. C., Levy, R. B., Louzada, M. L., & Jaime, P. C. (2018). The UN decade of nutrition, the NOVA food classification and the trouble with ultra-processing. *Public Health Nutrition, 21*(1), 5–17.
17. Monteiro, C. A., Levy, R. B., Claro, R. M., Castro, I. R., & Cannon, G. (2010). A new classification of foods based on the extent and purpose of their processing. *Cadernos de Saúde Pública, 26*(11), 2039–2049.
18. Monteiro, C. A., Moubarac, J. C., Cannon, G., Ng, S. W., & Popkin, B. (2013). Ultra-processed products are becoming dominant in the global food system. *Obesity Reviews, 14*(Suppl 2), 21–28.
19. Moubarac, J. C., Parra, D. C., Cannon, G., & Monteiro, C. A. (2014). Food classification systems based on food processing: Significance and implications for policies and actions: A systematic literature review and assessment. *Current Obesity Reports, 3*(2), 256–272.
20. O'Connor, A., & Steckelberg, A. (2023). Melted, pounded, extruded: Why many ultra-processed foods are unhealthy. *The Washington Post*.
21. Pagliai, G., Dinu, M., Madarena, M. P., Bonaccio, M., Iacoviello, L., & Sofi, F. (2021). Consumption of ultra-processed foods and health status: A systematic review and meta-analysis. *The British Journal of Nutrition, 125*(3), 308–318.
22. Sadler, C. R., Grassby, T., Hart, K., Raats, M., Sokolović, M., & Timotijevic, L. (2021). Processed food classification: Conceptualisation and challenges. *Trends in Food Science & Technology, 112*, 149–162.
23. Touvier, M., da Costa Louzada, M. L., Mozaffarian, D., Baker, P., Juul, F., & Srour, B. (2023). Ultra-processed foods and cardiometabolic health: Public health policies to reduce consumption cannot wait. *BMJ, 383*, e075294.

24. Valicente, V. M., Peng, C. H., Pacheco, K. N., Lin, L., Kielb, E. I., Dawoodani, E., et al. (2023). Ultraprocessed foods and obesity risk: A critical review of reported mechanisms. *Advances in Nutrition, 14*(4), 718–738.

25. Visioli, F., Marangoni, F., Fogliano, V., Del Rio, D., Martinez, J. A., & Kuhnle, G. (2023). The ultra-processed foods hypothesis: A product processed well beyond the basic ingredients in the package. *Nutrition Research Reviews, 36*(2), 340–350.

Chapter 2

Effects of High Pressure on Food Nutritional Qualities

Contents

INTRODUCTION

A non-thermal method of food preservation known as "high-pressure processing" (HPP) involves applying extremely high pressures to food—typically between 100 and 800 MPa—to inactivate bacteria and enzymes that can lead to spoiling. Since this method does not rely on high temperatures, which can degrade nutrients and change flavors, textures, and colors, it is especially appealing for maintaining the nutritional and sensory qualities of food. Consequently, the food industry is using HPP more and more to extend the shelf life of perishable goods while preserving their freshness. The ability of HPP to preserve heat-sensitive vitamins and minerals is one of its main advantages. For example, high-pressure conditions improve the retention of water-soluble vitamins like B vitamins and C, which can be greatly reduced during traditional thermal processing. In a similar vein, fat-soluble vitamins like A and E also exhibit better retention when contrasted with meals that have been heated. Because the vitamins are preserved, the food's nutritional value is maintained, giving customers healthier options.

HPP affects some nutrients' bioavailability favorably in addition to vitamins. The food matrix may change as a result of the high pressure, improving the accessibility of nutrients for absorption and digestion. For instance, high-pressure processing can improve the bioavailability of carotenoids in fruits and vegetables. This is especially crucial for nutrients that are inherently bound to intricate food structures because the pressure aids in dissolving the structures and liberating the nutrients. High-pressure processing also affects food's proteins and amino acids. This method can modify the structure of proteins without leading to the widespread denaturation that heat produces. This implies that it is possible to enhance the solubility and digestibility of proteins, among other useful qualities. Moreover, high pressure can enhance the nutritional value of foods high in protein by neutralizing some anti-nutritional factors, such as protease inhibitors in legumes. Notwithstanding these benefits, HPP has certain drawbacks. Under high pressure, some delicate bioactive compounds might still be prone to breakdown. Furthermore, high-pressure processing can only be used on high-value food items due to the high cost of the equipment needed. Ongoing technological developments and falling equipment prices, however, are opening up HPP to a wider variety of food producers.

2.1. THE EVOLUTION OF HIGH-PRESSURE PROCESSING OF FOODS

Foods lose quality at some point after being harvested, killed, or manufactured, with very few exceptions. The type of food, as well as its formulation, processing, and storage conditions, all affect the kind of quality loss. Microbiological, enzymatic, chemical, and physical reactions are the most significant quality-loss reactions and, consequently, the most significant targets for preservation.

When preservation fails, there are many different outcomes: from severe danger (e.g., when harmful bacteria are present and have the ability to proliferate), to commercially significant spoiling (which causes the loss of priceless products), to comparatively minor quality loss (e.g., alterations in texture, flavor, or color). The most severe types of food quality loss are caused by pathogenic bacteria that are either infectious (Salmonella species, *Campylobacter jejuni*, enteropathogenic, and verocytotoxigenic [verotoxin producing, toxic to cells] strains of *Escherichia coli*, *Listeria monocytogenes*, *Yersinia enteroco/itica*, *Vibrio parahaemolyticus*, *Aeromonas hydrophila*, *Plesiomonas shigelloides*, enterotoxin-producing strains of *Bacillus cereus*, *Clostridium perfringens*) or toxinogenic bacteria (*Staphylococcus aureus*, emetic toxin-producing strains of *Bacillus cereus*, *Clostridium botulinum*).

A limited range of techniques are available to combat pathogenic and spoilage microorganisms and to delay other types of quality loss. Most of the antimicrobial techniques (e.g., chill storage, frozen storage, drying, curing, conserving, vacuum packing, modified atmosphere packing, acidifying, fermenting, adding preservatives) act by inhibiting, slowing down, or preventing the growth of microorganisms. In contrast, few available techniques act primarily by inactivating microorganisms in foods (e.g., heat pasteurization; heat sterilization; and, to a much smaller extent, irradiation). Generally speaking, inactivation techniques are preferable to inhibitory ones, especially when trying to lower the risk of food poisoning. Removing pathogens from food is likely to be a more successful intervention than growth inhibition, which necessitates some initial contamination. Therefore, it is encouraging that the majority of methods being developed or put into use act through inactivation (e.g., the addition of bacteriolytic enzymes, the addition of bacteriocins, high hydrostatic pressure, electroporation, manothermosonication, high-intensity laser and noncoherent light pulses, high-strength magnetic field pulses, and surface decontamination methods for carcasses).

In light of this, consumers around the world are calling for more practical foods (i.e., fresher, simpler to prepare and store; e.g., chilled); more organic; and consequently, less intensively processed (e.g., less heat-processed, less intensively preserved (e.g., reduced in acidity, salinity, and sugar content), and less dependent on additive preservatives (e.g., benzoate, propionate, sorbate, nitrite, sulfite, and parabens) than items that were previously sold. A decrease in food's inherent preservation as well as its stability and safety could be the result of these trends. New technologies must maintain or enhance the effectiveness of preservation and guarantee safety that might otherwise be lost as the food industry responds to these trends.

Foods that satisfy a large number of these consumer demands may be produced through the use of physical methods, such as high pressure, in food preservation. Physical techniques are mostly nonthermal or involve reduced heat treatments, they are essentially additive-free, they have minimally negative effects on product quality, and they are typically thought of as natural. Their foundation is the inactivation of microorganisms rather than their inhibition.

2.1.1. Preservation Technologies

2.1.1.1. Traditional Technologies

A small number of preservation technologies, like drying, have been used for a long time to preserve food, while others, like modified atmosphere packaging, are relatively new. Using these methods in novel or enhanced combinations has become popular as a way to reduce the excessive use of any one technology and, as a result, raise the caliber of the final goods. This tactic served as the foundation for Leistner and Gorris's (1995) successful hurdle technology, which has progressively encouraged the creation of new pathways for food preservation all over the world. New technologies are being developed with the help of mathematical models of microbial growth and survival, whereas traditional hurdle technologies were developed empirically (Baranyi and Roberts, 1995). These models enable the confident prediction of the effects of parameters like pH, water activity (a_w), temperature, gas phase, and preservatives on the growth, survival, and thermal death of particular microorganisms in foods. They are derived from large multifactorial experiments and validated in food (Baranyi and Roberts, 2000). The effects of high pressure or other novel and developing technologies on microorganisms have not been well modeled to date. But as these technologies become more widely used, modeling will become increasingly important.

2.1.1.2. Lowered Temperature

A chilled food's decreased temperature inhibits microorganisms' ability to proliferate and grow. The lower limits of growth for the strict anaerobes *Clostridium perfringens* and *Clostridium botulinum* (type A and some types of B) are approximately 12°C, and the lower limits of growth for the nonproteolytic strains of *Clostridium botulinum* (types E and F and some types of B) are approximately 3°C. These are two particularly important temperatures. Since *Yersinia enterocolitica* and *Listeria monocytogenes* can grow at temperatures below 1°C, recommended shelf lives and sell-by dates are crucial for guaranteeing food safety, especially in situations where temperature control cannot be reliably guaranteed (e.g., at home) (Herbert and Sutherland, 2000). Up to roughly –7°C, a variety of spoilage microorganisms can still proliferate. At the temperature that frozen foods should be kept, which is often –18°C, microorganism growth is totally inhibited. However, enzyme activity, chemical reactions, and physical alterations can cause a gradual loss of quality.

2.1.1.3. Reduced Water Activity

Values for water activity (a_w) are frequently used to forecast how stable food is against microbial growth and enzymatic, chemical, and physical alterations that cause quality degradation (Christian, 2000). The values correspond to equilibrium relative humidities (ERR) on a scale from 100% to 0%, with values ranging from I (pure water) to 0 (no water). Foods can be made a_w-reducing by drying them out, adding solutes like sugar or salt (like in cured products), or using a combination of these methods. Minimal decreases (to approximately 0.98–0.97) are adequate to stop the growth of certain significant spoilage microorganisms.

For instance, fresh foods like red meats kept in the presence of air quickly deteriorate due to the high growth rates of several Pseudomonas species. Curried meats typically have longer shelf lives because their a_w has been sufficiently reduced to guarantee that they are free of pseudomonas. Rather, they experience gradual souring due to lactic acid bacteria that is more tolerant of aw. Even the lactic acid bacteria are inhibited if the a_w is less than 0.95, which is the case with some salamis and dry-cured meat products. These goods are spoiled more slowly by microorganisms that have a lower a_w tolerance. The storage stability and microbiological safety of food are frequently predicted and explained using these and related relationships. *Staphylococcus aureus* is the food-poisoning bacteria that is most tolerant to a_w; to limit its multiplication in the presence of oxygen, a_w must be reduced to approximately 0.86, and to limit its multiplication anaerobically, to 0.91. *Staphylococcus aureus* can therefore proliferate and yield enterotoxin in comparatively low-a_w foods provided that the temperature and other environmental factors are favorable.

Microorganisms that grow at low a_w values are able to do so because they divert energy to the synthesis or accumulation of high levels of intracellular solutes such as glycine, betaine, proline, and polyols. In this way, they "osmoregulate" so as to avoid water loss by osmosis and maintain turgor (and therefore functionality) in their cytoplasmic membranes. Few bacteria, and none that should be of concern to the public health, can multiply at a_w values below 0.86. Yeasts and molds, some of which can grow slowly at a_w values as low as

0.6, are the culprits behind food spoilage in products with aw values below this threshold. Dried foods that are shelf-stable are typically made with a_w values of approximately 0.3, which is near the minimal point at which lipid oxidation and other enzymatic and chemical changes occur. Foods with low acidity (a_w) can effectively preserve food, but they can also retain bacteria, allowing the bacteria to live longer in low-a_w foods than in higher-a_w ones. Additionally, at low a_w values, the resistance of vegetative and spore forms of microorganisms to heat may be significantly increased, most likely due to osmotic dehydration of their cytoplasmic contents (Murrell and Scott, 1966). Pressure is one more physical antimicrobial treatment that might exhibit the same behavior.

2.1.1.4. Vacuum Packaging and Modified Atmosphere Packaging

Because they eliminate oxygen, vacuum packaging and modified atmosphere packaging (MAP) are successful in preventing the growth of microorganisms that are exclusively oxidative. In the absence of oxygen, fermentation-producing microorganisms keep growing. But they usually do so more slowly, and they have less unpleasant effects on food quality for certain kinds of foods. Precautions must be taken to prevent the growth of food poisoning microorganisms that are strictly anaerobic, like *Clostridium botulinum*. Because of this, it is advised that foods like sous vide items—which are vacuum-packed and heat-pasteurized as opposed to sterilized—be distributed with stringent temperature controls and minimal heat treatments (Notermans et al., 1990). Because carbon dioxide has a particular antimicrobial activity and functions as a preservative that specifically evaporates when the food package is opened, it is frequently used in MAP foods. For instance, a lot of meats from supermarkets are packaged in gas mixtures that are roughly 70% oxygen and 30% carbon dioxide. The carbon dioxide slows down the growth of Gram-negative spoiling bacteria to roughly double the product's shelf life, while the oxygen keeps the meat's pigments in the bright-red oxymyoglobin state—the color that consumers prefer.

2.1.1.5. Acidification

At pH values below 3.5 or so, very few bacteria that are significant in food can grow. Such species are acclimated to acidic conditions (e.g., lactic acid bacteria). Some of these bacteria are used in a variety of acid-producing food and feed fermentations in addition to being spoilage microorganisms. Since *Clostridium botulinum* cannot grow below this pH, 4.5 is a particularly crucial pH for food safety. It is therefore not necessary to treat foods that are more acidic than this to the same extent as higher pH, low-acid foods when thermal processing or other microorganism-inactivating procedures, such as pressure, are applied. Other food poisoning and spoiling bacteria are largely suppressed when the pH falls below approximately 4 points. The spore-forming bacteria *Alicyclobacillus acidoterrestris*, which can grow at pH levels as low as 2, has, however, recently been linked to spoiling issues with certain acidic foods. Furthermore, the flora of rotting acidic foods is primarily composed of yeasts and molds, which are able to proliferate at pH values well below 2.

2.1.1.6. Preservation

Table 2.1 shows that the majority of preservatives used in food are acids, including sulfite and nitrite, as well as weak lipophilic organic acids like sorbate, benzoate, and propionate. These acids are all more potent at lower pH values. In fact, there aren't any broad-spectrum antimicrobial food preservatives that work well at pH values close to neutral, possibly except parabens, which are the alkyl esters of p-hydroxybenzoate. The effectiveness of weak acids and their combination with hydrogen ions (i.e., synergy) are well-established (i.e., in an environment with a low pH). They are easily able to equilibrate across the microbial cell membrane and enter the cell cytoplasm in their unionized forms, which are favored by low pH.

The typical weak acid preservatives have pH values between 4.2 (benzoic) and 4.87 (propionic). Activity is significantly decreased at pH values significantly above these ranges. Microorganisms maintain an internal pH higher than that of their surroundings at the pH of most foods. As a result, the undissociated acids tend to dissociate upon entering the cytoplasm, delivering hydrogen ions in addition to the specific anion. The microorganism may export the extra hydrogen ions, but this requires energy, which limits cell growth. The pH of the cytoplasm eventually drops to a point where growth is impossible if the energy supply is insufficient. Furthermore, the accumulated anions might possess particular antimicrobial properties.

Table 2.1. *Application of a Smart Cold Chain*

Method	Food	Research Results
Digital twinning technology	Strawberry	Presents a novel, holistic approach to selecting the optimal ventilated packaging of strawberries from farm to retailer. This approach can help reduce food loss and contribute towards making supply chains smart and efficient.

Kinetically modelled, enzymatic TTIs	Ready-to-eat chilled smoked fish	Appropriate TTIs were selected for shelf life monitoring of smoked fish in the cold chain.
Blockchain technology	Livestock products	Real-time risk point detection for food safety can reduce food fraud and contamination while also strengthening the mechanism for recalling affected batches of products.
Proportion-integral-derivative (PID) control algorithm, 4G	Aquatic products	Established a mathematical model for the system based on the step response method, realized the gas dynamic volumetric based on the embedded PID control algorithm and mass flow controller; used 4G network for remote data transmission; realized remote monitoring function of mobile client through cell phone terminal applet software.
A diffusion-based time–temperature-indicator (TTI), kinetically modelled.	Juice	TTI systems are efficient and economical tools for monitoring, recording, and translating the overall effect of temperature history on food safety and quality at a product unit level.

Source: Bai L, Liu M, Sun Y. Overview of Food Preservation and Traceability Technology in the Smart Cold Chain System. Foods. 2023; 12(15):2881. https://doi.org/10.3390/foods12152881 Combination Preservation and "Hurdle" Technologies.

A popular and successful combination preservation method involving two hurdles (hydrogen ions and the specific acid) is combining lowered pH with a weak acid preservative. Additionally, the method shows how adding more hurdle technologies can enhance preservation. For instance, from the perspective of practical food preservation, it makes sense to incorporate a weak acid whenever possible, to acidify a food product to the extent that it is acceptable from an organoleptic standpoint in order to take advantage of the synergy between weak acid and low pH, and to vacuum pack in order to limit the energy available for the extrusion of hydrogen ions. Then, it makes sense to lower the a_w value as much as possible, since doing so will increase the energy needs of the cell for osmoregulation, the synthesis of compatible solute, and other processes. In this way, numerous preservation technologies are employed, and new ones that make sense are sought after. The incorporation of new physical preservation technologies, like high hydrostatic pressure, in such combinations has received minimal research attention up to this point. Physical preservation techniques will almost certainly inspire the development of novel and practical barrier technologies.

2.1.1.7. Heat

Pasteurization and heat are sometimes used by some of the biggest food industries in the world to inactivate vegetative microorganisms, and sterilization and heat are sometimes used to inactivate bacterial spores (Pflug and Gould, 2000). Heat is still the only effective way to inactivate microorganisms in food because most countries are reluctant to accept irradiation for food preservation. The fact that the majority of the novel and developing technologies that have been studied and supported recently function through inactivation without requiring significant heating is encouraging.

2.1.2. New and Emerging Preservation Technologies

Instead of inhibiting microorganisms, the majority of new and developing food preservation technologies work by rendering them inactive. A variety of new technologies are being developed, including chemical, enzymatic, and physical ones (Gould, 1995a).

While heat is still the most popular method of inactivation, new technologies are emerging. The majority of these serve as nonthermal or low-heat substitutes for traditional heat processing. A few of these methods are currently in widespread use. For instance, more than 100 tons of hen egg white lysozyme, a bacteriolytic enzyme, are used annually to stop cheeses from forming gas (also known as blowing). The way the enzyme functions is by lysing *Clostridium tyrobutyricum* vegetative cells that are proliferating from contaminant spores. In nations where refrigerated transport systems are underdeveloped and pasteurization soon after milking is not feasible, bulk milk's shelf life can be extended by activating the milk-derived lactoperoxidase system. Particularly in nations with high ambient temperatures, the bacteriocin nisin is being used more and more to stop food spoiling caused by thermophilic, spore-forming bacteria as well as to stop some cheeses from going bad. There are now over 40 different bacteriocins known to exist, some of which are being considered for use in food. Numerous substances that are extracted from herbs and spices have been found and demonstrated in lab experiments to possess antimicrobial properties. While some of these compounds have been demonstrated to be useful in food settings, their effectiveness is frequently diminished there due to protein binding, partitioning into lipids, and other factors.

Food scientists will probably continue to discover new applications for chemical- and enzyme-based methods, especially those that come from natural sources, but the most exciting alternatives to heat for food preservation will probably come from new and developing physical processes.

2.1.2.1. Ionizing Radiation

It was first suggested to use ionizing radiation to preserve food in the early 1900s, and it is currently permitted in about 40 countries at pasteurization doses of up to 10 kGy. The World Health Organization (WHO) has recently advised against imposing an upper-dose limit because an overwhelming amount of research has demonstrated that there is no safety concern. Instead, the WHO suggests that sterilization treatments utilizing ionizing radiation (e.g., 30 to 50 kGy, contingent on the food kind) ought to be allowed. Compared to many thermal processes, food irradiation has much simpler processing requirements and straightforward inactivation kinetics and geometry, making dosage control and application technologically relatively straightforward. Food irradiation has the potential to significantly reduce the risk of food poisoning by removing pathogens like Salmonella, *Escherichia coli* OI57:H7, and Campylobacter from certain animal-based foods and seafood. But the public's sluggish acceptance is still impeding its introduction in the majority of nations.

2.1.2.2. Electroporation

While bacterial spores are more resilient, high-voltage electric shock treatments, or electroporation, are the most effective at inactivating the vegetative cells of bacteria, yeasts, and molds. Because electroporation breaches the cell membrane, one of the most crucial structures regulating a large number of the homeostatic mechanisms of the vegetative cell, it has a fatal effect. It has been demonstrated that liquid foods like fruit juices and milk can be safely pasteurized and have vegetative food-poisoning microorganisms rendered inactive by high-voltage pulse machines.

2.1.2.3. Manothermosonication

It has long been known that sufficiently high intensities of ultrasonication can inactivate vegetative microorganisms and decrease spores' heat resistance; the effect is enhanced by raising the temperature. On the other hand, the relative magnitude of the amplification decreases with increasing temperature. This is believed to happen because the cavitation effect, which is what causes the killing effect, is lessened as the vapor pressure increases. It has been reported that this decreased effectiveness can be overcome by applying a slight overpressure—ideally, 500 kPa—so that the amplification is maintained at the higher temperatures. Manothermosonication, a combination process, is said by researchers to have the ability to lower pasteurization and sterilization temperatures for foods that are pumpable and semisolid.

2.1.2.4. High-Intensity Light

In addition to potential uses in dentistry, high-intensity lasers and noncoherent light-pulse generators have been developed for food surface decontamination, potentially including transparent foods. In certain applications, the killing effect is caused by the UV content; in other applications, it is caused by localized, intense heating that eliminates microorganisms through ablation. Some researchers have claimed additional nonthermal and nonultraviolet effects.

2.1.2.5. High-Strength Magnetic Fields

It has been documented that exposure to intense, oscillating magnetic fields can affect biological systems in a number of ways, from the selective inactivation of cancerous cells to the inactivation of bacteria on food packaging materials. Treatment times are very short, usually less than 25 ms and field strengths are high, usually between 2 and 100 tesla at 5 to 500 kHz frequencies. The number of vegetative microorganisms inoculated into milk (*Streptococcus thermophilus*), orange juice (Saccharomyces), bread (mold spores), and other media was reduced by approximately 102 times, and no reports of bacterial spore inactivation have been made. Therefore, it appears that the technique's practical potential as it has been developed thus far is limited.

2.1.3. Evolution of High Pressure Processing

All other physical technologies aside from irradiation, high pressure processing has not kept pace with advancements. This is partially attributable to the effectiveness of high pressure processing. If sufficient pressure is applied, it deactivates all vegetative and spore forms of microorganisms. Furthermore, for high-value niche market products at least, the engineering aspects of high pressure processing have progressed to the point where within the last 10 or so years, commercially viable processes are now possible. The original observations that served as the foundation for high pressure processing were made a long time ago, as is the case with many other new technologies. More than a century ago, the idea that high pressure could render microorganisms inactive was first documented.

2.1.3.1. Historical Survey

The military's need to improve the cannon drove the early development of high pressure technology, not food technologists. Crossland (1995) has provided an overview of the early cannon designs that tried to hold pressures higher and higher so that gunpowder charges could shoot ever-heavier missiles. This work produced vessel designs that were later made available for use in large-scale laboratory experiments, including food-related ones. The first measurements of the compressibility of water and other fluids were reported by Perkins in 1820. The first persuasive experiments with microorganisms were reported at the end of the 19th century by Hite (1899), and the effects of pressure on the physical properties of foods were reported soon after. For example, Bridgman (1914) studied the coagulation of egg albumin by high pressure and pointed out that the properties of the product differed from those resulting from coagulation by heat. Likewise, Basset et al. (1933) reported that pressure caused the gelation of horse serum proteins. Much later, Macfarlane (1973) reported the practical potential of high pressure technology with the demonstration of pressure-tenderization of meat. Although Hite (1899) had reported that some of the physical properties of milk changed on pressurization (e.g., it became more translucent), detailed investigations of the effects of pressure on the physicochemical properties of food biopolymers were not made until much later. For example, Payens, and Heremans (1969) studied the effects of pressure on 13-casein from milk, and Zipp and Kauzmann (1973) studied the effects of the ionic environment on the pressure-denaturation of metrnyoglobin.

The impact of pressure on food's physical characteristics is now recognized as significant and possibly exploitable. Nevertheless, the earlier research focused on rendering microorganisms inactive, specifically the vegetative cells of yeasts and bacteria, with spores being the focus later. Cruess (1924), more than 75 years ago, predicted that pressure would be used to preserve fruit juices (in which the low pH ensures that spore formers are not a problem), preempting some of the most successful modern developments. Although Giddings et al. (1929) reported the pressure inactivation of tobacco mosaic virus, there has been little interest in the pressure tolerance of the viruses that are of concern in foods.

2.1.3.2. Vegetative Microorganisms

Hite (1899), who reported the treatment and prevention of souring of milk, was the first investigator to show clearly that microorganisms can be inactivated by pressure. Larson et al. (1918) cited unsuccessful experiments by Certes as early as 1889 and successful experiments by Buchner that were reported in 1897.

In 1899, Hite pointed out some of the potential advantages of pressure treatment that we still recognize today: "antiseptics are always objectionable;" "heating milk at the temperature of boiling water. for about half an hour each day for three or four successive days" (Tyndallization);. or heating it to a somewhat higher temperature for a longer time should keep indefinitely. but it has a "cooked" odor and taste;" "pasteurization reduces the germ content-but as it does not kill all the germs the milk must be kept at a low temperature or it will sour in a very short time." He mentioned that, since late 1897, the chemist in his institute (West Virginia Experiment Station) had been studying the effects of pressure on the preservation of milk, meats, and fruit juices (which seems very relevant to modem interests, more than 100 years later). The first experiments, which involved driving a heavy screw into a cylinder to deliver a theoretical pressure that was something tremendous and holding it for periods of days, failed to postpone the souring of milk. Higher pressures were obtained when the researchers acquired a piston-cylinder device from the father of a student in Hite's agricultural chemistry class, who was connected with the National Transit Company of Oil City, Pennsylvania. Following many trials and errors and machine development, pressure-treated milk samples stayed sweet "24 hours longer than untreated controls, some of them 48 to 60 hours longer. Two samples that had received 90 tons for one hour, remained sweet for a week. Thus encouraged, an attempt was made to submit other samples to 90 tons, when with a loud report the tool steel bush split from one end to the other." Despite such difficulties, this seminal publication clearly showed that pressure inactivated microorganisms in milk and improved its keeping qualities. Pressures greater than 30 tons per square inch applied for one hour generally postponed the souring of milk at ambient temperature for at least 24 hours. Occasionally, a sample did not sour even after two weeks, but there was much experiment-to-experiment and sample-to-sample variation.

It's intriguing that concerns about microbiological safety were recognized at such an early stage. In fact, Hite (1899) reported that pure cultures of typhoid germs were obtained in experiments that today would require a high-category containment laboratory. However, in the initial attempt to subject an inoculated milk sample to a pressure of 90 tons, the steel cylinder, the lead tube, the tin tube, and one of the pistons broke, dispersing the germs throughout the room. It was decided to stop this aspect of the work after a member of the station staff became seriously ill with typhoid before another cylinder could be made. Later on, though, experiments with anthrax, typhoid, tuberculosis, Proteus vulgaris, and bubonic plague were conducted with new presses.

It's amazing that so many novel observations were made in this early study, even though Hite's report did not consistently result in the full preservation of milk. Hite's 1899 report's latter section even foresaw numerous contemporary studies of the impact of pressure combined with a modest temperature increase. A 67-page report covering these studies was published approximately 15 years after they were conducted. For concerns that persist to this day, many of the earlier studies on the effects of high pressure on milk were abandoned. According to

Hite (1899), some slow changes in the milk persisted because of enzymes that the pressure was unable to eliminate, not even when 100,000 pounds per square inch of pressure was applied for seven days at room temperature or for an hour a day for more than a week. Remarkably, these results anticipated contemporary attempts to take advantage of pressure cycling. No sample of milk that has been treated in this way has yielded bacterial cultures.

In contrast to Hite's 1899 publication, the 1914 work featured counts of microbes. Pressures in the range of 700 MPa applied at room temperature for 10 minutes were shown to reduce the number of microorganisms in the vicinity of 10^7 mL^{-1} to a few hundred or less. Apple juice, grape juice, peaches, pears, plums, blackberries, raspberries, tomatoes, peas, beans, beets, and other vegetables were among the foods used in the experiments. Similar to the previous milk experiments, the researchers observed a pattern of mixed results. While some pressure-treated samples spoiled, others were successfully preserved. (Today, we would most likely blame the remaining spores for a large number of the failures.). As a result of the variability, Hite, and colleagues decided to undertake a systematic study using pure cultures of microorganisms; unfortunately, however, actual kinetic studies were not performed. Most data were derived from "end point-like" experiments. That is, results were recorded as "no growth," "several colonies," "good growth," etc., on agar plates. Most data were obtained with *Bacillus prodigiosus* (not a spore-forming bacillus, now named Serratia marcescens). Other, less extensive studies were done with lactic acid bacteria, *Bacillus jluorescens* (now *Pseudomonas jluorescens*), vegetative cells of *Bacillus subtilis*, yeasts (*Saccharomyces cerevisiae* and Saccharomyces [now *Candida] albicans*], as well as the pathogens *Bacillus typhosus* (*Salmonella typhimurium*) and *Bacillus diphtheriae* (*Corynebacterium diphtheriae*).

Hite and associates (1914) determined that the growth-no-growth limits for the vegetative cells under investigation were within the 400–600 MPa range when applied for 10 minutes; however, significant fluctuations transpired. The authors thought about how the decompression rate could affect the outcomes. After conducting experiments to look into this, they came to the conclusion that there was no difference between a sudden and slow release of pressure. Hite et al. Chlopin and Tammann's (1903) paper, which is mentioned in (1914), asserted that pressures of approximately 300 MPa did not inactivate microorganisms, but rather put them in a lethargic state where they might lose their ability to reproduce (what we now refer to as a viable but nonculturable state). Still, Hite et al. found that pressurized microorganisms (Serratia marcescens) that did not grow on the first day of incubation did not resurrect even after 89 days of incubation, based on extensive incubation experiments. Although the impact of medium composition was recognized, it wasn't thoroughly investigated until Timson and Short (1965) demonstrated the synergistic effects of extreme pH values and the significant protective effect of high concentrations of solutes (such as NaCI, K_2HPO_4, K citrate, and glucose) on microorganisms pressurized in milk (a finding that has been repeatedly confirmed since).

In order to reduce the possibility of spores being present, Rite, and colleagues (1914) used young (12- to 18-hour) cultures of *Bacillus subtilus*. They did, however, note that this could not be guaranteed. Hite and associates did not intentionally investigate spores.

This comprehensive study by Hite and colleagues set the scene for later work. Most of the phenomena described have stood the test of time, and some have been rediscovered many times since. Soon after the comprehensive Hite et al. (1914) paper was published, Larson et al. (1918) reported investigations of the effect of pressure on microorganisms with the aim of improving vaccines. They pressurized *Bacillus typhosus* (*Salmonella typhimurium*), Bacillus tuberculosis (*Mycobacterium tuberculosis*), *Bacillus proteus* (Proteus), *Bacillus subtilis*, staphylococci, streptococci, and pneumococci at pressures from about 300 to 1200 MPa. Later, Basset, and Macheboeuf (1933) similarly reported the pressure inactivation and antigenicity of *Clostridium diphtheriae* and *C. tetani*. Larson et al. used long pressurization times (up to 14 hours). Although the initial number of microorganisms is unclear, they found that 300 MPa was insufficient to inactivate any of the bacteria studied, whereas 600 MPa inactivated them all. However, Larson et al. (1918) were the first to clearly show that spores (of *Bacillus subtilis*), in contrast to the vegetative cells, were not "regularly killed" at the highest pressure used (1200 MPa). Further experiments reported in this paper concentrated on the use of high gas pressures (carbon dioxide, hydrogen, and nitrogen) and the effects of decompression during the sudden release of pressure. In particular, they reported that carbon dioxide under pressures as low as 5 MPa inactivated vegetative cells in 1.5 to 2.5 hours. Gram-negative microorganisms tended to disintegrate, while Gram-positive ones kept their shape. Hydrogen and nitrogen were relatively ineffective.

It's interesting to note that pressure-treated Salmonella typhi filtrate produced significantly higher antibody titers in inoculated rabbits than did living cell inoculation. Similarly, the carbon dioxide-pressed inactivated cells exhibited a high degree of antigenicity (Crelier, 1995; Baranyi & Roberts, 2000).

2.2. EFFECTS OF HIGH PRESSURE ON VEGETATIVE MICROORGANISMS

Food scientists are investigating alternative physical preservation techniques to replace conventional heat treatments in response to growing consumer demand for minimally processed, additive-free, shelf-stable products. Among these methods, ultra-high pressure appears to be the most promising. Foods produced under pressure should ideally be stable and safe for microbes under the broadest range of circumstances. Pressure treatments that inactivate bacterial spores, as well as vegetative cells, are therefore desperately needed. We are still in the exploratory stage of high pressure application as a sporicidal treatment. But pressure treatment has been used as a preservation method for over a century, ever since Hite (1899) proved that applying high pressure could postpone the microbial deterioration of milk. There are several pressure-treated products on the market right now. These are typically pressure-treated foods that are somewhat acidic, like fruit juices, in order to lower the amount of molds, yeasts, and vegetative bacteria. There are currently a plethora of additional potential uses for dairy and meat products being researched.

Like other physical preservation techniques, high pressure relies on a variety of internal and external factors to determine how effective it is. When optimizing pressure treatments for the production of safe, high-quality foods, these factors—which include the effect of pressure on water, adiabatic heat and heat dissipation, the composition of the food, and the physiological state of the microorganisms to be inactivated—must be considered.

2.2.1. Mode of Action of Temperature and Pressure on Microorganisms

The role of temperature in pressure treatments is almost universal, so here we briefly discuss the heat's mode of action. The synergistic impact of pressure and heat has been extensively researched. The primary foundation for heat and pressure processes is empirical data regarding the inactivation of microbes. The microbial physiology of the combined process would help predict the effects of particular food components on heat and pressure inactivation, even though the design of combined heat and pressure processes could be based on these data. Nevertheless, not much research has been done to compare the physiological effects of temperature and pressure on different cell functions. Pressure and elevated temperature work together to hit multiple targets rather than just one. In addition, the physiological effects are contingent upon the ambient temperature and atmospheric pressure. For example, damage to the membrane appears to be the primary target at 64°C, while at higher temperatures, there is a greater degree of enzyme inactivation. Anticipate that the intrinsic stability of macromolecules (ribonucleic acid [RNA], ribosomes, nucleic acids, enzymes, and proteins in the cell, membrane, and occasionally the cell wall) will contribute to the basal heat and pressure resistance. While the molecular changes for temperature and pressure inactivation appear to differ, many of the targets appear to be similar.

2.2.2. Morphological Changes Brought About by High Pressure

In general, no morphological changes can be observed in the cells of prokaryotes or lower eukaryotes by observation under a light microscope. Perrier-Comet et al. (1995) measured the diameter of yeast cells before, during, and after pressure treatment by image analysis. During pressure treatment, the average volume gradually decreased at 250 MPa. After pressure treatment, the cells did not totally recover their original volume. Shimada et al. (1993) observed a slight alteration in the outer shape of *Saccharomyces cerevisiae* with pressures below 400 MPa. However, the inner structure of the cell began to be affected as shown by transmission microscopy. Osumi et al. (1992) studied changes in the ultrastructure and the cytoskeleton of *Candida tropicalis* and found that the outer shape of the yeast was unaffected after treatment at 100 MPa. At 200 MPa the cell wall was damaged and the mitochondrial cristae were damaged. Kobori et al. (1995) observed changes in the nuclear apparatus of *Saccharomyces cerevisiae* by means of immuno electron microscopy. Hamada et al. (1992) noticed changes in colony form after pressure treatment of Saccharomyces cerevisiae.

2.2.3. Gene Expression, Protein Synthesis, and Stress Response

In order to survive later, potentially fatal temperatures, microorganisms have the ability to withstand and recover from the effects of heat and other adverse environmental conditions, such as starvation and low pH. It is unknown if they also have systems in place to resist pressure. The ideal temperature for growth typically lies in a narrow range between the maximum temperature at which stress proteins are produced and the cell gains more heat resistance. The development of thermotolerance and the synthesis of heat-shock proteins may or may not be related, but this is unclear.

Most food-spoilage and food-poisoning organisms grow best under conditions that are probably close to atmospheric pressure. Pressures of roughly 50 MPa have an impact on a number of bacterial biochemical processes, including protein synthesis, glucose utilization, and L-phenylalanine utilization. The types of proteins that are synthesized can be influenced by culture pressure, and transcription and translation are extremely sensitive to pressure. One of the cell processes that is most sensitive to pressure is known to be protein synthesis. Protein synthesis in *Escherichia coli* cell-free systems is totally blocked at 68 MPa, but this inhibition is fully reversible upon pressure release. Gene and protein expression can be influenced by high hydrostatic pressures ranging from 30 to 50 MPa. Following an 80 MPa pressure treatment, *Yarrowia lipolytica* exhibited increased proteolytic activity, most likely due to the high pressure activating the proteolytic enzymes rather than lysis. Sato et al. (1996) demonstrated that in *Escherichia coli*, promotors malK-lam and mal-EFG were the source of gene expression. They discovered that high pressure significantly decreased the expression of the genes encoding osmoregulatory membrane proteins, most likely at the transcriptional level. Hydrostatic pressure generally causes *Saccharomyces cerevisiae* to become tetraploid, suggesting that high pressure can obstruct DNA replication.

Three organisms that represent archaeobacteria, yeast, and bacteria—*Methanococcus thermolyticus, Rhodotorula rubra*, and *Escherichia coli*—have been found to contain pressure-inducible proteins. *Escherichia coli* may produce proteins at 53 MPa that are comparable to those found at high temperatures. Sublethally low pH can cause resistance to heat, and vice versa. Additionally, sublethal heat can increase bacteria's and yeasts' resistance to pressure. It has not yet been noted that sublethal pressure increases resistance to heat. Yeasts or *Lactobacillus plantarum* cells did not exhibit enhanced heat resistance or pressure resistance when exposed to a variety of sublethal pressures. Exponentially growing cells are more sensitive to pressure than stationary phase cells, which is not surprising given the previously mentioned findings that starvation or sublethal pH increases resistance. Even greater resistance to pressure can be achieved with an extended stationary phase. Additionally, some membrane protein synthesis is inhibited by hydrostatic pressure.

2.2.4. Ribosomes as a Target for Heat and Pressure Damage

Pressure and heat have an impact on ribosomes. It makes sense that a single cell would die when its quantity of functioning ribosomes falls below a certain point, above which it is impossible for the cell to survive. Ribosomal RNA (rRNA) in microbial cells maintains the specific secondary and tertiary structure of ribosome proteins, allowing ribosomes to function during protein synthesis. Heat damage in cells has been linked to RNA modification in its early stages. Pressure damage appears to be similar. Gross and al. (1993) looked into the effects of high pressure and found that at 40 to 60 MPa, ribosome dissociation begins.

Hurst (1984) reported that heating produces membrane damage, resulting in Mg^{++} depletion in the cells. This, in turn, can lead to destabilization of 70S ribosomes. This phenomenon might be similar to high pressure; Hauben et al. (1997c) reported that divalent cations such as Ca^{++}, Mg^{++}, Mn^{++}, and Fe^{++} reduced inactivation by high pressure. It has been demonstrated that ribosomes dissociate *in vitro* under high pressure. Mackey et al. (1991) demonstrated in *in vivo* studies with differential scanning calorimetry that the 50S subunit is more heat stable than the 30S subunit; however, both subunits are more heat labile than the complete 70S particle. Mackey and colleagues (1991) showed that heating in the range of 50°C to 80°C initiated denaturation of the 30S subunit by comparing the behavior of the 30S, 50S, and 70S subunits in isolation and in whole cells. A correlation between the ribosomes' melting points and maximum temperature would imply a relationship between ribosome stability and thermal resistance. Niven et aI. (1999) carried out a similar study of the change of conformation of ribosomes by pressure *in vivo* and found a correlation between cell death and ribosome damage.

2.2.5. Susceptibility of Proteins and Enzymes to Heat and Pressure

Enzymes and other proteins are significantly impacted by temperature and pressure. Proteins are best able to withstand pressure at a specific temperature. Microorganisms that have been rendered inactive have shown a similar pattern. Sadly, the majority of enzymes have only been studied in isolation—that is, not in whole cells. The enzymes of extreme thermophiles are more resistant to pressure and heat than those of mesophilic microorganisms, according to research by Hei and Clark (1994). Enzymes and transport system carriers may be directly impacted by hydrostatic pressure. Pressures of 200 and 100 MPa, respectively, were used to inactivate glyceraldehyde-3-phosphate dehydrogenase from baker's yeast and lactic dehydrogenase from rabbit muscle. This offers some hints that the pressure inactivation of microorganisms is significantly aided by enzyme inactivation. On the other hand, there are no kinetic data available regarding the inactivation of different sites, such as membrane-bound enzymes, glycolytic enzymes, ribosomes, or membranes. Furthermore, it is easier to study single molecules in isolation, such as enzymes and nucleic acids, but most biopolymers exhibit distinct cellular behaviors. Because the quaternary structure is primarily maintained by hydrophobic interactions, which are pressure sensitive, pressure changes the quaternary structure rather than the tertiary structure. Because of this, monomeric enzymes typically withstand pressure better than multimeric enzymes. Schmid et al. (1975) studied the effect of pressure on glycolytic enzymes with coenzymes *in vitro* and found that rabbit muscle lactic dehydrogenase was more stable against pressure than yeast glyceraldehyde-3-phosphate dehydrogenase.

However, little attention has been paid to the extent to which the weakest link in a system is affected. Degraeve et al. (1996) studied the effects of pH and temperature on pressure inactivation of *Escherichia coli* j3-galactosidase. No inactivation took place below 250 MPa. Cioni and Strambini (1997) studied the dissociation of yeast glyceraldehyde-3-phosphate dehydrogenase. The main effect was due to dissociation of the tetramer. Simpson and Gilmour (1997b) studied the effect of high hydrostatic pressure on the activity of 13 metabolic enzymes of *Listeria monocytogenes*. They isolated the enzymes before and after pressure treatment of whole cells. Phosphoglucomutase and aconitase were particularly susceptible to pressure, and even pressures of around 200 MPa were sufficient to inactivate these enzymes. Because *Listeria monocytogenes* was hardly affected by this amount of pressure, these enzymes cannot be considered as critical sites.

Thermophiles and deep-sea bacteria may have unique adaptations that reveal the locations within the organism that are most susceptible to pressure. Similar to heat inactivation, it is reasonable to assume that, depending on the pressure level, multiple distinct targets may have an impact on the entire organism.

2.2.6. Nucleic Acids

Particularly in DNA, nucleic acids are very resistant to pressure and comparatively resistant to heat. However, it has been discovered that *Salmonella typhimurium*, Lactobacillus plantarum, and *Listeria monocytogenes* exhibit an extreme condensation of their nuclear material. According to the theory, endonucleases come into contact with DNA at high pressures and cleave it. This condensation is reversible and likely involves an enzyme that is also in charge of renaturation, as it has been observed in numerous other cases. The cell can no longer divide if high pressure deactivates this enzyme.

2.2.7. Membrane-Related Events

The bacterial membrane is almost always disturbed during pressure treatment. Heat treatment works similarly, though usually to a lesser degree. Presumably, the stiffness of the membrane has an inverse relationship with pressure resistance. Cells should be expected to respond to stress by changing the composition of the membrane, since pressure and temperature changes reduce the fluidity of the membrane. There is a lot of data linking bacterial growth at low temperature or high pressure to the synthesis of higher proportions of membrane unsaturated acids. As growth pressure increases, the fatty acids in barophilic microorganisms become more polyunsaturated. Wirsin et aI. (1987) reported that in a psychrophilic and barophilic bacterium of the genus Alteromonas, the ratio of unsaturated to saturated fatty acid increased with growth pressure between 1 and 45 MPa at 2°C. The ratio of unsaturated fatty acids was higher when cells were grown at 2°C than when grown at l0°C. According to Allen et al. (1999), monounsaturated rather than polyunsaturated fatty acids would be required for the growth of the deep-sea bacterium *Photobacterium profundum*.

When *Lactobacillus plantarum* cells were grown at suboptimal temperatures, the cells grew exponentially and were more resistant to pressure. Under these circumstances, cells grown at optimal temperatures had a higher ratio of unsaturated fatty acids. Lanciotti et al. (1996) found that while the results for Yarrowia lipoiytica and *Listeria monocytogenes* were similar, the trends for *Escherichia coli* were occasionally different. Regretfully, it is unclear from their study if they employed stationary-phase or exponentially growing cultures. This could account for *Escherichia coli*'s enhanced resistance to pressure when grown at 30°C. Additionally, it has been demonstrated that the presence of cholesterol causes eukaryotic cells' cell membranes to become less fluid and more susceptible to pressure.

After being incubated at temperatures below optimal, the stationary phase cells' resistance to pressure does not increase. The fact that cells grown in the stationary phase are exposed to various stress conditions, including starvation and low pH, may be the cause of this discrepancy. We found that greater resistance to pressure is produced by longer incubation in the stationary phase. When comparing stationary phase cells grown at various temperatures, this factor needs to be considered.

The following is a list of the carbohydrates' protective effects on the membrane, ranked from highest to lowest: trehalose, sucrose, fructose, glucose, and glycerol. For these carbohydrates' ability to fend off pressure, the same sequence of events was discovered. Nucleic acids are bound by propidium iodide and ethidium bromide, but these substances can only enter a cell when the membrane is breached. Pressure-treated bacteria, in contrast to untreated cells, exhibit membrane damage that can be detected by staining with either propidium iodide or ethidium iodide. Another indication of membrane leakage is the rise in extracellular ATP (adenosine triphosphate) that coincides with pressure inactivation. When the membrane bilayer is sufficiently disrupted by pressure, integral, and peripheral proteins separate from the plasma membrane.

2.2.8. Events Related to Maintenance of Intracellular pH

Microbial cells require a constant internal pH to sustain vital cell functions. Reduced intracellular pH can be the result of high pressure treatment, and acid efflux may be hampered by membrane damage. In Lactobacillus plantarum, pressure treatment has been found to lower intracellular pH. Glycolysis impairment may also be the cause of the intracellular pH shift, which is accompanied by impaired acid efflux. Observations on yeasts and other conditions including membrane damage, protein inactivation, and lowered intracellular pH imply that membrane-bound enzymes involved in proton efflux may be a target in high pressure inactivation. Given that pressure can cause membrane-bound FoFj ATPase to become inactive or dislocate, it could be one of these targets. Similar to heat, extremely high pressure stress damages cells significantly and makes them more vulnerable to unfavorable environmental factors. Vacuolar pH in yeasts was found to be lowered by pressures between 40 and 60 MPa by Abe and Horikoshi (1995). They came to the conclusion that the yeast vacuole prevented the cytosol from becoming acidic by acting as a sequestrant for protons. Iwahashi et al. (1997) discovered that *Saccharomyces cerevisiae* mutants with reduced barotolerance and thermotolerance were unable to accumulate trehalose and/or heat-shock protein 104. They discovered, though, that trehalose appears to be more crucial for barotolerance (Crelier, 1995; Baranyi & Roberts, 2000).

2.3. THE EFFECT OF HIGH PRESSURE ON ENZYMES RELATED TO FOOD QUALITY

Seyderhelm et al. (1996) conducted a screening study based on the pressure resistance of food quality-related enzymes in buffer solutions. Ordered the enzymes lipoxygenase, lactoperoxidase, pectinmethylesterase, lipase, phosphatase, catalase, polyphenoloxidase, and peroxidase in ascending order of increasing pressure stability. However, because the pressure stability of enzymes is heavily reliant on multiple intrinsic (e.g., composition) as well as extrinsic (e.g., temperature) variables, a review of the literature may yield inconsistent findings.

2.3.1. Polyphenoloxidase

Enzymatic browning of harmed fruits and vegetables is brought on by polyphenoloxidase (PPO) activity, and this is thought to be one of the primary reasons for quality deterioration (brown coloration and accompanying changes in appearance and organoleptic properties) during postharvest handling, storage, and processing. Inactivation of PPO is highly desirable because it poses a significant issue for the food industry.

Even though numerous studies on the impact of pressure on PPO from various sources have been conducted (Table 2.2), the information that is currently available is frequently incomplete, and very few detailed kinetic data—such as data from a wide pressure-temperature domain—have been established. Thus far, scientists have discovered that PPO can exhibit either inactivation or activation (increased catalytic activity) in response to pressure. It is clear from Table 2.2 that the pH of the medium and the source of the PPO have a significant impact on the pressure level required for PPO inactivation at room temperature. Furthermore, it has been demonstrated that microenvironmental factors like the presence of sugars, salts, and other additives affect the pressure stability of PPO.

Table 2.2. *Effect of Pressure at Ambient Temperature on PPO From Various Origins*

Origin	Variety	Medium	Pressure Treatment	Activation/ Inactivation	Reference
Apple	NM	PB (pH 4.5, 0.5 M)	150–300 MPa/25°C/1 min	Activation	Anese et al., 1995
			≥500 MPa/25°C/1 min	Inactivation	Anese et al., 1995
	NM	PB (pH 5.4, 0.5 M	<300 MPa/25°C/1 min	Activation	Anese et al., 1995
			>700 MPa/25°C/1 min	Inactivation	Anese et al., 1995
	NM	PB (pH 7, 0.5 M)	≥100 MPa/25°C/1 min	Inactivation	Anese et al., 1995
	Golden	Crude extract (pH 7)	500 MPa/25°C/10 min	Activation	Jolibert et al., 1994
		Crude extract (pH 3–6)	500 MPa/25°C/10 min	Inactivation	Jolibert et al., 1994
		Cell-free extract (pH 3.9)	>250 MPa/25°C/10 min	Inactivation	Jolibert et al., 1994
		Puree	≥1000 MPa/20°C/5 min	Inactivation	Rovere et al., 1996
		PB (pH 6, 0.1 M)	≥600 MPa/25°C/5–60 min	Inactivation	Weemaes et al., 1998a
	Granny Smith	Cell-free extract (pH 3.5)	≥125 MPa/25°C/10 min	Inactivation	Jolibert et al., 1994
	Jonagolden	Cell-free extract (pH 3.9)	≥250 MPa/25°C/10 min	Inactivation	Jolibert et al., 1994
	Red Chief	Cell-free extract (pH 4.1)	>250 MPa/25°C/10 min	Inactivation	Jolibert et al., 1994
Apricot	NM	Cell-free extract (pH 3.9)	≥125 MPa/25°C/15 min	Inactivation	Jolibert et al., 1994
Avocado	NM	PB (pH 7, 0.1 M)	≥800 MPa/25°C/>60 min	Inactivation	Weemaes et al., 1998b
Grape	Trebbiano	Must (pH 3.2)	>600 MPa/15°C/10 min	Inactivation	Castellari et al., 1997
		Synthetic must (pH 2.9)	900 MPa/15°C/2 min	Inactivation	Castellari et al., 1997
	NM	McIlvaine (pH 6)	≥700 MPa/25°C/5–30 min	Inactivation	Weemaes et al., 1998a
Guava	Chung shan	Puree (pH 3.8)	≥600 MPa/25°C/15 min	Inactivation	Yen and Lin, 1996
Onion	Stuttgarter Riese	TB (pH 6.5, 0.01 M)	≤700 MPa/25°C/10 min	Activation	Butz et al., 1994
Pear	Bartlett	Slices	400 MPa/25°C/10 min	Activation	Asaka and Hayashi, 1991
		Cell-free extract	200–600 MPa/25°C/10 min	Activation	Asaka and Hayashi, 1991
	La France	PB (pH 7, 0.02 M)	400–700 MPa/20°C/≥3 min	Activation	Asaka et al., 1994
	Durondeau	PB (pH 7, 0.1 M)	900 MPa/25°C/200 min	Inactivation	Weemaes et al., 1998a
Plum	Mirabelle	Cell-free extract (pH 4.6)	≤500 MPa/25°C/15 min	No inactivation	Jolibert et al., 1994
	Red Beauty	PB (pH 7, 0.1 M)	900 MPa/25°C/180 min	No inactivation	Weemaes et al., 1998a
Potato	NM	Cubes	400 MPa/25°C/15 min	No inactivation	Knorr, 1993
	NM	Crude extract	≥200 MPa/25°C/10 min	Inactivation	Gomes and Ledward, 1996
	Desirée	Cell culture	100–200 MPa/25°C/5–10 min	Activation	Dörnenburg and Knorr, 1997
	Desirée	Water surrounded cubes	400 MPa/20°C/15 min	Inactivation	Esthiaghi and Knorr, 1993
Strawberry	NM	Crude extract (pH 2–5)	500 MPa/25°C/10 min	Inactivation	Jolibert et al., 1994
		Crude extract (pH 6)	500 MPa/25°C/10 min	Activation	Jolibert et al., 1994
		Cell-free extract (pH 3.8)	≥375 MPa/25°C/15 min	Inactivation	Jolibert et al., 1994
	Pajaro	Puree (pH 3.9)	50–285 MPa/20°C/15 min	Inactivation	Cano et al., 1997
Tomato	Pera	Puree (pH 4.1)	<200 MPa/20°C/15 min	Activation	Hernández and Cano, 1998
			200–500 MPa/ 25 °C/15 min	Inactivation	Hernández and Cano, 1998
Mushroom	NM	PB (pH 6.5, 0.1 M)	≥750 MPa/25°C/20 min	Inactivation	Weemaes et al., 1997
	NM	PB (pH 6.5, 0.05 M)	>100 MPa/25°C/>1 min	Inactivation	Gomes and Ledward, 1996
	NM	Crude extract	200 MPa/25°C/10 min	Inactivation	Gomes and Ledward, 1996
			400 MPa/25°C/10 min	Activation	Gomes and Ledward, 1996
			≥600 MPa/25°C/10 min	Inactivation	Gomes and Ledward, 1996
	NM	NM	300–700 MPa/20°C/>10 min	Activation	Aoyama et al., 1994
	NM	Intact tissue	600 MPa/25°C/5 min	Activation	Matser et al., 1998
			800 MPa/25°C/5 min	Inactivation	Matser et al., 1998

NM = not mentioned; PB = phosphate buffer; TB = Tris buffer

Source: Mare E.G., Hendriekx, Dietrieh Knorr: Ultra High Pressure Treatments of Foods: ISBN 978-1-4613-5211-2.

They investigated whether the pressure stability of avocado and mushroom PPO could be modified by adjusting the pH, adding antibrowning agents, and raising the temperature. A first-order kinetic model could adequately describe the pressure inactivation of avocado and mushroom PPO at room temperature and at their optimal pHs of 6.5 and 7, respectively (Figure 2.1).

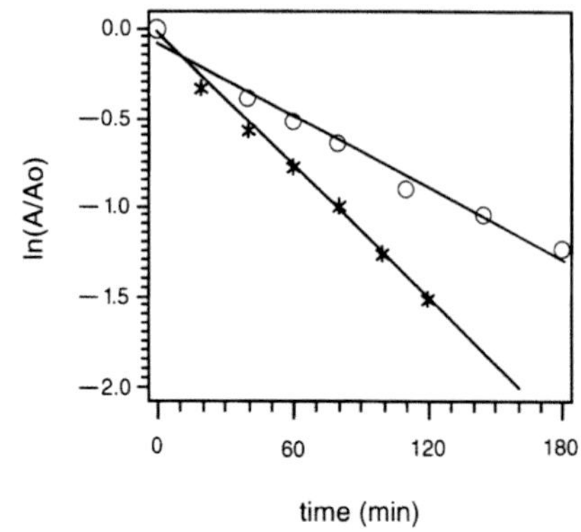

Figure 2.1. *First-order pressure inactivation (850 MPa) of avocado (*) and mushroom (0) PPO at 25°C.*

Source: Mare E.G., Hendriekx, Dietrieh Knorr: Ultra High Pressure Treatments of Foods: ISBN 978-1-4613-5211-2.

The threshold pressure for inactivation for avocado PPO and mushroom PPO dropped as the pH dropped below the ideal level (Figure 2.2). Furthermore, the pressure sensitivity of the inactivation rate constant increased as a result of acidification. Stated differently, the activation volume increased in absolute value as pH decreased (Figure 2.2).

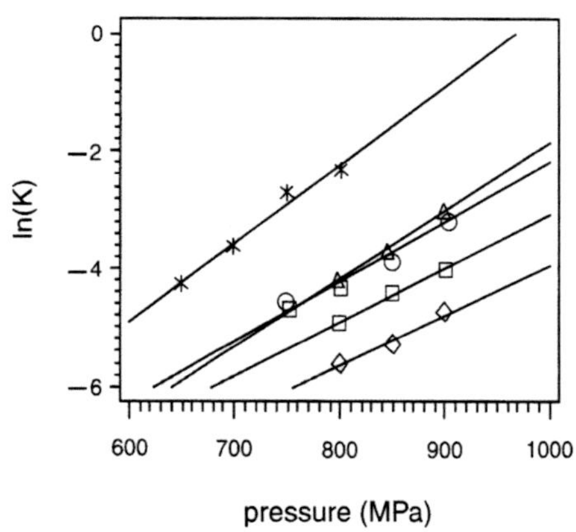

Figure 2.2. *Pressure dependence of the inactivation rate constant of avocado PPO at pH 5 (*), pH 6 [McIllvaine buffer (0), phosphate buffer (Δ)], pH 7(□) and pH 8 (◊).*

Source: Mare E.G., Hendriekx, Dietrieh Knorr: Ultra High Pressure Treatments of Foods: ISBN 978-1-4613-5211-2.

The pressure inactivation behavior of avocado PPO changed when the pH was lowered to 4, and a first-order model could no longer explain the pressure inactivation. In this instance, it appeared that the inactivation course could be described by a fractional conversion model (Figure 2.3). Two avocado PPO isozymes were found to have different pressure stability at this low pH, which may be the cause of the change in inactivation kinetics.

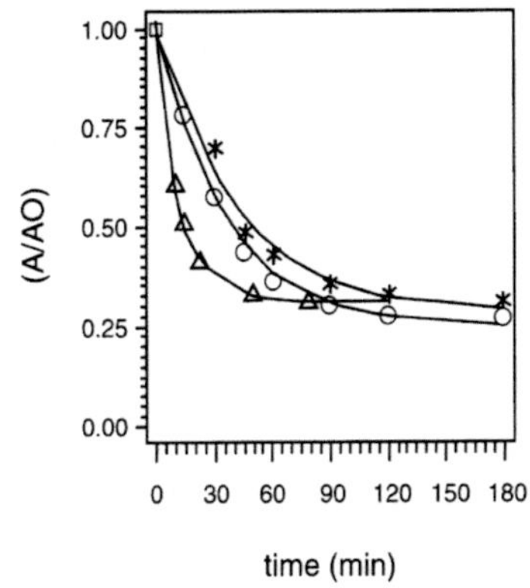

Figure 2.3. *Pressure inactivation of avocado PPO in McIllvaine buffer (PH 4) at 25°C and 550 (*), 600 (○), or 700 (Δ) MPa.*

Source: Mare E.G., Hendriekx, Dietrieh Knorr: Ultra High Pressure Treatments of Foods: ISBN 978-1-4613-5211-2.

The enzymes were impacted differently by the presence of antibrowning agents (benzoate, ethylenediaminetetraacetic acid [EDTA], glutathione, sodium chloride [NaCI], and 4-hexylresorcinol), despite the fact that the pH effect on avocado and mushroom PPO was fairly similar. When 4-hexylresorcinol was present, the pressure stability of mushroom PPO at pH 6.5 was significantly decreased; in contrast, it was either not affected at all or only marginally affected when EDT A, glutathione, NaCl, or benzoate were present. The addition of EDTA decreased pressure stability for avocado PPO at pH 5, and this effect was solely due to the pH drop caused by EDTA. However, a higher pressure was needed to cause inactivation after benzoate, NaCI, or glutathione were added, suggesting a higher pressure stability. The pressure affected 4-hexylresorcinol's effects. Pressure stability was found to be lower at pressures below 700 MPa and higher at pressures above 700 MPa. Lastly, a thorough investigation of the combined impact of temperature and pressure was conducted for avocado PPO dissolved in phosphate buffer (0.1M) at a pH of 7. First-order inactivation rate constants were used to characterize over 100 combinations of constant temperature (25°C to 77.5°C) and pressure (0.1 to 900 MPa) as a function of time.

At high temperatures (>62°C) and low pressures (<250 MPa), an antagonistic relationship between temperature and pressure was observed, suggesting that low pressure provides a buffer against thermal inactivation. The pressure-temperature kinetic diagram (Figure 2.4) for the inactivation of avocado PPO illustrates this antagonistic effect. It is a two-dimensional diagram that shows combinations of constant temperature and pressure that produce the same inactivation rate constant. The inactivation rate constant's temperature dependence (E_a) was measured at various constant pressures and was found to decrease with increasing pressure, going from 320 kJ/mol at 0.1 MPa to 60 kJ/mol at 900 MPa. However, for the whole pressure range under investigation, the Eyring equation was unable to adequately characterize pressure dependence. Different types of PPO exhibit different pressure-temperature behavior, according to a screening study on the combined pressure-temperature inactivation of plum, pear, and mushroom PPO. Pressure and temperature seemed to act antagonistically in the low temperature-high pressure domain in the case of pear PPO, a tendency that was not seen in the case of avocado PPO.

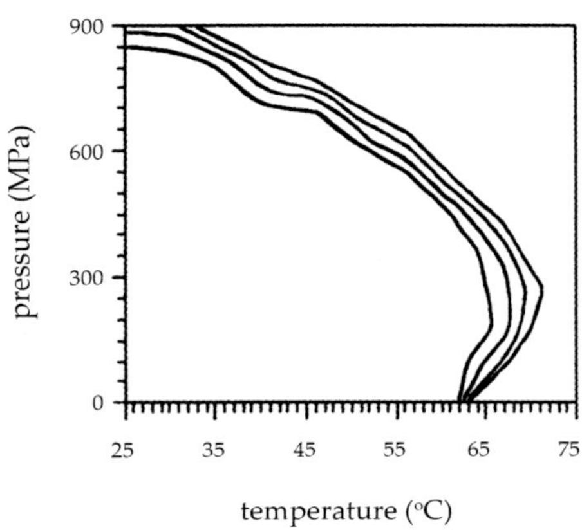

Figure 2.4. *Pressure-temperature kinetic diagram for inactivation of avocado PPO in phosphate buffer at pH 7 (lower line: k_{obs} = 0.005 min^{-1}; upper line: k_{obs} = 0.015 min^{-1}).*

Source: Mare E.G., Hendriekx, Dietrieh Knorr: Ultra High Pressure Treatments of Foods: ISBN 978-1-4613-5211-2.

2.3.2. Pectinmethylesterase

The demethylation of carboxylate groups on pectic substances is catalyzed by pectinmethylesterase (PME), which enables the substances to participate in calcium-mediated cross-linking reactions. PME causes orange juice to become unstable in the cloud, concentrates to gel, and tomato products to lose consistency, all of which are significant issues for the food industry. It has been reported that threshold pressures for PME inactivation from various sources range from 150 to 1200 MPa, contingent upon the source and the inactivation medium. Van den Broeck et al. (2000b) conducted a thorough kinetic analysis on the pressure activation and inactivation of PME from oranges, looking at the effects of environmental variables like pH, Ca^{++}, sucrose, and enzyme concentration. These authors investigated pressure inactivation under various conditions, ranging from 400 to 900 MPa. They found a pressure-resistant fraction of PME under every condition they looked at, which suggests that the kinetics adhere to a fractional conversion model (Figure 2.5). In general, orange PME's pressure stability increased in more concentrated solutions. On the other hand, enzyme concentration had no discernible effect on the inactivation rate constant in the case of a very low inactivation. Acidification significantly decreased orange PME's pressure stability. The threshold pressure at pH 3.7 (citric acid buffer) was 400 MPa instead of 600 MPa when compared to distilled water. Furthermore, there was a minor decrease in the inactivation rate constant's sensitivity to pressure. The concentration of Ca^{++} had a significant impact on the inactivation rate.

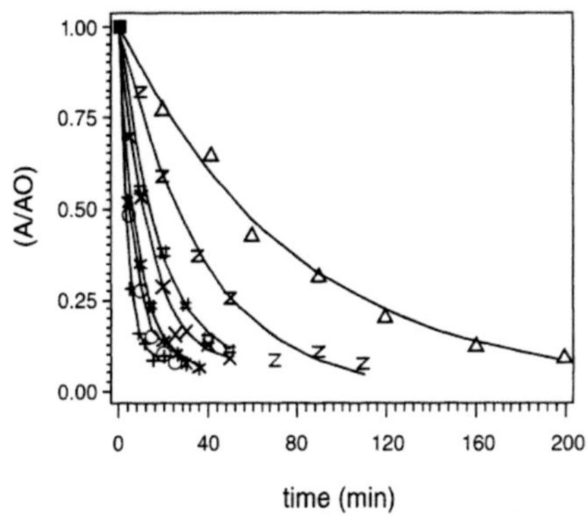

Figure 2.5. *Inactivation of commercial orange PME at different combinations of pressure (MPa) and temperature (°C): (#) 0.1/57; (O) 0.1/60; (Δ) 100/57; (z) 150/60; (x) 400/60; (+) 800/35; (*) 850/20.*

Source: Mare E. G. Hendriekx, Dietrieh Knorr: Ultra High Pressure Treatments of Foods: ISBN 978-1-4613-5211-2.

The inactivation rate constant rose at relatively low molarities (0.5 to 1.0 M); at higher molarity (1.5 M), a decrease in inactivation was observed. Pressure stability of PME and pressure sensitivity of the inactivation rate constants both increased in the presence of sucrose. An antagonistic effect at low pressure was revealed by a thorough kinetic study on the inactivation of orange PME by combined pressure (0.1 to 900 MPa) and temperature (15°e to 65°C). < 200 to 300 MPa) and high temperature (> 55°C) (*see* Figure 2:6).

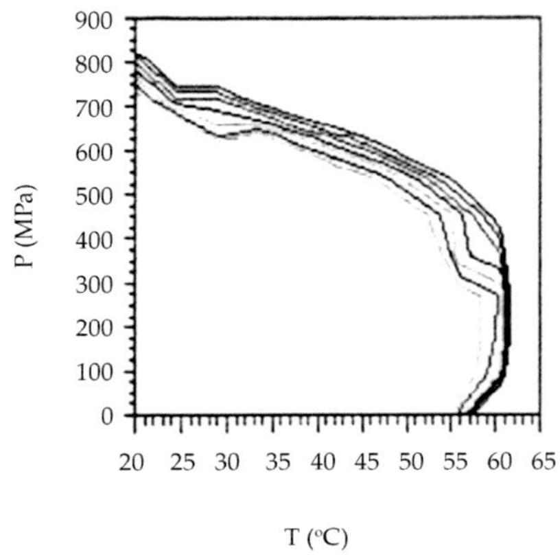

Figure 2.6. *Pressure-temperature kinetic diagram for commercial orange PME in deionized water (lower line: $k_{obs} = 0.02\ min^{-1}$; upper line: $k_{obs} = 0.08\ min^{-1}$).*

Source: Mare E. G. Hendriekx, Dietrieh Knorr: Ultra High Pressure Treatments of Foods: ISBN 978-1-4613-5211-2.

In contrast, tomato PME exhibited remarkable resistance to pressure, and a strong antagonistic relationship between pressure and elevated temperatures was observed. Tomato PME inactivation was totally inhibited at pressures between 100 and 500 MPa at 60°C (a temperature that causes thermal inactivation at atmospheric pressure). The inactivation rate was lower at 60°C and 900 MPa than it was at atmospheric pressure (Figure 2.7).

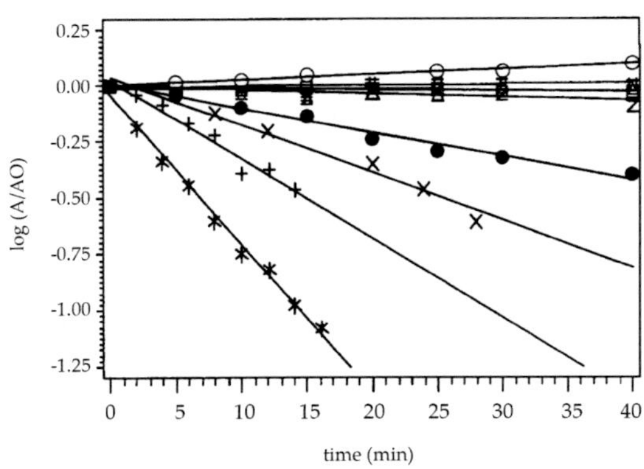

Figure 2.7. *First-order inactivation of commercial tomato PME in deionized water at 60°C and pressures of 0.1 (*), 100 (0), 200 (#), 300 (), 500 (Δ), 700 (z), 800 (●), 850(x), and 900 (+) MPa.*

Source: Mare E. G. Hendriekx, Dietrieh Knorr: Ultra High Pressure Treatments of Foods: ISBN 978-1-4613-5211-2.

The same researchers looked at the effects of high pressure on pure tomato PME in solution as well as high pressure in the presence of pectin, the PME's substrate. It was observed that under pressure, tomato PME continues to function. The ideal activity was found at 55°C and neutral pH at atmospheric pressure. The activity of the enzyme increased with pressure up to 300 MPa. The range of temperatures between 60°C and 65°C was found to have the maximum activity (Figure 2.8). By adding Ca^{++} ions (60 mM), enzyme activity could be increased even more, moving the parameters for maximal activity toward higher pressures (200–300 MPa) and temperatures (65–70°C).

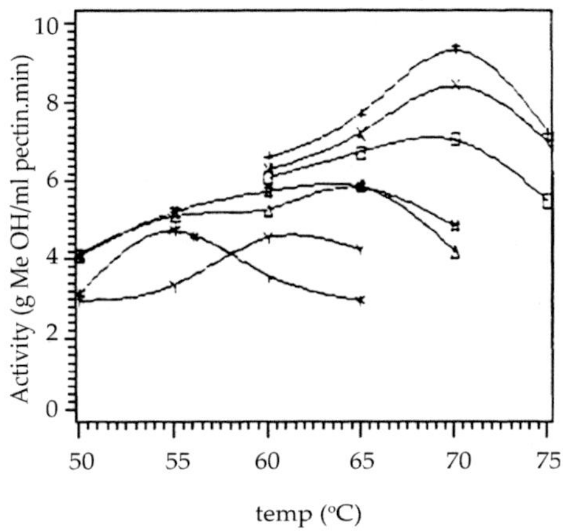

Figure 2.8. *Tomato PME activity as a function of temperature in absence [0.1 (*), 100 (Δ), 200 (#), and 300 (Y) MPa] and presence of 60 mM $CaCl_2$ [100 (), 200 (+), and 300 (x) MPa].*

Source: Mare E. G. Hendriekx, Dietrieh Knorr: Ultra High Pressure Treatments of Foods: ISBN 978-1-4613-5211-2.

2.4. EFFECTS OF HIGH PRESSURE ON CHEMICAL REACTIONS RELATED TO FOOD QUALITY

The food industry is interested in high pressure technology as a potential food processing/preservation method that has minimal impact on sensory and nutritional quality attributes, in response to the growing consumer demand for high-quality, nutritious food products. The majority of food quality characteristics can be preserved while pathogenic/spoiled microorganisms and food-spoiling enzymes are rendered inactive by high pressure. This benefit is explained by the fact that only nonvalent bonds—such as hydrogen, ionic, and hydrophobic bonds—are impacted by high pressure, which preserves covalent bonds. Texture, flavor, color, and nutritional value are all crucial aspects of fine food. The first three attributes establish the consumer's acceptance of the product and are associated with the quality of purchase and consumption. Calorie value (i.e., minerals, vitamins, and other nutrients) is an unnoticed characteristic. During preservation/processing treatments and subsequent storage, chemical or biochemical reactions that occur in food products can cause unfavorable changes or deterioration of these food quality attributes. Nonenzymatic and enzymatic browning, lipid hydrolysis or oxidation, protein denaturation, hydrolysis or cross-linking, polysaccharide hydrolysis or synthesis, and the breakdown of natural pigments are among the reactions that result in these unwanted alterations. A substantial amount of research has been conducted to examine the effects of different processing variables (e.g., temperature, time, heating rate, pH, and water activity) on the reactions that cause food products to deteriorate to plan processes that will both induce the necessary enzymatic and microbial inactivation and delay or inhibit particular deteriorative reactions. Understanding the effects of pressure is essential for the food industry to adopt high pressure technology as a substitute for traditional preservation and processing techniques. Since bacterial spores and certain enzymes are very pressure stable, the food industry's most likely course of action will be to combine high pressure processing with other processing techniques, particularly moderately elevated temperature (Crelier, 1995; Baranyi & Roberts, 2000).

2.4.1. The Effect of High Pressure on the Color of Food Products

2.4.1.1. Fruit and Vegetable Products

High pressure treatment keeps most fruits and vegetables looking mostly like they did when they were first produced. The color of pressure-processed orange juice was analyzed chemically and spectrophotometrically, and the results indicated that neither the pressure treatment nor the subsequent storage period significantly changed the color values. According to color parameters, fruit jam that had been pressure-treated was brighter (L-value) and redder (a-value) than jam that had been treated traditionally. Strawberries treated with pressure retained a significant amount of their red color. When compared to traditional hot break methods, pressure treatment improved the color of tomato juice (as measured by the a-value/b-value ratio); the highest alb ratio was noted at pH 4.5, regardless of the pressure used. The homogenizing and compacting effects were solely physical in nature, and the variations in the lycopene content of the pressure-treated and heat-treated samples were not the reason for this effect.

Guava puree under pressure (600 MPa, 25°C, 15 minutes) kept its original color after being freshly extracted. Lightness and greenness consistently decreased in all samples during a 60-day storage period at 4°C, particularly in the untreated puree. Samples of pressurized puree had far more stable color. The color of avocado puree treated at pressures ranging from 345 to 689 MPa (10 to 30 minutes) was the same as that of the puree when it was first prepared, (i.e., a shade of green having colorimetric a-values ranging from –7.9 to –7.3). But a few things changed while it was being stored. After 20 minutes of treatment at 689 MPa and storage at 5°C, samples at pH 4.1 showed less than 1% variation in lightness (L-value). Purees treated at higher pH or lower pressure, or stored at higher temperature, showed a greater decrease in lightness. More noticeable changes were seen for hue angle, net color difference, and a-value. Green's contribution to color gradually reduced during storage (i.e., due to severe browning brought on by lingering polyphenoloxidase (PPO) activity, the a-value turned positive. Lower storage temperature, high pressure leading to greater PPO inactivation, small initial pH, and longer acceptability storage time (time to reach the limits of acceptance) could all be used to achieve this. Banana puree produced comparable outcomes. The original color of the banana puree was maintained by high pressure treatment (689 MPa for 10 minutes); however, multiple color changes transpired during storage as a result of residual PPO activity, which caused the puree to enzyme brown.

Researchers have conducted extensive quantitative kinetic studies on the impact of high temperature and pressure on the amount of chlorophyll and green color in broccoli juice. When it comes to green color, one common transformation seen in processed green vegetables is the conversion of green chlorophyll to gray-brown pheophytin, a process that is strongly linked to quality loss and therefore unfavorable to consumers. Food processors therefore place a high value on reducing greenness loss and chlorophyll degradation. Extreme chlorophyll stability was noted when pressure-processed green vegetables at a low temperature. For other pigments like annatto, carotene, anthocyanins, and hibiscus extract, the same phenomenon has been observed. No decrease in the amount of chlorophyll a and b was observed, even following a 4-hour treatment at 800 MPa and 40°C. This observation can be explained by the fact that the covalent structure of chlorophyll is stable under high pressure due to the covalent bonds' minimal compressibility. According to a first-order model, a significant reduction in chlorophyll content is only noticeable when temperatures surpass 50°C and high pressure is present. Chlorophyll a is more temperature-sensitive and less pressure-stable than chlorophyll b, according to calculations of degradation rate constants and activation energies. Furthermore, a synergistic effect of temperature and pressure was observed for both chlorophylls separately as well as for the total chlorophyll content (Figure

2.9). On the other hand, a pressure increase of 100 MPa had only a slight impact on the rate of chlorophyll degradation, whereas a temperature increase of lODe significantly increased it. A related study was then conducted to measure the a-color value of broccoli juice post-treatment using spectrophotometry. Even after treatment at 800 MPa for 180 minutes, no discernible decrease in green color was seen at low temperatures (less than 40°C). A certain amount of greenness was lost when temperatures between 50°C and 60°C were combined with high pressure. At all pressures examined (0.1 to 850 MPa), greenness loss was observed at temperatures between 70°C and 80°C.

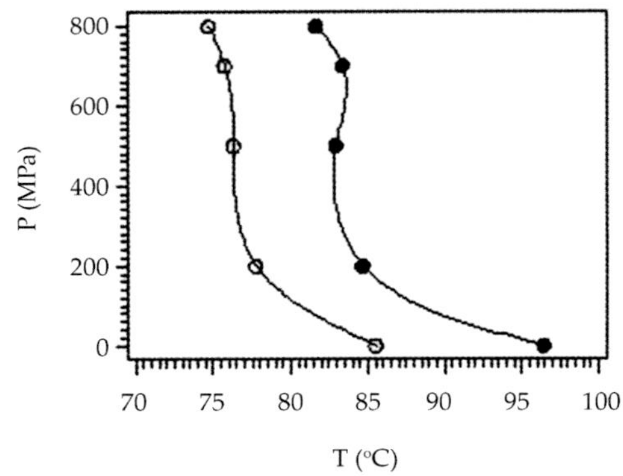

Figure 2.9. *Pressure-temperature combinations resulting in 90% loss of total chlorophyll content after process time of 50 minutes (●) and 100 minutes (○) at the selected pressure-temperature combination.*

Source: Mare E. G. Hendriekx, Dietrieh Knorr: Ultra High Pressure Treatments of Foods: ISBN 978-1-4613-5211-2.

Two successive first-order degradation steps could explain this loss, suggesting that both pyropheophytin and pheophytin formation took place. Temperature and pressure responded antagonistically to the second degradation step and cooperatively to the first at pressures lower than 150 MPa. Therefore, it is advised to apply high pressure treatment at temperatures below 50°C to reduce the amount of chlorophyll degradation and the ensuing loss of greenness in broccoli juice during processing.

Conversely, it has been discovered that high pressure causes onions and mushrooms to become discolored due to the action of the browning-causing enzyme PPO. PPO is a soluble enzyme that is mostly found attached to particulate cell fractions or in the cytosol of plant cells. Phenolic compounds are confined to vacuoles and are spatially separated from the enzyme in intact cells. Brown complexes are created when the cell and tonoplast are damaged, bringing the substrate and enzyme together and producing phenolic oxidation products. Microscopic analysis showed that onion epidermis cells treated at 300 MPa and 25°C had significant damage to their vacuoles. Enzymatic browning can occur as a result of PPO's ability to catalyze the oxidation of the released phenolic compounds at this low pressure because it is highly pressure-resistant.

2.4.1.2. Milk and Dairy Products

The color of milk is not significantly impacted by pressure. Researchers have found that while a- and b-values stayed constant, L-values consistently changed, suggesting a gradual change in brightness. A first-order model could be used to explain color degradation overall, showing that more significant color changes were produced by higher pressure and longer treatment times. However, only slight color changes were seen at pressures high enough to bring the microbial load down to a manageable level. The L-value, or lightness, of milk dropped from 85.6 to 67.0 after an hour of pressurization at 300 MPa. Higher pressure did not significantly accelerate this decline. This increased translucency indicates that pressure treatment has changed the micelle structures and caused them to break up into smaller pieces. However, the increase in translucence is not as strong as when calcium-chelating agents cause micelle disintegration. Because pressurized samples have less light-scattering ability than unpressurized ones, the mean values for lightness were lower in both cases. Pressurization reduced the a-values (greener) and b-values (bluer) in unconcentrated milk. Only in concentrated milk did pressure treatment result in a greener hue.

2.4.1.3. Meat

Since high pressure processing may offer potential to preserve and restructure meat—so long as the red color is maintained—many authors have examined the effects of high pressure on meat and meat products. Meat's visual color intensity is derived from its total myoglobin content, while the presence of myoglobin in its ferrous oxygenated form determines the meat's stability. The varying hues of meat can be attributed to the relative amounts of metmyoglobin (gray-brown), oxymyoglobin (bright red), and myoglobin (dark red). The concentration of multivalent ions, partial oxygen pressure, oxygen consumption by the tissues, temperature, light, pH, microflora, and pressure are among the factors that accelerate the rate of myoglobin oxidation and, consequently, the discoloration of meat. In fact, applying pressure causes red meat to change dramatically in color even at chilled temperatures (5°C to 10°C). The color of pressure-processed meat samples became paler above 200 MPa and gray-brown above 400 MPa (decrease in a-value). Similar results were obtained at 20°C. Shigehisa et aI. (1991) reported that L-values of pork slurries started to increase between 100 and 200 MPa, reached a maximum between 300 and 400

MPa, and did not change with further increase in pressure up to 600 MPa. On the other hand, the decrease in a-value was moderate between 100 and 200 MPa and progressed until 600 MPa. Similar findings were noted with regard to the redness of beef meat.

Therefore, it appears that (1) a whitening effect (increase in L-value) in the pressure range of 200 to 350 MPa, possibly due to globin denaturation or heme displacement or release, and (2) a loss of red color (decrease in a-value) at or above 400 MPa, caused by oxidation of ferrous myoglobin into ferric myoglobin, are the main causes of meat discoloration caused by pressure processing. Critical pressure thresholds seem to have a greater influence on both degradation stages than processing time.

Untreated control samples developed a gray-brown color during storage as a result of partial oxymyoglobin conversion to metmyoglobin. L-, a-, and b-values for samples pressurized to 500 MPa did not alter over storage, indicating a stable gray-brown color. However, meat samples treated at 350 MPa were pink when first processed, and as they were stored, they gradually lost their reddish color and turned a gray-brown color that was similar to samples treated at 500 MPa. When there is no oxygen present, the color red (i.e., the oxidation of myoglobin) did not occur, but it was still impossible to stop the meat from turning white. Furthermore, whitening, and color alterations were noted in meat emulsions with and without fat content. Increased lightness was brought on by pressurization in both situations, and this effect grew stronger with higher pressure and longer treatment times. Pressure treatment at 300 MPa (10–20 minutes) decreased the a- and b-values in low-fat emulsions. Similar outcomes were seen in high-fat emulsions treated for 10 minutes at pressures between 100 and 300 MPa; when pressurization time was increased to 20 minutes, the results were the opposite. Pressurization up to 150 MPa had no effect on the L-, a-, and b-color values of minced meat samples. Higher pressure did, however, once more cause less redness and more lightness.

Therefore, it can be said that pressure processing fresh red meat is a feasible option only if further cooking is done before the products are offered for sale or consumption (like with ready-to-eat meals). On the other hand, processing cured or white meats under high pressure is not likely to result in significant color issues.

2.4.1.4. Maillard Browning in Model Systems

The development of brown color in processed food products is a result of the Maillard reaction, which is known to be highly dependent on pH and temperature. Nevertheless, the effect of pressure has been the subject of relatively few studies. Tamoaka et al. (1991) reported the development of brown color to be suppressed when pressures of 200 to 400 MPa were applied to a xylose-lysine system at 50°C in sodium carbonate buffer at pH 8.2. Using a glucose-lysine system at pH 8 and 10.1, Hill et al. (1996) observed that pressure treatment at 600 MPa enhanced Maillard browning; however, at pH 5.1 and 6.5, application of pressure depressed the rate of browning. It was evident that the browning rate constants rose with increasing pH for both pressurized and unpressurized solutions. For every pH unit, the rate increased by about 40 times at atmospheric pressure and by about 1000 times at 600 MPa. Pressure is thought to speed up the Maillard browning process, but when the pH drops below 7, the pressure-induced pH drop cancels out the enhancement. Solutions under pressure will have a lower pH than they would be at room pressure, which will also result in a less browning reaction (Crelier, 1995; Baranyi & Roberts, 2000).

2.4.2. The Effects of High Pressure on the Flavor of Food Products

2.4.2.1. Fruit and Vegetable Products

The potential of high pressure as a new processing technology for the production of fruit juices partly arises from the fact that fresh flavor can be maintained during pressure treatment. Ogawa et al. (1990) and Takahashi et al. (1993) reported that Satsuma mandarin juice maintained freshness and original flavor during high pressure processing. Results of extensive sensory testing of orange, grapefruit, and apple juices allowed Mermelstein (1999) to conclude that tasters could not differentiate between the flavor of pressure-processed and unprocessed juice made from the same raw material.

Parish (1994) found that an expert panel of sensory analysts was unable to distinguish between orange juice that had been pressurized for 30 seconds at 500 MPa and juice that had not been pressurized. Furthermore, the pressure-treated juice stayed free from fermentation for five weeks, whereas the control juice began to exhibit signs of fermentation after two weeks at 6°C and turned into an organoleptically unacceptable state. Parish (1998) examined the effects of pressure in the 500–700 MPa range on Valencia orange juice in a more thorough investigation. Although some notable differences in flavor between pressurized and control juices were noted, sensory panels generally assessed the flavor of pressure-treated juices to be significantly closer to that of fresh/frozen control juices than those of thermally treated juices. Pressure-treated juices kept their flavor advantages over heat-treated ones during storage as well. When even very little heating (between 50 and 60°C) was added, the overall sensory quality suffered. According to Bignon (1996), sensory qualities of pressure-processed orange juice resemble those of freshly squeezed juice and remain so for more than 30 days. Thus, the retention of fresh flavor is a major advantage of cold, high pressure processing.

With fruit jams (e.g., strawberry jam), high pressure processing was found to significantly preserve fresh flavor over conventional thermal processing.

Pressure has a major impact on the flavor of tomato juice and onions, unlike orange juice. Pressure-treated onions smelled more like cooked or fried onions, according to sensory panels, and less strongly than fresh onions. Samples of tomato juice treated at various pressure, temperature, and time combinations turned out to be completely unpalatable due to a strong rancid flavor.

There are studies in the literature that provide comprehensive information about specific aroma compounds in addition to reports on how pressure affects the overall flavor of fruit and vegetable products. Donsi et al. (1996) conducted a gas chromatographic analysis of orange juice's distinct flavor constituents. The least stable aroma component, limonene, is significantly altered by heat and is present in pressure-treated orange juice at a level comparable to fresh juice. In addition, other components of aroma like terpinene, cymene, linalool, and myrcene were not significantly altered by pressure treatment. A pressure-treated fruit jam's head space gas chromatographic analysis revealed that the majority of the original flavor compounds (e.g., methyl butyric acid, trans-2-hexenol, linalool, ethyl butyrate, and so forth) were retained to a far higher extent than with conventional heat treatment. This outstanding quality was kept unchanged for two to three months at 5°C in storage. However, because of the dissolved oxygen and residual enzyme activity, storage at 25°C led to a quick decline in quality.

According to reports, pressure treatment of onions results in an increase in transpropenyldisulfide and 3,4-dimethylthiophene concentrations and a decrease in dipropylsulfide concentration, which is the compound that gives fresh onions their pungent flavor. This alteration in flavor makes the onions taste like braised or fried onions. Due to pressure-induced cell injury and decompartmentalization, 2-methyl-pent-2-enal, the primary aldol condensation product of propanal resulting from alliinase action, increased simultaneously. Pressure-processing tomato juice reduced the amount of hexanol and significantly increased the amounts of trans-2-hexenal and n-hexanal, the latter of which, at concentrations of roughly 1 to 2 mg/kg, gives fresh tomato flavor. Increased concentrations give off a rancid taste.

Several alcohols and aldehydes that were absent from fresh peaches were found in pressurized peaches; these were most likely created by enzyme-induced oxidation of unsaturated fatty acids. Benzaldehyde concentration in particular increased significantly. Prunasin is thought to release benzaldehyde enzymatically when fruit tissue is disrupted during pressurization. Emulsin was found to retain its enzyme activity even after pressure treatment, and the enzymatic reaction may be sped up by the adiabatic temperature increase. This benzaldehyde plays a major role in the pressure-treated peaches' flavor. Additionally, it was discovered that pressure treatment had no appreciable effect on the concentration of -y-decalactone, which also contributes to the distinctive peach flavor.

Yen and Lin (1999) studied the changes in volatile components of guava juice during pressure processing (25°C, 600 MPa, 15 minutes) and subsequent storage at 4°C and 25°C, using gas chromatography/mass spectrometry. Esters were the most volatile fraction in guava juice and alcohols were the second most volatile, with ethyl-l-hexanoate being the most important flavor component. When compared to fresh juice, volatile compounds were not significantly changed by pressure; however, during storage at 4°C and 25°C, a decrease was noted.

Total alcohols and esters as well as the contents of the individual components were relatively close to that of fresh juice after storage at 4°C for 30 days. Pressure-treated juice stored at 4°C maintained flavor stability better than untreated juice, probably because enzymes are partially inhibited. After storage for 60 days, alcohol concentration was markedly increased, caused not to post-processing contamination but by enzymatic activity. Significant changes were observed when the juice was stored at 25°C for 30 days. Methanol, ethanol, ethyl acetate, methyl-I-propionate, and 2-ethylfurfuran concentration increased while that of other compounds decreased.

2.4.2.2. Milk and Dairy Products

There hasn't been much research done on how pressure affects the flavor of milk and dairy products. The shelf-life of high pressure processed milk, at 400 to 500 MPa for 3 to 15 minutes, followed by refrigeration, was found to be similar to that of thermal pasteurization; however, the sensory qualities of the pressure-processed milk were not mentioned in this report. Drake et al. (1997) found that trained dairy panels' flavor ratings for pasteurized and pressurized milk cheeses were comparable.

Both kinds of cheeses were criticized for having flavor defects that were high in acidity and bitterness, which were linked to their consistently low salt content. Conversely, raw milk cheese received a much lower flavor score due to defects in its fermented and fruity flavor, which are brought on by the presence of unwanted bacteria in raw milk.

2.4.2.3. Meat

There isn't much information on how high pressure affects the flavor of meat, but pressurized beef has reportedly been found to taste sweeter than control meat. Suzuki et al. (1994) investigated how pressure affected the water-soluble substances that give food its taste and meaty flavor. Up to 300 MPa (5 minutes at 2°C), there appeared to be an increase in the amount of peptides and amino acids estimated as phenol-reagent positive materials with increasing pressure. Peptides and amino acids increased in the extracts of both untreated and pressurized muscles after the samples were kept at 2°C for 7 days. Pressurization decreased the amount of inosinic acid, a flavor enhancer. Soluble peptides were subjected to high-performance liquid chromatography (HPLC) analysis. The results showed a significant decrease in the peptide fraction from muscles pressurized at 300 to 400 MPa and no significant changes in any fraction from pressurized muscles up to 200 MPa. We conclude that the content of taste-related amino acids and peptides is altered like meat conditioning under pressure (up to 300 MPa). These changes could be explained by the release of proteolytic enzymes from lysosomes and pressure-induced denaturation of muscle proteins (Crelier, 1995; Baranyi & Roberts, 2000).

2.4.3. The Effects of High Pressure on Texture of Food Products

2.4.3.1. Fruit and Vegetable Products

It's common knowledge that high pressure treatment makes fruits and vegetables softer and more malleable. On the other hand, plant systems can withstand pressures of up to 350 MPa without experiencing significant changes in texture or structure. Kasai and associates discovered that the amount and duration of the applied pressure significantly increased the relative hardness (the ratio of the hardness of pressure-treated vegetables to untreated vegetables). After 30 minutes of cooking, the relative hardness of the vegetables treated at 400 MPa for 120 minutes ranged from 2 to 3. Pressure-induced tissue softening for potato cubes was similar to that caused by hot water blanching. They found that the pressurization time and pressure level had an impact on the altered firmness of the treated samples. The softening curves generally showed that texture changes brought on by pressure happened in two stages: (1) an abrupt loss due to the pressure's pulse action, and (2) additional loss or a slow recovery during the pressure-holding phase. Figures 2.10 and 2.11 show that the sensitivity ranking depended on pressure intensity, while the instantaneous pressure softening was clearly dependent on the product type and pressure level.

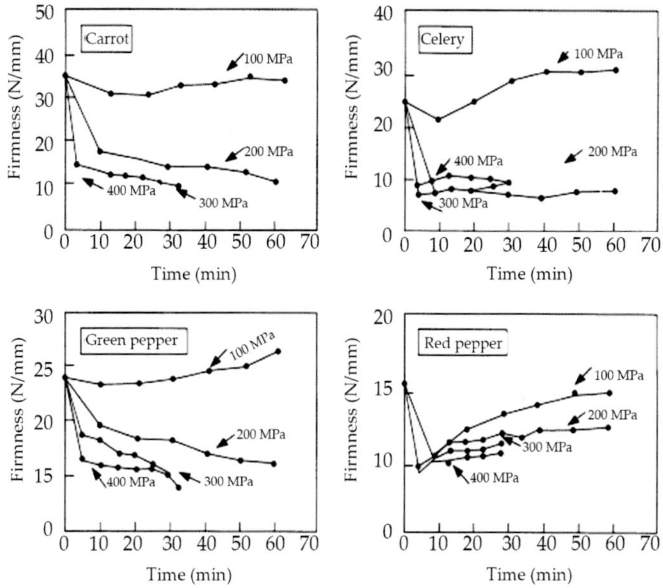

Figure 2.10. *Effect of pressure level and treatment time on the firmness of fruits.*

Source: Mare E. G. Hendriekx, Dietrieh Knorr: Ultra High Pressure Treatments of Foods: ISBN 978-1-4613-5211-2.

Pressure sensitivity was highest for pears at 100 MPa and highest for celery at pressures over 200 MPa. A small amount of firmness was lost at low pressure (100 MPa) due to the compaction of cellular structures without any disruption. Severe texture loss happened at pressures higher than 200 MPa due to cellular membrane rupture and the ensuing drop in turgor pressure. The entire texture of processed products was obviously impacted by the pressure holding time. After being held for 30 to 60 minutes at 100 to 200 MPa, the majority of the instantaneous texture loss was gradually restored in many fruits and vegetables, and some even became firmer than when they were fresh. However, at 200 MPa, additional texture loss was noted for green pepper and carrots.

Pectinmethylesterase (PME) is responsible for the fact that pressure-treated vegetables frequently did not soften when cooked further. Pressure treatment causes structural disruption and partial inactivation of PME, which brings the enzyme and substrate into contact. Thus, PME-initiated deesterification of cell wall pectin, which results in cross-linking with divalent ions, is primarily responsible for the firming effects. There may also be a role for increased cellular structure compactness after tissue degassing. Products may sustain severe damage and irreversible texture loss at higher pressures.

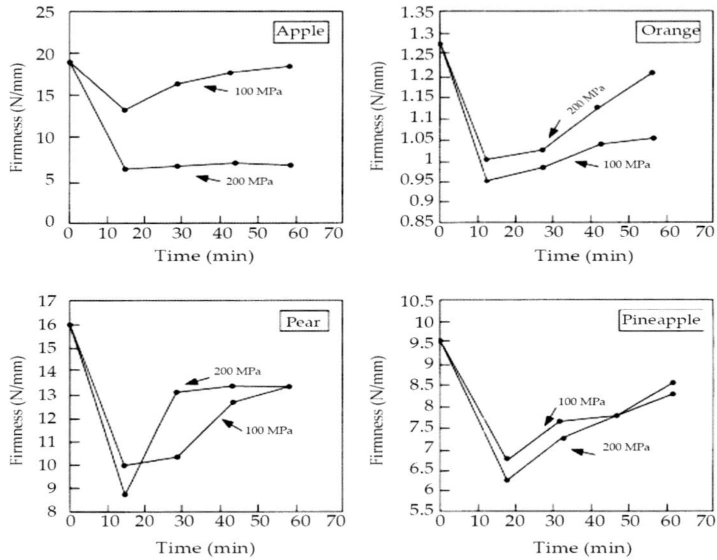

Figure 2.11. *Effect of pressure level and treatment time on the firmness of vegetables.*

Source: Mare E. G. Hendriekx, Dietrieh Knorr: Ultra High Pressure Treatments of Foods: ISBN 978-1-4613-5211-2.

Consistency of tomato products refers to their viscosity and the ability of their solid portion to remain in suspension throughout the shelf life of the product. A high-consistency product has almost no syneresis (no separation of juice pulp and serum). Poretta et al. (1995) reported that high pressure treatment clearly affected the viscosity of tomato juice and gave rise to a jelly-like, translucent structure caused by protein-tissue coagulation and compacting. For pressures above 700 MPa, viscosity increased with increasing pH, indicating that certain pH ranges enhance enzyme activity and result in viscosity changes. For guava puree, turbidity, viscosity, and cloud content were not significantly affected by pressure treatment (400 to 600 MPa). However, some decrease in turbidity, viscosity, and cloud content was observed during storage for 20 days.

2.4.3.2. Milk and Dairy Products

The viscosity of milk was only marginally affected by high pressure treatment. A first-order reaction can be used to explain how milk viscosity builds up. The small negative activation volume suggests that at higher pressure levels, the rate at which milk viscosity builds up accelerates. The disintegration of casein micelles has been attributed to an increase in viscosity. Since milk's viscosity is far less sensitive to pressure changes than the microorganisms present in it, the viscosity increase may not be a significant disadvantage of pressure processing milk. Compared to raw or pasteurized milk cheeses, pressurized milk cheeses received substantially lower texture ratings. They were assessed to have a weak, pasty texture and a flavor tainted by whey. The higher moisture content and wet weight yield of pressurized cheeses are responsible for these flaws. However, this increased moisture content could be advantageous when making cheese with less fat.

2.4.3.3. Meat

Raw meat is produced that is extremely firm and contracted when different muscles are pressed. However, due to reduced contraction and less drip loss, pressurized meat has lower shear values, a higher moisture content, and is more tender after cooking than nonpressurized meat. According to sensory analysis, pressure-treated meat is more tender but less juicy than control meat. The sarcomere structure is damaged when tenderization rather than toughening takes place. According to numerous authors, tenderization of pre-rigor meat can occur when it is briefly exposed to pressures between 100 and 200 MPa. Even though post-rigor application of pressure is far more important from a commercial standpoint, it has no positive effect on beef tenderness when done below 30°C. Furthermore, cooking meat afterward does not improve tenderness, juiciness, or shear strength. On the other hand, toughening brought on by cold-shortening can be effectively countered by applying pressure (150 MPa) along with heat (55°C to 60°C). The reason for this tenderization effect was thought to be due to changes in myofibrillar structure rather than connective tissue. Without the need for additional heating, meat could be tenderized by applying pressure up to 500 MPa, but this changed the meat's color and appearance. In conclusion, pressure does appear to have some potential for tenderizing meat, particularly for white and cured meats or ready-to-eat foods where the color defects caused by pressure are minimal (Crelier, 1995; Baranyi & Roberts, 2000).

2.4.4. The Effects of High Pressure on Nutritive Value and Health Components of Food Products

2.4.4.1. Fruit and Vegetable Products

Pressure treatment has no effect on the levels of vitamin C, A, B_1, B_2, E, or folic acid. Pressure-processed strawberry jam retained approximately 95% of the vitamin C content of fresh strawberries, which is approximately 1/7 of what was found in heat-processed jam. After 30 minutes at 20°C and 400 MPa of pressure treatment, the ascorbic acid content of strawberry coulis/nectar was preserved. Furthermore, pressure did not accelerate or slow down the kinetics of degradation during storage, as evidenced by the similar profiles of decreased nutritional value observed in pressurized and untreated strawberry coulis. Pressure (400 to 600 MPa for 15 minutes) or heat treatment (88°C to 90°C for 24 seconds) had no effect on the guava puree's initial ascorbic acid content. Whereas the ascorbic acid content in heated and 600-MPa treated puree stayed constant during 30 and 40 days of storage, it started to decrease in untreated and 400-MPa treated puree after 10 and 20 days, respectively. For freshly squeezed orange juice and mandarin orange juice, high pressure (up to 600 MPa) did not change the initial vitamin C content. Donsi et al. (1996) studied the effect of pressure (200 to 500 MPa) on several vitamins (vitamin C, B_6, B_2, B_1, and niacin), sugars (sucrose, fructose, and glucose), and organic acids (malic acid, citric acid, and isocitric acid) in orange juices and found no substantial modifications at any of the pressure levels tested.

Van den Broeck et al. (1998) conducted a more thorough kinetic analysis on the pressure-temperature stability of ascorbic acid in buffer systems and natural environments (orange and tomato juice). Ascorbic acid degraded according to first-order kinetics at temperatures between 65°C and 80°C when combined with a pressure of 850 MPa; this process advanced more quickly at higher temperatures, suggesting that pressure and temperature work in concert.

Clearly, orange juice's ascorbic acid was more sensitive to pressure than tomato juice's. Furthermore, these authors observed that the pressure level had no bearing on the ascorbic acid degradation's thermal sensitivity. These tests showed that ascorbic acid is only unstable when extremely high pressure and high temperature are present. Thus, the authors concluded that ascorbic acid is unaffected by mild treatments, which are significant for the food industry.

There is currently a lot of interest in researching the connections between dietary practices and human cancer incidence. Fruit and vegetable-rich diets have been linked to a lower incidence of many types of cancer in humans. Nonetheless, heat treatment significantly diminishes the antimutagenic properties of the majority of fruits and vegetables. Consequently, Butz et al. (1997) a thorough investigation into the impact of pressure on fruits' and vegetables' antimutagenic properties was conducted. Even when subjected to heat or pressure, strawberries, and grapefruits demonstrated a moderate level of antimutagenic activity. Strong antimutagenic properties that were heat-sensitive but not pressure-sensitive were found in carrots, cauliflower, kohlrabi, leeks, and spinach. Pressure only had an impact on beets' and tomatoes' antimutagenic properties under extremely harsh circumstances (600 MPa at 50°C or 800 MPa at 35°C). Vegetable juices that have undergone pressure processing generally lose much of their antimutagenicity compared to those that have undergone thermal processing.

2.4.4.2. Milk and Dairy Products

One food allergen found in milk is β-lactoglobulin. The objective of adding milk concentrates to modified milk for infants is to remove f)-lactoglobulin without reducing the amount of ex-lactalbumin. β-lactoglobulin can be selectively removed from whey concentrates through thermolysin hydrolysis under high pressure. Because α-Lactalbumin contains four disulfide bonds, it is resistant to hydrolysis. High pressure (400 to 1000 MPa) did not alter the vitamin C content of egg yolks. When thermolysin digestion is carried out under pressure, β-lactoglobulin is hydrolyzed much faster and more completely than at atmospheric pressure, while exlactalbumin is unaffected. Ascorbate concentrations in egg yolks were mostly unaffected when they were pressurized for 30 minutes at 20°C and 200, 400, or 600 MPa (92.6%, 101.3%, and 102.7%, respectively). Similar results regarding the thiamin content of egg yolks have been reported. Butz et aI. (1997b) studied the effect of high pressure on aspartame in milk, using aspartame solutions of about 0.5 giL, corresponding to the concentration in commercial diet chocolate milk. A treatment of 3 minutes at 60°C and 600 MPa induced 50% loss of active substance, while the non-sweet aspartylphenylalanine and diketopiperazine formed; the molar sum of these compounds is the symmetrical reflection of the aspartame degradation curve (Figure 2.12).

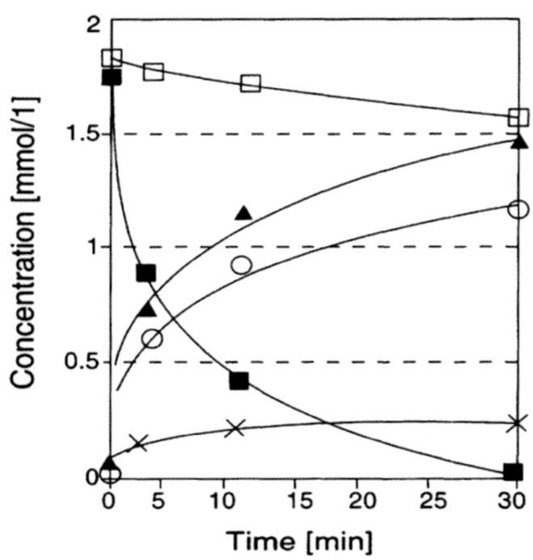

Figure 2.12. *Pressure treatment of aspartame solutions of 0.5 gIL in full-cream milk at 600 MPa and 60°C: (▲) aspartame; () aspartame control at 60°C; (■) molar sum of diketopiperazine and aspartylphenylalanine; (O) diketopiperazine; (X) aspartylphenylalanine.*

Source: Mare E. G. Hendriekx, Dietrieh Knorr: Ultra High Pressure Treatments of Foods: ISBN 978-1-4613-5211-2.

Smaller effects were seen in the buffer solution, suggesting that the milk matrix had another effect. Because aspartame is stabilized and water undergoes self-ionization, the effects in an aqueous solution were still less pronounced. Therefore, high pressure treatment may cause issues, particularly in dairy products with a neutral pH that contain aspartame (like chocolate milk and ice cream). After five minutes of pressure treatment at 600 MPa, 60°C, and pH 7, 1.15 mmol of diketopiperazine might be detected. Thus, the maximum amount of diketopiperazine would be consumed by a 40 kg human consuming 1 L of pressure-treated chocolate milk. Higher initial concentrations would compensate for pressure-related aspartame loss, leading to even higher final product concentrations of diketopiperazine.

2.4.4.3. Meat

Information on the effect of high pressure on the nutritional value and health components of meat is very scarce. Sancho et al. (1999) found that thiamin in pork meat was not degraded by pressure treatment at 125 to 600 MPa for 10 minutes to 18 hours at 20°C. Although it is unlikely that partial denaturation of muscle proteins and structures could reduce protein digestibility or biological value, this should be confirmed experimentally. It would also be useful to evaluate whether the nutritional availability of vitamins and iron is modified after pressure treatment (Crelier, 1995; Baranyi & Roberts, 2000).

REFERENCES

1. Baranyi, J., & Roberts, T. A. (2000). Principles and application of predictive modeling of the effects of preservative factors on microorganisms. In B. M. Lund, A. C. Baird-Parker, & G. W. Gould (Eds.), *The Microbiological Safety and Quality of Food* (pp. 342–358). Gaithersburg, MD: Aspen.
2. Barbosa-Cánovas, G. V., Pothakamury, U. R., & Swanson, B. G. (1995). State of the art technologies for the sterilization of foods by nonthermal processes: Physical methods. In G. V. Barbosa-Cánovas & J. Welti-Chanes (Eds.), *Food Preservation by Moisture Control: Fundamentals and Applications* (pp. 493–532). Lancaster, PA: Technomic.
3. Booth, I. R. (1998). Bacterial responses to osmotic stress: Diverse mechanisms to achieve a common goal. In D. S. Reid (Ed.), *The Properties of Water in Foods* (pp. 456–485). London: Blackie Academic & Professional.
4. Cano, M. P., Hernandez, A., & De Ancos, B. (1997). High pressure and temperature effects on enzyme inactivation in strawberry and orange products. *Journal of Food Science, 62*, 85–88.
5. Castellari, M., Matricardi, L., Arfelli, G., Rovere, P., & Amati, A. (1997). Effects of high pressure processing on polyphenoloxidase enzyme activity of grape musts. *Food Chemistry, 60*, 647–649.
6. Cheftel, J. C. (1992). Effects of high hydrostatic pressure on food constituents: An overview. In C. Balny, R. Hayashi, K. Heremans, & P. Masson (Eds.), *High Pressure and Biotechnology* (Vol. 224, pp. 195–209). Montrouge: John Libbey Eurotext.
7. Christian, I. H. B. (2000). Drying and reduction in water activity. In B. M. Lund, A. C. Baird-Parker, & G. W. Gould (Eds.), *The Microbiological Safety and Quality of Food* (pp. 146–174). Gaithersburg, MD: Aspen.
8. Clouston, I. G., & Wills, P. A. (1969). Initiation of germination and inactivation of *Bacillus pumilus* spores by hydrostatic pressure. *Journal of Bacteriology, 97*, 684–690.
9. Clouston, I. G., & Wills, P. A. (1970). Kinetics of germination and inactivation of *Bacillus pumilus* spores by hydrostatic pressure. *Journal of Bacteriology, 103*, 140–143.
10. Crelier, S., Robert, M. C., & Juillerat, M. A. (1998). Effect of a high pressure treatment on the texture and enzyme activities of selected vegetables. Poster presentation at the *International Conference on High Pressure Bioscience and Biotechnology*, Heidelberg, Germany.
11. Crelier, S., Tüche, M. C., Raemy, A., Renken, A., & Raetz, E. (1995). High

pressure for the inactivation of enzymes in food products: Thermal and HHP treatment of tomato pectin methyl esterase. Poster presentation at the *International Symposium on High Pressure Effects on Foods*, Budapest, Hungary.

12. Damodaran, S. (1988). Amino acids, peptides, and proteins. In O. R. Fennema (Ed.), *Food Chemistry* (3rd ed., pp. 321–430). New York: Marcel Dekker.
13. Denys, S., Ludikhuyze, L., Van Loey, A., & Hendrickx, M. (2000a). Modelling conductive heat transfer and process uniformity during batch high-pressure processing of foods. *Biotechnology Progress, 16*(1), 92–101.
14. Denys, S., Van Loey, A., & Hendrickx, M. E. (2000b). A modeling approach for evaluating process uniformity during batch high hydrostatic pressure processing: Combination of a numerical heat transfer model and enzyme inactivation kinetics. *Innovative Food Science and Emerging Technologies, 1*(1), 5–19.
15. Dunford, N. T., & Temelli, F. (1996). Effect of supercritical CO_2 on myrosinase activity and glucosinolate degradation of canola. *Journal of Agricultural and Food Chemistry, 44*, 2372–2376.
16. Ekstrand, B. (1994). Lactoperoxidase and lactoferrin. In Y. M. Dillon & R. G. Board (Eds.), *Natural Antimicrobial Systems and Food Preservation* (pp. 15–63). Wallingford, Oxon: CAB International.
17. Eshtiaghi, M. N., & Knorr, D. (1993). Potato cubes' response to water blanching and high hydrostatic pressure. *Journal of Food Science, 58*, 1371–1374.
18. Gatfield, L. L., & Sand, T. A. (1983). Coupled enzymatic procedure for the determination of myrosinase activity. *Lebensmittel-Wissenschaft & Technologie, 16*, 73–75.
19. Giovane, A., Quagliuolo, L., Castoldo, D., Servillo, L., & Balestrieri, C. (1990). Pectin methyl esterase from *Actinidia chinensis* fruits. *Phytochemistry, 29*, 2821–2823.
20. Goodner, J. K., Braddock, R. J., & Parish, M. E. (1998). Inactivation of pectinesterase in orange and grapefruit juices by high pressure. *Journal of Agricultural and Food Chemistry, 46*, 1997–2000.
21. Gould, G. W. (1995). Homeostatic mechanisms during food preservation by combined methods. In G. V. Barbosa-Canovas & J. Welti-Chanes (Eds.), *Food Preservation by Moisture Control: Fundamentals and Applications* (pp. 397–410). Lancaster, PA: Technomic.

22. Gross, M., & Jaenicke, R. (1994). Proteins under pressure: The influence of high hydrostatic pressure on structure, function, and assembly of proteins and protein complexes. *European Journal of Biochemistry, 221*, 617–630.
23. Hauben, K. J. A., Wuytack, E. Y., Soontjens, C. C. F., & Michiels, C. W. (1996). High-pressure transient sensitization of *Escherichia coli* to nisin and lysozyme by disruption of outer-membrane permeability. *Journal of Food Protection, 59*, 350–355.
24. Heinisch, O., Kowalski, E., Goossens, K., Frank, J., Heremans, K., Ludwig, H., & Tauscher, B. (1995). Pressure effects on the stability of lipoxygenase: Fourier Transform Infrared spectroscopy and enzyme activity study. *Zeitschrift für Lebensmitteluntersuchung und –Forschung, 201*, 562–565.
25. Herbert, R. A., & Sutherland, J. P. (2000). Chill storage. In B. M. Lund, A. C. Baird-Parker, & G. W. Gould (Eds.), *The Microbiological Safety and Quality of Food* (pp. 101–121). Gaithersburg, MD: Aspen.
26. Heremans, K. (1993). The behavior of proteins under pressure. In R. Winter & J. Jonas (Eds.), *High Pressure Chemistry, Biochemistry, and Material Science* (pp. 443–469). Dordrecht: Kluwer Academic.
27. Hite, B. H., Giddings, N. J., & Weakley, C. W. (1914). The effect of pressure on certain microorganisms encountered in the preservation of fruits and vegetables. *Bulletin of the West Virginia University Agricultural Experiment Station, 146*, 1–67.
28. Hoover, D. G. (2000). Microorganisms and their products in the preservation of foods. In B. M. Lund, A. C. Baird-Parker, & G. W. Gould (Eds.), *The Microbiological Safety and Quality of Food* (pp. 251–276). Gaithersburg, MD: Aspen.
29. Indrawati, I., Ludikhuyze, L. R., Van Loey, A. M., & Hendrickx, M. E. (2000). Lipoxygenase inactivation in green beans (*Phaseolus vulgaris* L.) due to high pressure treatment at subzero and elevated temperatures. *Journal of Agricultural and Food Chemistry, 48*(5), 1850–1859.
30. Indrawati, I., Van Loey, A., Ludikhuyze, L., & Hendrickx, M. (1999). Soybean lipoxygenase inactivation by pressure at subzero and elevated temperature. *Journal of Agricultural and Food Chemistry, 47*, 2468–2474.
31. Indrawati, I., Van Loey, A., Ludikhuyze, L., & Hendrickx, M. Lipoxygenase inactivation in green peas (*Pisum sativum*) due to high pressure treatment at subzero and elevated temperatures. *Journal of Food Science*. In press.
32. Irwe, S., & Olsson, I. (1994). Reduction of pectinesterase activity in orange juice by high pressure treatment. In R. P. Singh & F. A. R. Oliveira (Eds.), *Minimal Processing of Foods and Process Optimization–An Interface* (pp.

35–42). Boca Raton: CRC Press.

33. Kalchayanand, N., Sikes, T., Dunne, C. P., & Ray, B. (1994). Hydrostatic pressure and electroporation have increased bactericidal efficiency in combination with bacteriocins. *Applied and Environmental Microbiology, 60*, 4174–4177.
34. Kermasha, S., Bisakowski, B., Ramaswamy, H., & Van de Voort, F. R. (1993). Thermal and microwave inactivation of soybean lipoxygenase. *Lebensmittel-Wissenschaft und Technologie, 26*, 215–219.
35. Lambrecht, H. S. (1995). Sulfite substitutes for the prevention of enzymatic browning in foods. In C. Y. Lee & J. R. Whitaker (Eds.), *Enzymatic Browning and its Prevention* (pp. 313–323). Washington, DC: American Chemical Society.
36. Leistner, L. (1995). Use of hurdle technology in food: Recent advances. In G. V. Barbosa-Canovas & J. Welti-Chanes (Eds.), *Food Preservation by Moisture Control: Fundamentals and Applications* (pp. 377–396). Lancaster, PA: Technomic.
37. Lencki, R. W., Arul, J., & Neufeld, R. J. (1992). Effect of subunit dissociation, denaturation, aggregation, coagulation, and decomposition on enzyme inactivation kinetics: 1. First-order behavior. *Biotechnology and Bioengineering, 40*, 1421–1426.
38. Ludikhuyze, L. R., Claeys, W., & Hendrickx, M. (2000). Combined pressure-temperature inactivation of alkaline phosphatase in bovine milk: A kinetic study. *Journal of Food Science, 65*, 155–160.
39. Ludikhuyze, L. R., Van den Broeck, I., Weemaes, C. A., Herremans, C. H., Van Impe, J. F., Hendrickx, M. E., & Tobback, P. P. (1997). Kinetics for isobaric-isothermal inactivation of *Bacillus subtilis* α-amylase. *Biotechnology Progress, 13*, 532–538.
40. Ludikhuyze, L., Indrawati, I., Van den Broeck, I., Weemaes, C., & Hendrickx, M. (1998). Effect of combined pressure and temperature on soybean lipoxygenase: I. Influence of extrinsic and intrinsic factors on isobaric-isothermal inactivation kinetics. *Journal of Agricultural and Food Chemistry, 46*, 4074–4080.
41. Ludikhuyze, L., Indrawati, I., Van den Broeck, I., Weemaes, C., & Hendrickx, M. (1998). Effect of combined pressure and temperature on soybean lipoxygenase: II. Modeling inactivation kinetics under static and dynamic conditions. *Journal of Agricultural and Food Chemistry, 46*, 4081–4086.
42. Ludikhuyze, L., Ooms, V., Weemaes, C., & Hendrickx, M. (1999). Kinetic study of the irreversible thermal and pressure inactivation of myrosinase

from broccoli (*Brassica oleracea* L. cv. Italica). *Journal of Agricultural and Food Chemistry, 47*, 1794–1800.

43. Ludikhuyze, L., Rodrigo, I., & Hendrickx, M. (2000). The activity of myrosinase from broccoli (*Brassica oleracea* L. cv. Italica): Influence of intrinsic and extrinsic factors. *Journal of Food Protection, 63*(3), 400–403.
44. Matser, A. M., Knorr, E. R., & Bartels, P. Y. (1998). Effect of high isostatic pressure on polyphenoloxidase and colour of mushrooms. Poster presentation at the International Conference on High Pressure Bioscience and Biotechnology, Aug. 30–Sep. 3, Heidelberg, Germany.
45. Mertens, B. (1995). Hydrostatic pressure treatment of food: Equipment and processing. In G. W. Gould (Ed.), *New Methods of Food Preservation* (pp. 135–158). Glasgow: Blackie Academic & Professional.
46. Miyagawa, K., Sannoe, K., & Suzuki, K. (1964). Studies of Taka amylase A under high pressure: II. Recovery of enzymic activity of pressure inactivated Taka amylase A and its enhancement by retreatment at moderate pressure. *Archives of Biochemistry and Biophysics, 106*, 467–474.

Chapter 3

Metabolic Effects and Weight Management

Contents

INTRODUCTION

Ultra-processed foods (UPFs) have become a significant component of modern diets, characterized by their high content of added sugars, fats, preservatives, and artificial ingredients. These foods, which include items such as sugary beverages, snacks, fast foods, and pre-packaged meals, undergo extensive industrial processing that strips them of essential nutrients while enhancing their palatability, shelf-life, and convenience. The rising consumption of UPFs has been linked to various adverse health outcomes, particularly concerning metabolism and weight management.

Metabolically, UPFs can disrupt normal bodily functions. Their high glycemic index leads to rapid spikes in blood sugar levels, followed by sharp declines, which can promote insulin resistance over time. This insulin resistance is a precursor to type 2 diabetes and other metabolic disorders. Furthermore, the high caloric density and low satiety value of UPFs contribute to overeating and poor appetite regulation. The artificial ingredients and preservatives in these foods can also impact gut microbiota, potentially leading to dysbiosis, which is associated with metabolic disturbances and obesity.

From a weight management perspective, the consumption of UPFs is closely associated with increased body weight and a higher risk of obesity. Studies have shown that diets high in UPFs are typically more energy-dense but nutrient-poor, leading to an imbalance between calorie intake and expenditure. The convenience and addictive nature of these foods further exacerbate overeating, promoting a positive energy balance and weight gain. Additionally, the marketing and availability of UPFs make them more accessible and appealing, particularly in urban settings where lifestyle factors often prioritize convenience over nutrition.

The metabolic effects of ultra-processed foods significantly impact weight management and overall health. Understanding the mechanisms by which these foods influence metabolism and contribute to obesity is crucial for developing effective dietary guidelines and interventions aimed at reducing their consumption and promoting healthier eating habits.

3.1. ULTRAPROCESSED FOODS, WEIGHT GAIN, AND COMORBIDITY RISK

Any number of procedures used to change raw foods from their original state and prepare them for eating, cooking, or storage are considered food processing. This covers a variety of tasks like heating, freezing, cleaning, fermenting, grinding, and packing. Our ancestors have been using free to cook and heat food to maintain its nutritional value and organoleptic qualities since prehistoric times. The emphasis on food processing started to move away from home cooking and toward more industrialized processes during more recent historical events like the Industrial Revolution or the Second World War, with an emphasis on the safety, nutritional value, and preservation of food.

For centuries, food processing has been essential to the population's access to safe, palatable, and nourishing food. It helps to minimize losses and waste while extending food shelf life, improving nutrient availability, and improving food quality. Canning and pasteurization are two of the food processing technologies that have been around since the nineteenth century. Numerous physical, thermal, and chemical procedures, like the centrifugation and sterilization of dairy products and the bleaching of vegetable oils, came next. Food processing is a complicated topic with a variety of risks and benefits associated with each process type. For instance, heat processing lengthens raw milk's shelf life and reduces its pathogenic potential, but it also encourages the loss of some nutrients and the synthesis of chemicals that can cause cancer or mutagenesis in others. Therefore, there are advantages and disadvantages to various food processing techniques.

Many food products that are part of the so-called "Westernized diet" have been developed as a result of improvements in food processing and modifications to our agro-industrial systems. These food items that have been Westernized are typically highly processed, high in energy, and heavy in added sugar, saturated fat, and salt with minimal fiber content. Food classification systems are used to differentiate between various processed food categories in response to health concerns regarding the impact of industrial processing on diet quality and the risk of chronic diseases. The most widely used of those food classification schemes is the NOVA system (which is not an acronym), which coined the phrase "ultra-processed foods" to refer to foods that have undergone the most extreme processing. Ultra-processed foods (UPFs) are frequently sold and advertised in enticing ways, and they have a tendency to be very palatable, convenient, shelf-stable, and reasonably priced. Based on a PubMed search of title words, 72 articles from 2009 to 2016 and an additional 565 articles from 2017 to 2021 have been published on the topic since the term UPFs was first used. Numerous studies evaluating the relationship between UPFs and weight gain and/or co-morbidity risk have been conducted as a result of the increased focus on the health effects of UPFs. Recent research has also shown that during the shelter-in-place lockdowns that were put in place to stop the spread of the novel coronavirus (COVID-19), there was an increase in the consumption of UPFs. This was mostly caused by a rise in fast food and the intake of low-quality meals and snacks, including processed meat, chocolates, sweets, and beverages with added sugar. According to a recent review, only one study demonstrated an increase in the consumption of food quality (i.e., higher intake of fruits and vegetables) among participants while they were staying put.

This review's goal is to look at recent research because of the increased interest in and worries about the possible negative health effects of consuming UPFs; i.e., during the previous five years) on UPF use and the risk of co-morbidity and/or weight gain. We also go over the limitations of the NOVA classification system and possible mechanisms by which UPFs raise the risk of gaining weight and developing chronic diseases.

3.1.1. Defining Ultra-Processed Foods

The degree of food processing, as defined by food science and technology, depends on the type and quantity of operations used to improve shelf life, food safety, food quality, and the accessibility of edible portions of raw materials. Organizations such as the International Food Information Council (IFIC) and the International Agency for Research on Cancer—European Prospective Investigation into Cancer (IARC-EPIC) have provided several definitions regarding food processing. These classification schemes were generally created by researchers to examine the connections between industrial products and the risk of chronic diseases and/or nutritional intake. "Any raw agricultural commodity that has been subjected to washing, cleaning, milling, cutting, chopping, heating, pasteurizing, blanching, cooking, canning, freezing, drying, dehydrating, mixing, packaging, or other procedures that alter the food from its natural state" is what the United States Department of Agriculture (USDA) defines as processed food.

One of the most widely used systems for classifying foods and beverages in the public health literature is the NOVA system (Figure 3.1). According to one of the earliest systematic reviews on food processing that was printed in this publication, NOVA was the most precise, cohesive, understandable, all-encompassing, and practical definition. Food products are categorized into four groups according to the NOVA criteria: (1) minimally or fully processed foods; (2) processed culinary ingredients; (3) processed foods; and (4) ultra-processed foods (UPFs). The classification is based on the amount of processed ingredients in the food products. "Formulations mostly of cheap industrial sources of dietary energy and nutrients plus additives, using a series of processes" is how the UPF category is defined. Confectionery items, frozen pizzas, and reconstituted meats are a few examples. A group from the University of São Paulo first developed and coined the idea of UPFs in a 2009 commentary. The commentary's major contention was that the degree of food processing—rather than particular nutrients or food items—determines the relationship between nutrition and chronic illnesses. This work has been acknowledged in numerous prestigious reports, including those from the Food and Agriculture Organization (FAO) of the United Nations and the Pan American Health Organization (PAHO) of the World Health Organization (WHO), and it has now been formally adopted as part of Brazil's national dietary guidelines.

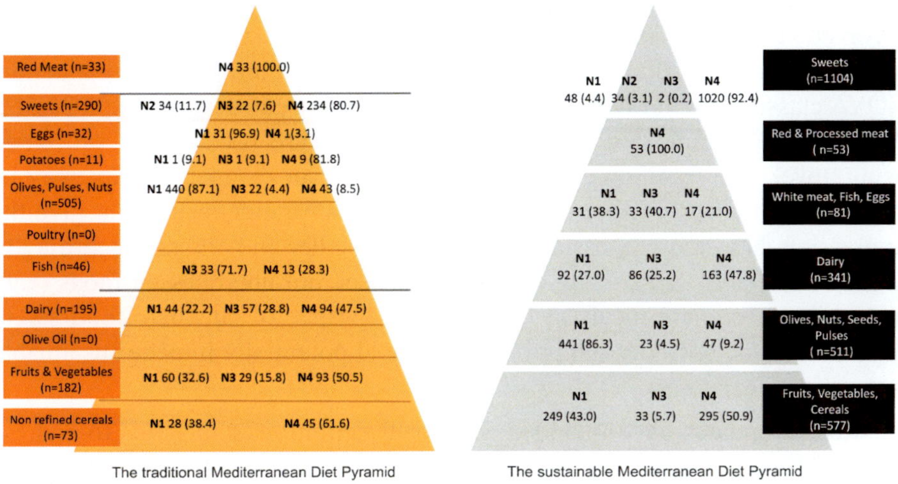

Figure 3.1. *Spectrum of processing of foods based on the NOVA classification.*

Source: Katidi A, Vlassopoulos A, Noutsos S, Kapsokefalou M. Ultra-Processed Foods in the Mediterranean Diet according to the NOVA Classification System; A Food Level Analysis of Branded Foods in Greece. Foods. *2023; 12(7):1520. https://doi.org/10.3390/foods12071520.*

It can be challenging for consumers to quickly identify UPFs because the food manufacturing industry is not required to disclose the procedures used in its products on food labels, and the information needed on food labels is not standardized across nations. Products such as unprocessed steel-cut oats, unflavored corn fakes, and shredded wheat, for instance, are categorized as minimally processed foods; however, the same foods are deemed processed if they additionally contain sugar, and Ultra-processed if they additionally contain colors or flavors. As a general guideline, one should identify food ingredients or additives whose main purpose is to increase the final product's palatability or appeal (i.e., "cosmetic adjuvants." This comprises, among other things, interesterifed oils, high-fructose corn syrup, and hydrolyzed proteins.

It is feasible to incorrectly classify some nutrient-rich foods as ultra-processed because classification schemes like NOVA primarily rely on classifying a processed food category based on the content of added sugars, saturated fat, and sodium. As in the case of Drewnowski et al. fortified ready-to-eat cereals, as well as beans and nuts (in the form listed in the FFQ), were classified as ultraprocessed from the NOVA definitions in's (2020) analysis of various foods using both the Nutrient Rich Food and NOVA criteria. Furthermore, when evaluating diets for scientific purposes (e.g., food frequency questionnaires, 24-h recalls), it is not typical to specify the level of processing involved in reported foods, and thus misclassification may also result here. For instance, commercially baked bread is categorized as ultra-processed under the NOVA classification system, but homemade bread is categorized as processed. Even though complex, non-traditional processes are used, foods processed using novel, non-traditional methods, like electric or magnetic fields, may be considered minimally processed. A few of the UPF guidelines might give the impression that food processing is a bad idea. Sadler and others. According to (2021), this might "encourage consumers to seek out unprocessed foods (e.g., raw milk) or prepare meals at home in the absence of adequate food safety regulations, and such consumer backlash could potentially impede sustainable innovations.

Recent updates have included additional definitions and terms to address some of the classification issues with the original NOVA criteria. By adding five additional, more focused subgroups to the initial four food processing categories, the Siga classification scheme for processed foods expands upon the NOVA classification system. The added sugar, fat, and salt contents; "at risk" additives; "matrix" effects; ultra-processed ingredients; and the quantity of markers of ultra-processing (MUPs) are all taken into consideration by this classification system. There haven't been many studies published using the Siga criteria, and the majority of the literature to date still uses the original NOVA criteria.

3.1.2. Ultra-Processed Food Consumption Levels

UPFs are becoming more prevalent in global food systems. Numerous articles about how UPF consumption affects daily total energy intake in various nations have been published. Selected recent articles on the topic are displayed in Table 3.1. High-income nations typically consume most of their calories from highly processed foods and drinks. For instance, it was estimated that UPF products contributed 45.0%, 50.4%, and 57.39% of total energy intake in the USA, the UK, and Canada, respectively. In other nations, such as Brazil, UPFs made up 22.7% of the total energy consumed. In other nations, such as Brazil, UPFs made up 22.7% of the total energy consumed. However, it should be noted that, depending on the sample, earlier research (not included in Table 3.1) has suggested that UPFs contributed anywhere from 21.5 to 51.2% of total energy intake. Two of the samples were limited to children, but the majority included both adults and children. Monteiro along with others. (2018) evaluated the availability of NOVA food groups in households across 19 European nations and examined the relationship between obesity prevalence and UPF availability. The utilization of population-based, real (non-modeled) estimates of the prevalence of obesity and standardized data were two of the study's strongest points. The prevalence of obesity increased by 0.25% points for every percentage point that UPF became more widely available in households after multiple confounding variables were taken into account.

In some of the countries, UPF consumption levels also seem to correspond with obesity rates. For instance, the highest rates of obesity and UPF consumption were found in the USA and the UK. There are disparities, though, with regard to other nations. Portugal has an obesity rate of 20.8%, which is comparable to other European nations, despite having a relatively low UPF consumption rate (10.2%) (Figure 3.2). Even though UPFs constitute a substantial portion of the energy intake in the diet, obesity rates are also influenced by other factors, and UPFs are only one food group among many that can be consumed.

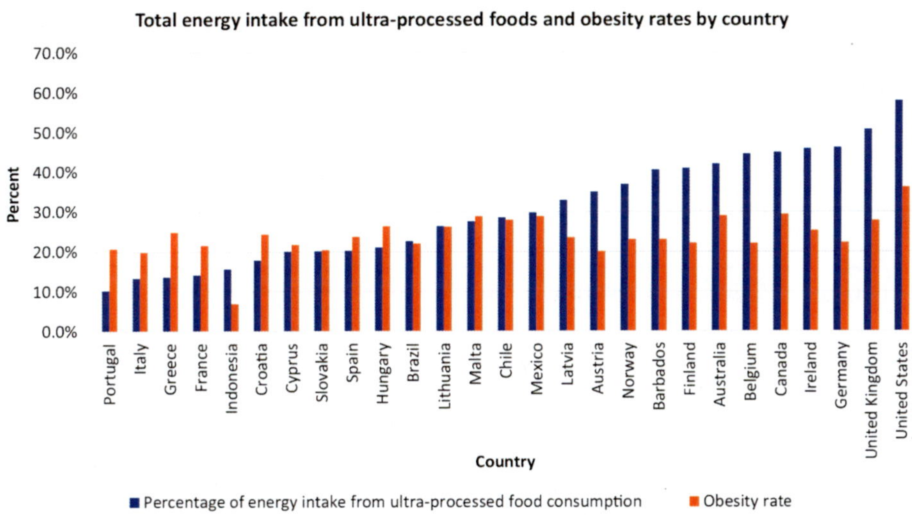

Figure 3.2. *Comparison of select countries totally energy intake from UPFs and obesity rates. Colombia and New Zealand were not included in the graphs, since those studies were based on children only. The obesity rates are based on 2016 data by the World Health Organization (WHO).*

Source: Crimarco A, Landry MJ, Gardner CD. Ultra-processed Foods, Weight Gain, and Co-morbidity Risk. Curr Obes Rep. 2022 Sep;11(3):80-92. doi: 10.1007/s13679-021-00460-y. Epub 2021 Oct 22. PMID: 34677812; PMCID: PMC8532572.

3.1.3. Sociodemographic Characteristics

Previous research has consistently demonstrated that UPF consumption varies by sociodemographic characteristic stratum. Depending on the sample, there are differences in consumption by nativity, age, gender, ethnicity, education level, and number of children living in the home. A person's or a household's socioeconomic status may also be a significant factor linked to consuming more ultra-processed foods, according to the literature; however, these associations differ depending on the income level of the nation. Global disparities in food costs, accessibility, and affordability can be linked to national variations. Higher UPF consumption is linked to lower income and education levels in the USA, and research shows that UPF accounts for a higher share of grocery spending in households that participate in the Supplemental Nutrition Assistance Program than in households that do not. Comparable outcomes have been observed in other affluent nations. In contrast, people with higher socioeconomic status, those who reside in urban areas, and those who have completed more education are linked to higher UPF intakes in middle- and low-income nations. Families that practice subsistence farming, where they consume a larger portion of the food they grow or raise, are more likely to do so in middle-class and lower-class nations. Higher earners in these nations stray from traditional eating habits because they can afford to buy more Westernized foods, which are more likely to be processed or overly processed.

Table 3.1. *Select Recent Articles on Ultra-Processed Food Consumption Levels From Various Countries*

Articles	Country	Sample	Key findings
Machado et al. (2020)	Australia	12,153 individuals from the National Nutrition and Physical Activity Survey (2011–2012) ages 2 years and above	Consumption of ultra-processed foods consisted of 42.0% of total energy intake
Harris et al. (2021)	Barbados	Nationally representative population-based sample of 364 adult Barbadians	Consumption of ultra-processed foods consisted of 40.5% of total energy intake
Simões et al. (2018)	Brazil	14,378 adults ages between 35 and 74 years sampled at multicenter cohort from 6 public universities	Consumption of ultra-processed foods consisted of 22.7% of total energy intake

Nardocci et al. (2019)	Canada	9363 adults ages 18 years or more from the 2004 Canadian Community Health Survey	Consumption of ultra-processed foods consisted of 45.0% of total energy intake
Cediel et al. (2018)	Chile	4920 individuals ages 2 years and above	Consumption of ultra-processed foods consisted of 28.6% of total energy intake
Cornwell et al. (2018)	Colombia	223 children ages 5–12 years	Consumption of ultra-processed foods consisted of 34.4% of total energy intake
Monteiro et al. (2018)	Multiple European countries	Households from the Living Costs and Food Survey (LCFS) or the Data Food Networking (DAFNE)	Consumption of ultra-processed foods consisted of 26.4% of total energy intake. The range of calories consumed from ultra-processed foods were 10.2% of total energy intake in Portugal to 50.4% of total energy intake from the UK
Setyowati et al. (2014)	Indonesia	Children and adults (n=145,360) grouped into the following age groups: 0–4, 5–12, 13–18, 19–55, and>55 years	Consumption of ultra-processed foods consisted of 15.7% of total energy intake
Marrón-Ponce et al. (2018)	Mexico	10,087 individuals from the 2012 Mexican National Health and Nutrition Survey	Consumption of ultra-processed foods consisted of 29.8% of total energy intake
Fangupo et al. (2021)	New Zealand	669 children ages 1–5 years born in Dunedin, New Zealand	Consumption of ultra-processed foods consisted of 45.0%, 42.0%, and 51.0% of energy intake to the diets of children at 12, 24, and 60 months of age, respectively
Steele et al. (2016)	USA	9317 individuals ages 1 year and above	Consumption of ultra-processed foods consisted of 57.9% of total energy intake

Source: Crimarco A, Landry MJ, Gardner CD. Ultra-processed Foods, Weight Gain, and Co-morbidity Risk. Curr Obes Rep. 2022 Sep;11(3):80-92. doi: 10.1007/s13679-021-00460-y. Epub 2021 Oct 22. PMID: 34677812; PMCID: PMC8532572.

3.1.4. Ultra-Processed Foods with Weight Gain and Disease Risk in Adults

There is mounting proof that UPFs raise the risk for some chronic diseases and contribute to weight gain, even though the majority of studies on their effects on health are observational in nature. The consumption of UPFs was linked to a higher risk of overweight (odds ratio: 1.36; 95% confidence interval [CI], 1.23–1.51; $P<0.001$), obesity (odds ratio: 1.51; 95% CI, 1.34–1.70; $P<0.001$), abdominal obesity (odds ratio: 1.49; 95% CI, 1.34–1.66; $P<0.0001$), all-cause mortality (hazard ratio: 1.28; 95% CI, 1.11–1.48; $P=0.001$), and metabolic syndrome (odds ratio: 1.81; 95% CI, 1.12–2.93; $P=0.015$) in adults. An association between UPF consumption and an increased relative risk for overweight/obesity (+39%), high waist circumference (+39%), low HDL-cholesterol levels (+102%), and metabolic syndrome (+79%) was also discovered in a recent systematic review of 23 studies (10 cross-sectional and 13 prospective cohort). Higher UPF consumption was linked to greater accumulation of visceral fat, android-to-gynoid fat ratio, and total body fat, according to an analysis conducted on a subsample of adult men and women from the PREDIMED-Plus cohort who had obesity and metabolic syndrome (Table 3.2).

Table 3.2. *Select Recent Studies Examining Sociodemographic Factors Associated With Greater Consumption of Ultra-Processed Foods*

Article	Country	Sociodemographic Factors Associated with Greater Consumption of UPF
Calixto Andrade et al. (2021)	France	• Younger age • Urban
Djupegot et al. (2017)	Norway	• Men • Native Norwegian • Lower educational attainment • $\geq$3 children within the home • Younger age
Nardocci et al. (2017)	Canada	• Men • Younger age • Lower educational attainment • Smokers • Physically inactive • Canadian-born individuals

Khandpur et al. (2020)	Columbia	• Younger age • Higher socioeconomic status • Area of residence/geographic region • Urban
Marrón-Ponce et al. (2018)	Mexico	• Younger age • Urban • Higher socioeconomic status • Lower educational attainment • Geographical region
Machado et al. (2020)	Australia	• Younger • Australian or English country • Physically inactive • Smoker • Lower educational attainment • Urban
Cediel et al. (2018)	Chile	• Younger • Urban • Geographic region • Higher income
Baraldi et al. (2018)	USA	• Non-Hispanic Whites and Non-Hispanic Blacks (compared to other race/ethnicity groups) • Higher educational attainment • Younger age • Lower income level

Source: Crimarco A, Landry MJ, Gardner CD. Ultra-processed Foods, Weight Gain, and Co-morbidity Risk. Curr Obes Rep. 2022 Sep;11(3):80-92. doi: 10.1007/s13679-021-00460-y. Epub 2021 Oct 22. PMID: 34677812; PMCID: PMC8532572.

Similar results have been reported by a number of other recent reviews that evaluated the impact of UPF consumption on weight gain or increased health risk. Since most of these studies are cross-sectional, there is no indication of a causal relationship, even though higher UPF consumption is typically seen in individuals with co-morbidities and higher levels of obesity. It is important to remember that obesity is a complex illness with numerous correlated lifestyle factors. Due to the observational nature of most UPF research, residual confounding may exist (Table 3.3).

Table 3.3. *Narrative Reviews, Systematic Reviews, and Meta-Analyses Summarizing Recent Evidence Evaluating the Association Between Ultra-Processed Food (UPF) Intake and Health Outcomes*

Reference	**Type of Review**	**Articles Reviewed and Study Design**	**Outcome**	**Key Findings**
Askari et al. (2020)	Systematic review and meta-analysis	13 cross-sectional, 1 prospective cohort	Excess body weight and obesity	UPF consumption was associated with increased risk of overweight and obesity
Chen et al. (2020)	Systematic review	8 cross-sectional, 12 prospective cohort	Any health outcome	UPF consumption was associated with increased risk of all-cause mortality, overall cardiovascular diseases, coronary heart diseases, cerebrovascular diseases, hypertension, metabolic syndrome, overweight, and obesity, depression, irritable bowel syndrome, overall cancer, postmenopausal breast cancer, gestational obesity, adolescent asthma and wheezing, and frailty. No association with cardiovascular disease mortality, prostate, and colorectal cancers, gestational diabetes mellitus, or gestational overweight

Costa et al. (2018)	Systematic Review	5 interventions, 6 cross-sectional, and 15 prospective cohorts	Body fat (during childhood and adolescence)	Consumption of UPF was positively associated with body fat during childhood and adolescence
de Miranda et al. (2021)	Narrative review	1 randomized controlled trial, 15 prospective cohort, 12 cross-sectional, 1 prospective and cross-sectional, and 1 ecological	Metabolic health	Consumption of UPF was positively associated with metabolic syndrome, body weight change and obesity indicators, blood pressure and hypertension, glucose profile, insulin resistance and type 2 diabetes, other metabolic risks and cardiovascular diseases and mortality
Elizabeth et al. (1995)	Narrative review	1 randomized controlled trial, 1 case–control, 19 prospective cohort, 19 cross-sectional, 3 ecological	All health outcomes	Consumption of UPF was positively associated with overweight, obesity, and cardio-metabolic risks; cancer, type 2 diabetes and cardiovascular diseases; irritable bowel syndrome, depression, and frailty conditions; and all-cause mortality in adults. Among children and adolescents, UPF consumption was associated with included cardiometabolic risks and asthma

Lane et al. (2021)	Systematic Review and Meta-Analysis	21 cross-sectional, 19 prospective, 2 case–control, 1 prospective and cross-sectional	Noncommunicable disease risk, morbidity, and mortality	Consumption of UPF was associated with increased risk of overweight, obesity, abdominal obesity, all-cause mortality, metabolic syndrome, cardiometabolic diseases, frailty, irritable bowel syndrome, functional dyspepsia, breast, and overall cancer, depression, and wheezing in adults. Associated with metabolic syndrome in adolescents and dyslipidemia in children. No association with asthma in adolescents
Moradi et al. (2021)	Systematic review and meta-analysis	9 cross-sectional and 3 prospective cohort	Overweight, obesity, and abdominal obesity	UPF consumption was associated with increase in the risk of overweight, obesity, and abdominal obesity
Pagliai et al. (2021)	Systematic review and meta-analysis	10 cross-sectional and 13 prospective cohort	Any health indicator	Highest UPF consumption associated with increase in the risk of overweight/obesity, high waist circumference, low HDL, metabolic syndrome, all-cause mortality, increased risk of CVD, cerebrovascular disease, and depression. No association with hypertension, hyperglycemia, or hypertriglyceridemia

Santos et al. (2020)	Systematic Review	9 cross-sectional and 2 prospective cohort	Cardiometabolic risk factors	UPF were positively associated with overweight and obesity, high blood pressure, and metabolic syndrome
Silva et al. (2020)	Systematic Review	13 cross-sectional, 6 prospective cohort, and 2 ecological	Noncommunicable chronic diseases	UPF was positively associated with excess body weight, hypertension, dyslipidemia, and metabolic syndrome

Source: Crimarco A, Landry MJ, Gardner CD. Ultra-processed Foods, Weight Gain, and Co-morbidity Risk. Curr Obes Rep. 2022 Sep;11(3):80-92. doi: 10.1007/s13679-021-00460-y. Epub 2021 Oct 22. PMID: 34677812; PMCID: PMC8532572.

Additional reports from prospective cohort studies indicate that UPFs have a positive correlation with several measures of obesity (i.e., BMI, waist size, and percentage of body fat). Diets high in UPFs were linked to a 30% increased risk of abdominal obesity (HR 1.30; 95% CI 1.14—1.48) and a 79% increased risk of obesity (HR 1.79; 95% CI 1.06—3.03). This information was found in a retrospective cohort study. Furthermore, during the follow-up period (median of 5.6 years), higher consumption of UPFs increased the risk of a gain in BMI, waist circumference, and body fat of 5% or more. According to a Brazilian civil servant cohort study, using UPF was linked to a higher relative risk of weight gain of 27% (95% confidence interval: 1.07–1.50) and a higher relative risk of waist circumference gain of 33% (95% confidence interval: 1.12–1.58). Fazzino along with others. (2021) found that a higher consumption of UPFs during a buffet meal was linked to a greater increase in weight gain over the course of the following 12 months in a prospective study involving 82 people without obesity. These cohort study results support the cross-sectional study findings by showing a causal relationship between UPF consumption and weight gain.

Consumption of UPFs is linked to a higher risk of hypertension, cardiovascular disease, type 2 diabetes, metabolic syndrome, overall cancer risk, and all-cause mortality in adults, according to results mainly from observational studies. Several of these studies suggested that high UPF diets raise the risk for co-morbidities regardless of body weight by adjusting for BMI in the main analyses and/or including sensitivity analyses to adjust for BMI, weight gain, physical activity levels, or a family history of the particular health condition. The individuals in the majority of these studies who consumed the most UPF also had diets that were of lower overall quality. In comparison to individuals who consumed fewer UPF products, those who consumed the most UPFs had higher intakes of sugar, saturated fats, and salt, but lower and/or inadequate intakes of fiber and micronutrients. Given that poor diet quality is linked to all of the aforementioned health conditions, it is conceivable that UPF consumption is linked to many of the major chronic diseases of today. Apart from inadequate diet quality, there exist several plausible biological mechanisms specific to UPF consumption that could account for some of their effects on elevated risk of weight gain and/or chronic disease (Crimarco et al., 2022).

3.1.5. Ultra-Processed Foods and Health Outcomes in Children and Adolescents

While numerous systematic reviews and meta-analyses have provided substantial evidence linking UPFs to adult health outcomes, there has been less research conducted on pediatric populations. More than two-thirds of the energy consumed by US children and adolescents in 2017–2018 came from UPFs, a 5.6% increase over the previous 20 years. Additionally, studies have shown a link between overweight children's frequent UPF consumption and food addition. When combined, a poor diet and an excessive calorie intake can cause children to become overweight or obese. Higher childhood UPF consumption was linked in several prospective studies to faster increases in BMI, FMI, weight, and waist circumference in adolescence and the early years of adulthood. According to a 2018 systematic review, eating a lot of ultra-processed food as a child or teenager was positively linked to obesity. On the other hand, many studies have not discovered any link between eating a lot of ultra-processed food and being overweight or obese. According to research, factors like physical activity, genetics, and family lifestyle that are linked to the etiology of obesity but were not examined in the studies may be the cause of null findings. These results suggest that cutting back on children's intake of ultra-processed foods may lower the prevalence of childhood obesity and overweight; nevertheless, a child population clinical trial is required.

Numerous studies have discovered a positive correlation between higher UPF consumption and blood lipids in addition to weight and adiposity outcomes. In one Brazilian cohort, children who consumed the most UPFs between the ages of three and four showed higher levels of total and triglycerides four years later. In a different Brazilian cohort, elevated levels of LDL and triglycerides were observed several years later. Furthermore, high UPF consumption was linked to the prevalence of metabolic syndrome, a group of risk factors that raises the risk for diabetes, stroke, and cardiovascular disease, according to a study of 210 teenagers in Brazil. A cross-sectional study conducted in Spain, however, did not discover any connections between the consumption of ultra-processed foods and levels of HDL or triglycerides. This null result might be explained by the fact that only a portion of the study population had their lipid levels measured, which could have an effect on lipid levels.

All things considered, the evidence currently available indicates that consuming UPFs may have short-term negative effects on blood lipid levels, weight gain, and excessive calorie intake, as well as long-term health consequences in adulthood. Considering that childhood nutrition shapes lifelong eating habits that carry into adulthood, it is important to limit children's exposure to and consumption of ultra-processed, high-energy foods that are also low in nutrients.

3.1.6. Potential Mechanisms of How Ultra-Processed Foods May Increase Weight Gain and Chronic Disease Risk

UPFs have a low potential for satiety but cause high glycemic responses. Comparatively speaking to whole foods, a well-controlled randomized crossover study showed that consuming UPFs increased energy intake and caused weight gain. For two weeks each, 20 adults in the study consumed a diet high in unsaturated polyphenols (UPFs; approximately 80% of calories) and a different diet high in whole grains and unprocessed foods. The researchers matched the diets for total energy intake and macronutrients, but the UPF diet produced higher levels of added total sugar (approximately 54% versus 1%, respectively), insoluble fiber (approximately 16% versus 77%, respectively), and saturated fat (approximately 34% versus 19%). One of the study's main conclusions was that, in the UPF phase, participants consumed 500 kcal/day more than they did on the alternate diet. They also gained 0.9 ± 0.3 kg ($P = 0.009$) and lost 0.9 ± 0.3 kg ($P = 0.007$) in the unprocessed diet. The elevation in ad libitum energy intake during the UPF dietary phase could be explained by changes in the participants' hunger-related hormones, pancreatic peptide YY, and ghrelin. Additionally, it has been proposed that certain aspects of food processing, such as the addition of additives and changes to the food matrix, give foods a softer texture that requires less chewing and enhance sensory qualities, which postpones the signal that indicates fullness and leads to overindulgence in food. UPFs are more palatable due to their higher sugar, fat, and salt content. This could lead to the replacement of a nutrient-dense, healthful diet with empty calories and a lower-quality diet that causes weight gain.

Additionally, it has been suggested that UPFs influence the gut microbiota, favoring microbes linked to inflammatory illnesses. Food matrix modification frequently results in changes to the foods' fiber and fat content, which affects the makeup of the microbiota and the interactions between bacteria and hosts. Natural or minimally processed foods have intact fibrous cell walls, which guarantee a gradual release of nutrients throughout the digestive tract and serve as a substrate for bacteria in the colon that break down fiber. However, since most of the nutrients in UPFs are cellular, the environment they create encourages an inflammatory gut microbiota, which is linked to a number of cardiometabolic disorders. As a result, not only do UPF diets typically have low dietary fiber content, but food processing also affects how effective dietary fiber is at fostering a favorable environment for the gut microbiome. Pigs fed a high-fiber diet consisting of processed, extruded grains had lower bacterial diversity than those fed a diet consisting of unprocessed whole grains, according to an animal model study. Higher UPF diets were most frequently linked to a decrease in the abundance of microbes associated with positive health outcomes and an increase in the abundance of microbes associated with negative health outcomes, according to a review of seven trials.

UPF products' complex packaging materials frequently contain endocrine-disrupting chemicals like bisphenol A (BPA), which could have an additional effect on biology or metabolism when consumed. Although BPA's exact mechanisms are still unknown, there is evidence that it increases insulin resistance, inflammation, oxidative stress, and adipogenesis. These factors raise our risk of developing major cardiovascular diseases (CVDs), such as diabetes, hypertension, and both abdominal and general obesity.

According to Gibney (2019 and 2020), nutritional factors, not the level of food processing, are to blame for any negative effects linked to UPFs. Results from our lab group's SWAP-MEAT crossover intervention study showed that after 8 weeks of consuming alternative plant-based meat products instead of organic animal meats, participants improved in several cardiovascular disease risk factors. This resulted from the plant-based meats' concurrent reduction in saturated fats and increase in dietary fiber. Since most UPFs are naturally high in energy and low in nutritional quality, it is unclear if the degree of food processing raises our risk for weight gain and other chronic diseases independently from the nutritional makeup of the foods themselves.

However, there was a slight but statistically significant drop in weight in the plant-based meat phase compared to the control group in our SWAP-MEAT study. phase of animal meat. However, the purpose of our study was to concentrate on just one instance of replacing animal meat with plant-based meat. Researchers found that UPFs were linked to an increased risk of cardiovascular disease even after adjusting for several nutritional quality factors in a cohort study that looked at UPF consumption overall across all food types. Similarly, after adjusting for diet quality and energy intake, a different study discovered a stronger correlation between type 2 diabetes and the consumption of ultra-processed foods among members of the NutriNet-Santé cohort.

3.1.7. Policies to Reduce Ultra-Processed Food Consumption

Recognizing the detrimental effects of UPF on health, several nations have started enacting policies to limit UPF consumption. These policies include taxes on sugar-sweetened beverages and snacks, front-of-package (FOP) warning labels, limiting the amount of trans fat and sodium in food products, laws that limit or prohibit UPF marketing, and limiting the availability and promotion of UPFs in educational settings. One of the first nations to thoroughly assess its tax laws on sweetened beverages was Mexico, which discovered that taxed beverage purchases decreased by 6%, with lower socioeconomic class households experiencing the largest reductions from pre-tax trends. Taxes on UPF foods are a useful tool for lowering sales of these goods, but one area of fiscal policy that needs improvement is the lack of incentives or subsidies for buying healthier food. There is proof that improving the population's diet quality and health results involves taxing unhealthy foods and providing subsidies for the purchase of healthier foods. However, enforcing stricter regulations is still difficult because the food industry and other interested parties oppose cutting back on UPF consumption or work to thwart public health initiatives (Floros et al., 2010; Augustin et al., 2016; Monteiro et al., 2019).

3.2. INSULIN RESISTANCE AND TYPE 2 DIABETES

Ultra-processed foods (UPFs) are becoming more and more associated with the emergence of type 2 diabetes and insulin resistance. Here's a detailed look at how UPFs contribute to these metabolic issues:

1. High Glycemic Index and Load
   i. Rapid Blood Sugar Spikes: UPFs often contain refined carbohydrates and sugars that cause rapid increases in blood glucose levels. This forces the pancreas to produce large amounts of insulin to manage the sudden surge.
   ii. Chronic High Insulin Levels: Over time, constant high insulin levels (hyperinsulinemia) can lead to insulin resistance, where the body's cells become less responsive to insulin, making it harder for glucose to enter cells.
2. Excessive Caloric Intake and Weight Gain
   i. Energy Density: UPFs are typically high in calories and low in nutritional value, leading to excessive caloric intake without satisfying hunger, promoting overeating and weight gain.

ii. Obesity: Increased body fat, particularly visceral fat, is strongly associated with insulin resistance. Adipose tissue secretes inflammatory cytokines that interfere with insulin signaling pathways.

3. Poor Nutritional Profile

i. Lack of Fiber: Fiber helps regulate blood sugar levels and promotes satiety. UPFs are usually low in dietary fiber, which can lead to poor glycemic control.

ii. Deficiency in Micronutrients: Essential nutrients like magnesium, chromium, and vitamin D are often lacking in UPFs, and these micronutrients play critical roles in glucose metabolism and insulin sensitivity.

4. High Fructose Content

i. Fructose Metabolism: Many UPFs contain high fructose corn syrup and other forms of added fructose, which is metabolized differently from glucose. Excessive fructose intake can lead to fatty liver and insulin resistance.

ii. Liver Fat Accumulation: Fructose is metabolized primarily in the liver, where it can be converted to fat, contributing to non-alcoholic fatty liver disease (NAFLD), which is associated with insulin resistance.

5. Inflammatory Additives

i. Emulsifiers and Preservatives: Certain additives in UPFs can disrupt gut microbiota, leading to increased gut permeability and systemic inflammation. Chronic inflammation can impair insulin signaling.

ii. Artificial Sweeteners: Some studies suggest that artificial sweeteners can negatively affect gut bacteria and glucose metabolism, contributing to insulin resistance.

6. Oxidative Stress

i. Pro-oxidant Ingredients: UPFs often contain ingredients that can induce oxidative stress, such as trans fats and advanced glycation end products (AGEs). Oxidative stress damages cells and impairs insulin signaling pathways.

ii. Reduced Antioxidant Intake: A diet high in UPFs and low in fruits and vegetables reduces the intake of antioxidants, exacerbating oxidative stress.

7. Disruption of Gut Microbiota

i. Microbiome Imbalance: UPFs can negatively impact the gut microbiome, which plays a crucial role in metabolic health. A disrupted microbiome can contribute to inflammation and insulin resistance.

ii. Low Prebiotic Content: UPFs often lack prebiotics (the fiber that feeds beneficial gut bacteria), further compromising gut health and metabolic function.

iii. Insulin Resistance Pathway: The chronic consumption of UPFs leads to constant high blood sugar levels, excessive insulin production, and eventually, the cells' decreased responsiveness to insulin.

iv. Beta-cell Dysfunction: The pancreas' beta cells, which produce insulin, can become overworked and start to fail, leading to decreased insulin production over time.

v. Inflammation and Oxidative Stress: These conditions further exacerbate insulin resistance and beta-cell dysfunction, creating a vicious cycle.

To prevent and treat insulin resistance and type 2 diabetes, consuming fewer UVPs is crucial. Stressing the importance of eating a diet high in whole, minimally processed foods—like fruits, vegetables, whole grains, lean meats, and healthy fats—can support healthy insulin sensitivity and glucose metabolism. Addressing the escalating type 2 diabetes epidemic requires raising awareness of the risks associated with UPFs and encouraging better eating habits.

3.2.1. Ultra-Processed Foods Linked to Increased Risk of Type 2 Diabetes

Consuming ultra-processed foods raises the risk of Type 2 diabetes, according to a growing body of scientific evidence.

Approximately 40 million individuals in the U.S. have diabetes, with Type 2 accounting for more than 90% of cases. Increased insulin resistance causes blood sugar levels to rise in people with diabetes, a chronic illness. If this condition is not treated, it can cause serious harm to the heart, brain, kidneys, gastrointestinal tract, eyes, gums, feet, nerves, and blood vessels.

The autoimmune disease known as type 1 diabetes usually manifests itself suddenly in childhood. Up until recently, adults over 45 were the main demographic for type 2 diabetes diagnoses. However, Type 2 diabetes is becoming more common in children and young adults today. It is predicted that the number of children affected by the disease will rise by almost 700% in the next 40 years. While obesity, waist circumference, and a sedentary lifestyle are the main risk factors for Type 2 diabetes, genetic factors can also contribute to the disease's development.

Ultra-processed food accounts for nearly 70% of the calories consumed by children and more than half of the calories consumed by adults in America today.

3.2.2. Lawmakers Discuss Diabetes Risks of Ultra-processed Foods

Congress has been notified of serious health risks associated with ultra-processed food. Sen. wrote the FDA a letter in February. Bernie Sanders cited the "skyrocketing" prevalence of obesity and Type 2 diabetes in the U.S. Then, he demanded that foods with high levels of processing bear explicit warning labels.

In December 2023, Sanders penned an opinion piece for USA Today titled, "We can't allow the food and beverage industry to destroy our kids' health." He stated, "It is past time that we take the obesity and Type 2 diabetes epidemic in America seriously. We need to have the guts to oppose the food and beverage industry's greed, which is systematically endangering our kids' health and wellbeing." Ashley Gearhardt, a food addiction specialist and psychology professor at the University of Michigan, gave testimony to the U.S. Senate that a "scientific body of evidence suggests that ultra-processed food intake patterns implicated in poor health, obesity, and diabetes play a significant role in contributing to addictive processes." If ultra-processed foods are triggering mechanisms for addiction, this could be a missed explanation for why it can be difficult to cut back on ultra-processed food intake even when dealing with health conditions like diabetes. "

3.2.3. Evidence Linking Ultra-processed Foods and Diabetes

In a 2019 observational study that was published in JAMA Internal Medicine, researchers looked into the possible link between ultra-processed foods (UPF) and Type 2 diabetes (T2D). The prospective analysis comprised over 104,000 individuals' data. The researchers discovered that "consumption of UPF was associated with a higher risk of T2D" after correcting for socioeconomic, lifestyle, and comorbidities. Even after accounting for weight change, other metabolic comorbidities, and a number of indicators of the diet's nutritional quality, these results were still statistically significant. Even after adjusting for the consumption of minimally or no-processed foods, the absolute amount of UPF consumption (grams per day) was consistently linked to the risk of type 2 diabetes."

Researchers looked at the relationship between the risk of different adverse health outcomes and the consumption of ultra-processed foods in an umbrella review that was published in 2024 in BMJ. By utilizing information from 45 analyses, the researchers discovered that "the strongest available evidence pertained to direct associations between higher risks of all-cause mortality, cardiovascular disease related mortality, common mental disorder outcomes, overweight, and obesity, and type 2 diabetes and greater exposure to ultra-processed foods." "We recommend urgent mechanistic research and the development and evaluation of comprehensive population-based and public health strategies, including government-led policy frameworks and dietary guidelines, aimed at targeting and reducing dietary exposure to ultra-processed foods for improved human health," they concluded, in conjunction with the population-based strategies already in place."

A meta-analysis that looked at studies that connected eating ultra-processed foods to an increased risk of Type 2 diabetes was published in the International Journal of Epidemiology in 2022. Researchers discovered that, "compared with non-consumption, moderate intake of ultra-processed food increased the risk of diabetes by 12%, whereas high intake increased risk by 31%" after screening more than 18 studies involving a total of more than 1 million people. The elevated risk had "a dose-response effect and moderate to high credibility of evidence," they continued. "As long as they are accompanied by a reduction in the consumption of ultra-processed foods, strategies like exchange for fresh or minimally processed foods and physical activity, in addition to alternative therapies, can be effective in preventing or treating diabetes," the study's authors concluded."

According to a mini-review that was published in the journal Nutrients, "understanding the factors that cause diabetes and conducting clinical trials has become a priority." The authors of the study observed a "significant" link between eating a lot of ultra-processed food and an increased risk of diabetes mellitus. Consuming sweetened beverages and salty processed foods was found to significantly increase the risk of prediabetes by 248% and 48%, respectively, and Type 2 diabetes by 219% and 600%, respectively, when compared to people who did not consume these products or had a rare consumption ratio, according to a notable study of participants."

An overview of 39 meta-analyses that were published in Clinical Nutrition in 2024 looked at the relationships between eating a lot of ultra-processed food and a number of bad health outcomes, such as Type 2 diabetes. "Five outcomes were reported with highly suggestive evidence, including diabetes mellitus," the study's authors wrote. The macronutritional dietary features of UPFs were linked to obesity and overweight, two conditions that increase the risk of diabetes, the researchers said. Because of the altered food structure, an experimental study confirmed that UPFs were, on average, more hyperglycemic than minimally and moderately processed foods. Carrageenan, an ingredient frequently found in UPFs, was found to increase insulin resistance and fasting blood glucose in mice in a study. Moreover, UPF is frequently packed in synthetic materials that could contain chemicals that disrupt hormones in the endocrine system, like bisphenol A (BPA), which is linked to diabetes."

More than 10,000 Brazilian adults participated in a long-term study that was published in the journal Diabetes and Metabolic Syndrome in 2023. The study evaluated the relationship between the consumption of ultra-processed foods (UPF) and Type 2 diabetes. After controlling for lifestyle and risk factors, they discovered that "relative risks (RR) was 1.24 when comparing the fourth with the first quartile of UPF distribution; every 150 g/day increment in UPF consumption resulted in an RR of 1.05." Reclassifying natural drinks that have sweeteners added as higher risk due to UPF. Consumption of processed meats and beverages with added sugar raised the risk of diabetes among UPF subgroups, whereas yogurt and dairy products reduced the risk. They said, "These results strengthen the body of research linking UVFs to the onset of diabetes and other chronic illnesses, endorsing the advice to limit UVF intake for the prevention and treatment of diabetes."

Researchers in the Netherlands sought to investigate the health effects of a variety of ultra-processed foods for a study that was published in BMC Medical in 2022. The information contained over 70,000 study participants' self-reported food intake, ranging in age from 35 to 70. "A 25% higher risk of developing type 2 diabetes was associated with a 10% increment in UPF consumption," according to a follow-up study conducted after a little over three years. Particularly, consuming a large quantity of savory snacks, both warm and cold, was linked to an elevated risk. Notably, participants tended to be younger, have higher BMIs, lower risk scores for type 2 diabetes, be less physically active, have worse diet quality overall, consume less alcohol, smoke less, be less highly educated, and spend more time watching TV when their UPF consumption increased in quartiles. The researchers came to the following conclusion: "Part of diabetes prevention strategies should be the active discouragement of unhealthy food products, such as savory UPF, in addition to encouraging consumption of healthy food products."

More than 200,000 adults from four different countries participated in a meta-analysis and systematic review published in the journal Nutrients in 2021. The results showed that consuming more ultra-processed foods was "significantly associated" with a higher risk of type 2 diabetes (T2DM). "Linear dose-response analysis indicated that among adults, a 10% increase in UPF consumption was associated with a 15% higher risk of type 2 diabetes mellitus," the researchers wrote. Adult T2DM and UPF consumption showed a positive linear correlation, according to non-linear dose-response analysis. Adult T2DM and UPF consumption showed a positive linear association as well, according to non-linear dose-response analysis."

Researchers looked at the relationship between ultra-processed foods and Type 2 diabetes in over 12,000 Chinese adults, whose mean age was 43.33 years, for a survey that was published in the journal Nutrients in 2022. They discovered that among Chinese adults, "both UPF consumption and the prevalence of diabetes increased during 1997–2011." Compared to non-consumers, higher UPF consumers had a noticeably higher risk of diabetes. One moderating factor in the relationship between UPF consumption and diabetes was overweight/obesity."

Researchers evaluated data from over 7000 Koreans between the ages of 40 and 69 in order to examine the relationships between ultra-processed food consumption and diabetes risk in Asian populations. Their findings were published in the Journal of Nutrition. The highest quartile of ultra-processed food intake was found to be positively associated with the risk of diabetes when compared to the lowest quartile. Even after additional adjustments for BMI and diet quality, the association remained unchanged. Particularly noteworthy were the correlations between the consumption of ice cream, instant noodles, ham, and sausage, and carbonated drinks and an elevated risk of type 2 diabetes. The study's authors responded to these findings by writing: "Public health strategies should also focus on lowering the cost and increasing accessibility of unprocessed or minimally processed foods and supporting industry efforts to create less processed food formulations. Food producers can reduce the amount of additives and hazardous ingredients in their products by reformulating their processing techniques. Food labels may also include more thorough information about the ingredients and preservatives used in food products, as well as information about food processing techniques, to help consumers make educated decisions."

Researchers looked at the relationship between emulsifiers—food additives that combine two or more substances—and Type 2 diabetes in a study that was published in The Lancet Diabetes and Endocrinology in 2024. They are frequently present in highly processed foods. More than 100,000 adult participants' data from the French NutriNet-Santé prospective cohort study were included in the study. The researchers discovered: "We discovered clear correlations between exposure to different food additive emulsifiers commonly used in industrial foods and the risk of type 2 diabetes." (Floros et al., 2010; Augustin et al., 2016; Monteiro et al., 2019).

3.3. THE ULTRA-PROCESSED FOODS & WEIGHT GAIN CONNECTION

We are taught that losing weight is all about burning off calories rather than consuming them and that eating less and exercising more is the easy fix.

The figures don't add up, which is the issue. According to official statistics, people's average calorie intake has decreased over the previous few decades. In the UK, the average daily energy intake is currently approximately 1972 kcal, compared to 2570 kcal in 1969.

The average body weight has increased over this period. According to government statistics, approximately 28% (1 in 4) of adults in the UK are obese, and an additional 36.2% are overweight.

Overindulgence in highly processed foods is one of the main causes of weight gain, but if calories alone aren't the problem, then what is?

3.3.1. 56.8% of Total Energy Intake in the UK is from Ultra-processed Foods

It's been observed that there is a direct correlation between the rise in obesity and the consumption of ultra-processed foods, such as instant soups, breakfast cereals, ready meals, sweets, and beverages sweetened with sugar.

They are extremely processed and have a significant amount of added sugar and salt, but very little fiber, vitamins, minerals, or other vital nutrients that your body needs to remain healthy. Ultra-processed foods are completely devoid of nutrients and disturb the body's natural equilibrium of body weight, known as body weight homeostasis.

3.3.2. Your "Healthy' Body Weight Range: Homeostasis

The body's natural mechanism for regulating weight is called body weight homeostasis. Simply put, the body initiates a variety of processes to bring a person back to their normal weight, also known as a "set point," when they begin to lose or gain weight.

Homeostatic control of body weight is thought to be influenced by many factors, including:

1. Hormones.
2. Genetic influences.
3. The gut microbiome.

However, the body's capacity to control weight can be readily undermined by a variety of factors, including stress, sleep patterns, dietary habits, and our surroundings when it comes to food. Crucially, eating a lot of processed food can have a negative impact on two important factors that regulate weight: hormones and gut health.

3.3.3. Ultra-processed Foods Negatively Affect Your Hormones

A recent study that compared the effects of an "ultra-processed diet" to an "unprocessed diet" on healthy adults found that ultra-processed foods had a deleterious effect on the hormones involved in body weight homeostatic control.

During the processed diet phase…

- Blood sugar levels were higher > this can make you hungrier.
- Insulin secretion was higher > this can increase fat storage.

During the unprocessed diet phase…

- The appetite-stimulating hormone, ghrelin, decreased > they felt less hungry.
- The appetite-suppressing hormone, PYY, increased > again, they felt less hungry.
- Inflammation and cholesterol reduced during the unprocessed diet > their general health improved.

It's not surprising that ultra-processed foods have such a detrimental long-term impact on weight and health as these effects were only noticed after two weeks!

3.3.4. Ultra-processed Foods Negatively Affect Your Gut Health

According to research, the bacteria in your gut that aid in food digestion and nutrient absorption may have a big impact on your capacity to lose weight.

Ultra-processed food diets are thought to be the cause of "gut dysbiosis." This occurs when an imbalance of gut bacteria causes an overgrowth of yeast and dangerous bacteria. In addition to raising the risk of cancer, heart disease, autoimmune diseases, and type 2 diabetes, gut dysbiosis also raises the risk of weight gain and obesity. Because the bacteria in your stomach can influence how fat is stored, how food is digested, and whether you feel hungry or full. When combined, these elements have a detrimental effect on body weight homeostasis, or your body's innate capacity to control your weight.

Ultra-processed foods may be the reason behind your failure to see long-term results from cutting back on calories to lose weight. But all it takes to see a positive change quickly is to eat more whole foods and less processed foods!

We are meant to consume, process, and assimilate whole foods. They cooperate with other systems that support body weight homeostasis, such as your gut microbiota and hormones that control hunger and satiety.

They naturally keep you fuller for longer periods of time and take longer to digest. That makes maintaining a healthy weight much simpler for you.

Getting started could be as simple as:

- Purition for breakfast.
- A healthy whole food snack or lunch.
- A Daily Feed meal for dinner (Floros et al., 2010; Augustin et al., 2016; Monteiro et al., 2019).

3.4. WEIGHT MANAGEMENT WITH ULTRA-PROCESSED FOODS

Consuming ultra-processed foods (UPFs) increases the difficulty of managing weight because of their high-calorie density, low nutritional content, and extremely appealing texture. UPFs are often high in refined carbohydrates, unhealthy fats, and added sugars, which can cause overindulgence in calories while offering little satisfaction. This promotes overindulging and frequent snacking, which helps people gain weight. Furthermore, UPFs frequently lack fiber and other nutrients that are necessary for sustaining a healthy metabolism and encouraging feelings of fullness. The confluence of these variables may interfere with regular appetite control and metabolic functions, making it more difficult to effectively regulate body weight. Moreover, the inclusion of additives and preservatives in UPFs may have adverse effects on gut health and inflammation, making attempts to control weight even more challenging. Reduce your intake of UPFs and concentrate on eating a balanced diet full of whole, minimally processed foods high in fiber and nutrients to reach and stay at a healthy weight.

3.4.1. Impact of Ultra-Processed Foods on Weight

i. **High Caloric Density**: UPFs often contain high levels of sugars, unhealthy fats, and refined carbohydrates, leading to high caloric intake.

ii. **Low Nutritional Value**: These foods typically lack essential nutrients like vitamins, minerals, and fiber.

iii. **Overconsumption**: UPFs are designed to be highly palatable, which can lead to overeating.

iv. **Altered Satiety Signals**: The additives and refined ingredients in

UPFs can disrupt normal hunger and satiety signals, making it easier to consume more than needed.

3.4.2. Strategies for Weight Management

1. **Mindful Eating**: Pay attention to hunger and fullness cues, and try to eat slowly to give your body time to signal when it's full.
2. **Portion Control**: Be mindful of portion sizes, especially with UPFs. Using smaller plates and pre-measuring portions can help.
3. **Balance with Whole Foods**: Include more whole foods like fruits, vegetables, lean proteins, and whole grains to balance the diet and provide necessary nutrients.
4. **Reading Labels**: Learn to read nutrition labels to make informed choices about the processed foods you consume. Look for lower-calorie options with fewer additives.
5. **Healthy Snacking**: Choose healthier snacks, such as nuts, yogurt, or fruit, instead of chips or candy.
6. **Hydration**: Sometimes thirst is mistaken for hunger. Drink plenty of water throughout the day.
7. **Regular Exercise**: Incorporate physical activity into your daily routine to help manage weight and improve overall health.
8. **Meal Planning**: Plan meals ahead of time to avoid relying on convenience foods, which are often ultra-processed.

3.4.3. Reducing UPFs in the Diet

1. **Cooking at Home**: Prepare meals from scratch using whole ingredients. This allows control over what goes into your food.
2. **Healthy Swaps**: Replace UPFs with healthier alternatives. For example, swap sugary cereals for oatmeal or whole-grain cereals.
3. **Gradual Changes**: Make gradual changes to your diet to reduce reliance on UPFs. Start by replacing one processed food item per week with a healthier option.
4. **Awareness and Education**: Educate yourself about the negative impacts of UPFs and become more aware of what you're eating.

Maintaining a balanced diet rich in whole, nutrient-dense foods and exercising mindfulness are essential to managing weight when following a UPF-based diet. Long-term weight management success can result from making modest, sustainable adjustments (Floros et al., 2010; Augustin et al., 2016; Monteiro et al., 2019).

REFERENCES

1. Alamri, E. S. (2021). Effects of COVID-19 home confinement on eating behavior: A review. *Journal of Public Health Research.*
2. Askari, M., Heshmati, J., Shahinfar, H., Tripathi, N., & Daneshzad, E. (2020). Ultra-processed food and the risk of overweight and obesity: A systematic review and meta-analysis of observational studies. *International Journal of Obesity (Lond)*, *44*(10), 2080–2091. https://doi.org/10.1038/s41366-020-00650-z.
3. Augustin, M. A., Riley, M., Stockmann, R., Bennett, L., Kahl, A., Lockett, T., et al. (2016). Role of food processing in food and nutrition security. *Trends in Food Science & Technology, 56*, 115–125. https://doi.org/10.1016/j.tifs.2016.08.005.
4. Baker, P., Machado, P., Santos, T., Sievert, K., Backholer, K., Hadjikakou, M., et al. (2020). Ultra-processed foods and the nutrition transition: Global, regional and national trends, food systems transformations and political economy drivers. *Obesity Reviews, 21*(12), e13126.
5. Baraldi, L. G., Steele, E. M., Canella, D. S., & Monteiro, C. A. (2018). Consumption of ultra-processed foods and associated sociodemographic factors in the USA between 2007 and 2012: Evidence from a nationally representative cross-sectional study. *BMJ Open, 8*(3).
6. Bawaked, R. A., Fernández-Barrés, S., Navarrete-Muñoz, E. M., González-Palacios, S., Guxens, M., Irizar, A., et al. (2020). Impact of lifestyle behaviors in early childhood on obesity and cardiometabolic risk in children: Results from the Spanish INMA birth cohort study. *Pediatric Obesity, 15*(3), e12590.
7. Bielemann, R. M., Motta, J. V. S., Minten, G. C., Horta, B. L., & Gigante, D. P. (2015). Consumption of ultra-processed foods and their impact on the diet of young adults. *Revista de Saúde Pública, 49*, 28.
8. Bleiweiss-Sande, R., Sacheck, J. M., Chui, K., Goldberg, J. P., Bailey, C., & Evans, E. W. (2020). Processed food consumption is associated with diet quality, but not weight status, in a sample of low-income and ethnically diverse elementary school children. *Appetite, 151*, 104696.
9. Bonaccio, M., Costanzo, S., Ruggiero, E., Persichillo, M., Esposito, S., Olivieri, M., et al. (2021). Changes in ultra-processed food consumption during the first Italian lockdown following the COVID-19 pandemic and major correlates: Results from two population-based cohorts. *Public Health Nutrition.*
10. Botelho, R., Araújo, W., & Pineli, L. (2018). Food formulation and not

processing level: Conceptual divergences between public health and food science and technology sectors. *Critical Reviews in Food Science and Nutrition, 58*(4), 639–650. https://doi.org/10.1080/10408398.2016.1209159.

11. Calixto Andrade, G., Julia, C., Deschamps, V., Srour, B., Hercberg, S., Kesse-Guyot, E., et al. (2021). Consumption of ultra-processed food and its association with sociodemographic characteristics and diet quality in a representative sample of French adults. *Nutrients, 13*(2), 682. https://doi.org/10.3390/nu13020682.
12. Canhada, S. L., Luft, V. C., Giatti, L., Duncan, B. B., Chor, D., Maria de Jesus, M., et al. (2020). Ultra-processed foods, incident overweight and obesity, and longitudinal changes in weight and waist circumference: The Brazilian Longitudinal Study of Adult Health (ELSA-Brasil). *Public Health Nutrition, 23*(6), 1076–1086.
13. Cediel, G., Reyes, M., da Costa Louzada, M. L., Steele, E. M., Monteiro, C. A., & Corvalán, C. (2018). Ultra-processed foods and added sugars in the Chilean diet (2010). *Public Health Nutrition, 21*(1), 125–133. https://doi.org/10.1017/S1368980017001161.
14. Chajès, V., Biessy, C., Byrnes, G., Deharveng, G., Saadatian-Elahi, M., Jenab, M., et al. (2011). Ecological-level associations between highly processed food intakes and plasma phospholipid elaidic acid concentrations: Results from a cross-sectional study within the European prospective investigation into cancer and nutrition (EPIC). *Nutrition and Cancer, 63*(8), 1235–1250. https://doi.org/10.1080/01635581.2011.617530.
15. Chang, K., Khandpur, N., Neri, D., Touvier, M., Huybrechts, I., Millett, C., et al. (2021). Association between childhood consumption of ultraprocessed food and adiposity trajectories in the Avon Longitudinal Study of Parents and Children Birth Cohort. *JAMA Pediatrics*, e211573.
16. Chen, X., Zhang, Z., Yang, H., Qiu, P., Wang, H., Wang, F., et al. (2020). Consumption of ultra-processed foods and health outcomes: A systematic review of epidemiological studies. *Nutrition Journal, 19*(1), 1–10. https://doi.org/10.1186/s12937-019-0518-3.
17. Cobiac, L. J., Tam, K., Veerman, L., & Blakely, T. (2017). Taxes and subsidies for improving diet and population health in Australia: A cost-effectiveness modelling study. *PLoS Medicine, 14*(2), e1002232.
18. Cornwell, B., Villamor, E., Mora-Plazas, M., Marin, C., Monteiro, C. A., & Baylin, A. (2018). Processed and ultra-processed foods are associated with lower-quality nutrient profiles in children from Colombia. *Public Health*

Nutrition, 21(1), 142–147. https://doi.org/10.1017/S1368980017000891.

19. Costa, C. S., Del-Ponte, B., Assunção, M. C. F., & Santos, I. S. (2018). Consumption of ultra-processed foods and body fat during childhood and adolescence: A systematic review. *Public Health Nutrition, 21*(1), 148–159. https://doi.org/10.1017/S1368980017001331.
20. Costa, C., Rauber, F., Leffa, P. S., Sangalli, C., Campagnolo, P., & Vitolo, M. R. (2019). Ultra-processed food consumption and its effects on anthropometric and glucose profile: A longitudinal study during childhood. *Nutrition, Metabolism & Cardiovascular Diseases, 29*(2), 177–184. https://doi.org/10.1016/j.numecd.2018.11.003.
21. Costard, S., Espejo, L., Groenendaal, H., & Zagmutt, F. J. (2017). Outbreak-related disease burden associated with consumption of unpasteurized cow's milk and cheese, United States, 2009–2014. *Emerging Infectious Diseases, 23*(6), 957. https://doi.org/10.3201/eid2306.151603.
22. Crimarco, A., Springfield, S., Petlura, C., Streaty, T., Cunanan, K., Lee, J., et al. (2020). A randomized crossover trial on the effect of plant-based compared with animal-based meat on trimethylamine-N-oxide and cardiovascular disease risk factors in generally healthy adults: Study With Appetizing Plantfood—Meat Eating Alternative Trial (SWAP-MEAT). *American Journal of Clinical Nutrition, 112*(5), 1188–1199. https://doi.org/10.1093/ajcn/nqaa203.
23. Davidou, S., Christodoulou, A., Fardet, A., & Frank, K. (2020). The holistico-reductionist Siga classification according to the degree of food processing: An evaluation of ultra-processed foods in French supermarkets. *Food & Function, 11*(3), 2026–2039. https://doi.org/10.1039/C9FO02271F.
24. de Miranda, R. C., Rauber, F., & Levy, R. B. (2021). Impact of ultra-processed food consumption on metabolic health. *Current Opinion in Lipidology, 32*(1), 24–37. https://doi.org/10.1097/MOL.0000000000000728.
25. DeSalvo, K. B., Olson, R., & Casavale, K. O. (2016). Dietary guidelines for Americans. *JAMA, 315*(5), 457–458. https://doi.org/10.1001/jama.2015.18396.
26. Djupegot, I. L., Nenseth, C. B., Bere, E., Bjørnarå, H. B. T., Helland, S. H., Øverby, N. C., et al. (2017). The association between time scarcity, sociodemographic correlates and consumption of ultra-processed foods among parents in Norway: A cross-sectional study. *BMC Public Health, 17*(1), 1–8. https://doi.org/10.1186/s12889-017-4408-3.
27. Drewnowski, A., Gupta, S., & Darmon, N. (2020). An overlap between "ultraprocessed" foods and the preexisting Nutrient Rich

Foods Index? *Nutrition Today, 55*(2), 75–81. https://doi.org/10.1097/NT.0000000000000400.

28. Eicher-Miller, H. A., Fulgoni, V. L., III, & Keast, D. R. (2012). Contributions of processed foods to dietary intake in the US from 2003–2008: A report of the Food and Nutrition Science Solutions Joint Task Force of the Academy of Nutrition and Dietetics, American Society for Nutrition, Institute of Food Technologists, and International Food Information Council. *The Journal of Nutrition, 142*(11), 2065S–2072S. https://doi.org/10.3945/jn.112.164442.
29. Elizabeth, L., Machado, P., Zinöcker, M., Baker, P., & Lawrence, M. (2020). Ultra-processed foods and health outcomes: A narrative review. *Nutrients, 12*(7), 1955. https://doi.org/10.3390/nu12071955.
30. Fagerberg, P., Langlet, B., Oravsky, A., Sandborg, J., Löf, M., & Ioakimidis, I. (2019). Ultra-processed food advertisements dominate the food advertising landscape in two Stockholm areas with low vs high socioeconomic status. Is it time for regulatory action? *BMC Public Health, 19*(1), 1–10.
31. Fangupo, L. J., Haszard, J. J., Taylor, B. J., Gray, A. R., Lawrence, J. A., & Taylor, R. W. (2021). Ultra-processed food intake and associations with demographic factors in young New Zealand children. *Journal of the Academy of Nutrition and Dietetics, 121*(2), 305–313. https://doi.org/10.1016/j.jand.2020.08.088.
32. Fardet, A. (2016). Minimally processed foods are more satiating and less hyperglycemic than ultra-processed foods: A preliminary study with 98 ready-to-eat foods. *Food Function, 7*(5), 2338–2346. https://doi.org/10.1039/C6FO00107F.
33. Fazzino, T. L., Dorling, J. L., Apolzan, J. W., & Martin, C. K. (2021). Meal composition during an ad libitum buffet meal and longitudinal predictions of weight and percent body fat change: The role of hyper-palatable, energy dense, and ultra-processed foods. *Appetite, 167*, 105592. https://doi.org/10.1016/j.appet.2021.105592.
34. Filgueiras, A. R., de Almeida, V. B. P., Nogueira, P. C. K., Domene, S. M. A., da Silva, C. E., Sesso, R., et al. (2019). Exploring the consumption of ultra-processed foods and its association with food addiction in overweight children. *Appetite, 135*, 137–145. https://doi.org/10.1016/j.appet.2018.11.005.
35. Fiolet, T., Srour, B., Sellem, L., Kesse-Guyot, E., Allès, B., Méjean, C., et al. (2018). Consumption of ultra-processed foods and cancer risk: Results from the NutriNet-Santé prospective cohort. *BMJ, 360*.

36. Floros, J. D., Newsome, R., Fisher, W., Barbosa-Cánovas, G. V., Chen, H., Dunne, C. P., et al. (2010). Feeding the world today and tomorrow: The importance of food science and technology: An IFT scientific review. *Comprehensive Reviews in Food Science and Food Safety, 9*(5), 572–599. https://doi.org/10.1111/j.1541-4337.2010.00127.x.
37. Food Agriculture Organization of the United Nations. (2015). Guidelines on the collection of information on food processing through food consumption surveys. FAO Rome.
38. Franckle, R. L., Moran, A., Hou, T., Blue, D., Greene, J., & Thorndike, A. N., et al. (2017). Transactions at a Northeastern supermarket chain: Differences by Supplemental Nutrition Assistance Program use. *American Journal of Preventive Medicine, 53*(4), e131–e138. https://doi.org/10.1016/j.amepre.2017.06.019.
39. Gibney, M. J. (2019). Ultra-processed foods: Definitions and policy issues. *Current Developments in Nutrition, 3*(2), nzy077.
40. Gibney, M. J. (2020). Ultraprocessed foods and their application to nutrition policy. *Nutrition Today, 55*(1), 16–21. https://doi.org/10.1097/NT.0000000000000393.
41. Grundy, M. M. L., Lapsley, K., & Ellis, P. R. (2016). A review of the impact of processing on nutrient bioaccessibility and digestion of almonds. *International Journal of Food Science & Technology, 51*(9), 1937–1946. https://doi.org/10.1111/ijfs.13192.
42. Hall, K. D., Ayuketah, A., Brychta, R., Cai, H., Cassimatis, T., Chen, K. Y., et al. (2019). Ultra-processed diets cause excess calorie intake and weight gain: An inpatient randomized controlled trial of ad libitum food intake. *Cell Metabolism, 30*(1), 67–77.e3.
43. Harris, R. M., Rose, A. M., Soares-Wynter, S., & Unwin, N. (2021). Ultra-processed food consumption in Barbados: Evidence from a nationally representative, cross-sectional study. *Journal of Nutritional Science, 10.*
44. Heinrich, V., Zunabovic, M., Varzakas, T., Bergmair, J., & Kneifel, W. (2016). Pulsed light treatment of different food types with a special focus on meat: A critical review. *Critical Reviews in Food Science and Nutrition, 56*(4), 591–613. https://doi.org/10.1080/10408398.2013.826174.
45. Horta, P. M., Rodrigues, F. T., & Dos Santos, L. C. (2018). Ultra-processed food product brands on Facebook pages: Highly accessed by Brazilians through their marketing techniques. *Public Health Nutrition, 21*(8), 1515–1519. https://doi.org/10.1017/S1368980018000083.
46. Jones, J. M. (2019). Food processing: Criteria for dietary guidance and public health? *Proceedings of the Nutrition Society, 78*(1), 4–18. https://

doi.org/10.1017/S0029665118002513.

47. Jones, J. M., & Clemens, R. A. (2017). Processed and ultraprocessed foods defined—An Alice in Wonderland question. *Cereal Foods World, 62*, 120. https://doi.org/10.1094/CFW-62-3-0120.
48. Juul, F., Vaidean, G., & Parekh, N. (2021). Ultra-processed foods and cardiovascular diseases: Potential mechanisms of action. *Advances in Nutrition*.
49. Khandpur, N., Cediel, G., Obando, D. A., Jaime, P. C., & Parra, D. C. (2020). Sociodemographic factors associated with the consumption of ultra-processed foods in Colombia. *Revista de Saúde Pública, 54*, 19. https://doi.org/10.11606/s1518-8787.2020054001176.
50. Kim, H., Hu, E. A., & Rebholz, C. M. (2019). Ultra-processed food intake and mortality in the United States: Results from the Third National Health and Nutrition Examination Survey (NHANES III 1988–1994). *Public Health Nutrition, 22*(10), 1777. https://doi.org/10.1017/S1368980018003890.
51. Konieczna, J., Morey, M., Abete, I., Bes-Rastrollo, M., Ruiz-Canela, M., & Vioque, J., et al. (2021). Contribution of ultra-processed foods in visceral fat deposition and other adiposity indicators: Prospective analysis nested in the PREDIMED-Plus trial. *Clinical Nutrition*.
52. Lane, M. M., Davis, J. A., Beattie, S., Gómez-Donoso, C., Loughman, A., O'Neil, A., et al. (2021). Ultra-processed food and chronic noncommunicable diseases: A systematic review and meta-analysis of 43 observational studies. *Obesity Reviews, 22*(3), e13146.
53. Lane, M., Howland, G., West, M., Hockey, M., Marx, W., Loughman, A., et al. (2020). The effect of ultra-processed very low-energy diets on gut microbiota and metabolic outcomes in individuals with obesity: A systematic literature review. *Obesity Research & Clinical Practice, 14*(3), 197–204. https://doi.org/10.1016/j.orcp.2020.04.006.
54. Lavigne-Robichaud, M., Moubarac, J. C., Lantagne-Lopez, S., Johnson-Down, L., Batal, M., Sidi, E. A. L., et al. (2018). Diet quality indices in relation to metabolic syndrome in an Indigenous Cree (Eeyouch) population in northern Québec, Canada. *Public Health Nutrition, 21*(1), 172–180. https://doi.org/10.1017/S136898001700115X.
55. Leffa, P. S., Hoffman, D. J., Rauber, F., Sangalli, C. N., Valmórbida, J. L., & Vitolo, M. R. (2020). Longitudinal associations between ultra-processed foods and blood lipids in childhood. *British Journal of Nutrition, 124*(3), 341–348. https://doi.org/10.1017/S0007114520001233.
56. Louzada, M. L. D. C., Martins, A. P. B., Canella, D. S., Baraldi, L. G.,

Levy, R. B., Claro, R. M., et al. (2015). Ultra-processed foods and the nutritional dietary profile in Brazil. *Revista de Saúde Pública, 49*.

57. Machado, P. P., Steele, E. M., da Costa Louzada, M. L., Levy, R. B., Rangan, A., Woods, J., et al. (2020). Ultra-processed food consumption drives excessive free sugar intake among all age groups in Australia. *European Journal of Nutrition, 59*(6), 2783–2792. https://doi.org/10.1007/s00394-019-02125-y.
58. Machado, P. P., Steele, E. M., Levy, R. B., da Costa Louzada, M. L., Rangan, A., Woods, J., et al. (2020). Ultra-processed food consumption and obesity in the Australian adult population. *Nutrition & Diabetes, 10*(1), 1–11. https://doi.org/10.1038/s41387-020-00141-0.
59. Mais, L., Horta, P., Martins, A., & Claro, R. (2020). Ultra-processed food and beverage advertising on Brazilian television by International Network for Food and Obesity/Non-Communicable Diseases Research, Monitoring and Action Support benchmark. *Public Health Nutrition, 23*(15), 2657–2662. https://doi.org/10.1017/S1368980020000518.
60. Marrón-Ponce, J. A., Sánchez-Pimienta, T. G., da Costa Louzada, M. L., & Batis, C. (2018). Energy contribution of NOVA food groups and sociodemographic determinants of ultra-processed food consumption in the Mexican population. *Public Health Nutrition, 21*(1), 87–93. https://doi.org/10.1017/S1368980017002129.
61. Marrón-Ponce, J. A., Tolentino-Mayo, L., Hernández-F, M., & Batis, C. (2019). Trends in ultra-processed food purchases from 1984 to 2016 in Mexican households. *Nutrients, 11*(1), 45. https://doi.org/10.3390/nu11010045.
62. Mendonça, R. d. D., Lopes, A. C. S., Pimenta, A. M., Gea, A., Martinez-Gonzalez, M. A., & Bes-Rastrollo, M. (2017). Ultra-processed food consumption and the incidence of hypertension in a Mediterranean cohort: The Seguimiento Universidad de Navarra Project. *American Journal of Hypertension, 30*(4), 358–366.
63. Messer, K. D., Costanigro, M., & Kaiser, H. M. (2017). Labeling food processes: The good, the bad and the ugly. *Applied Economic Perspectives and Policy, 39*(3), 407–427. https://doi.org/10.1093/aepp/ppx028.
64. Mialon, M., Crosbie, E., & Sacks, G. (2020). Mapping of food industry strategies to influence public health policy, research and practice in South Africa. *International Journal of Public Health, 65*(7), 1027–1036. https://doi.org/10.1007/s00038-020-01407-1.
65. Miclotte, L., & Van de Wiele, T. (2020). Food processing, gut microbiota and the globesity problem. *Critical Reviews in Food Science and Nutrition,*

60(11), 1769–1782. https://doi.org/10.1080/10408398.2019.1596878.

66. Mikkilä, V., Räsänen, L., Raitakari, O., Pietinen, P., & Viikari, J. (2005). Consistent dietary patterns identified from childhood to adulthood: The cardiovascular risk in Young Finns Study. *British Journal of Nutrition, 93*(6), 923–931. https://doi.org/10.1079/BJN20051418.
67. Misra, N., Koubaa, M., Roohinejad, S., Juliano, P., Alpas, H., Inácio, R. S., et al. (2017). Landmarks in the historical development of twenty first century food processing technologies. *Food Research International, 97*, 318–339. https://doi.org/10.1016/j.foodres.2017.05.001.
68. Moen, B., Berget, I., Rud, I., Hole, A. S., Kjos, N. P., & Sahlstrøm, S. (2016). Extrusion of barley and oat influence the fecal microbiota and SCFA profile of growing pigs. *Food Function, 7*(2), 1024–1032. https://doi.org/10.1039/C5FO01452B.
69. Monteiro, C. A. (2009). Nutrition and health. The issue is not food, nor nutrients, so much as processing. *Public Health Nutrition, 12*(5), 729–731.
70. Monteiro, C. A., Cannon, G., Lawrence, M., Costa Louzada, M. d., & Pereira Machado, P. (2019). Ultra-processed foods, diet quality, and health using the NOVA classification system. FAO.
71. Monteiro, C. A., Cannon, G., Levy, R. B., Moubarac, J. C., Louzada, M. L., Rauber, F., et al. (2019). Ultra-processed foods: What they are and how to identify them. *Public Health Nutrition, 22*(5), 936–941. https://doi.org/10.1017/S1368980018003762.
72. Monteiro, C. A., Cannon, G., Moubarac, J. C., Levy, R. B., Louzada, M. L. C., & Jaime, P. C. (2018). The UN Decade of Nutrition, the NOVA food classification and the trouble with ultra-processing. *Public Health Nutrition, 21*(1), 5–17. https://doi.org/10.1017/S1368980017000234.
73. Monteiro, C. A., Cannon, G., Moubarac, J. C., Martins, A. P. B., Martins, C. A., Garzillo, J., et al. (2015). Dietary guidelines to nourish humanity and the planet in the twenty-first century: A blueprint from Brazil. *Public Health Nutrition, 18*(13), 2311–2322.
74. Monteiro, C. A., Moubarac, J. C., Levy, R. B., Canella, D. S., da Costa Louzada, M. L., & Cannon, G. (2018). Household availability of ultra-processed foods and obesity in nineteen European countries. *Public Health Nutrition, 21*(1), 18–26. https://doi.org/10.1017/S1368980017001379.
75. Moradi, S., Entezari, M. H., Mohammadi, H., Jayedi, A., Lazaridi, A. V., Kermani, M. H., et al. (2021). Ultra-processed food consumption and adult obesity risk: A systematic review and dose-response meta-analysis. *Critical Reviews in Food Science and Nutrition*, 1–12.

76. Moran, A. J., Khandpur, N., Polacsek, M., & Rimm, E. B. (2019). What factors influence ultra-processed food purchases and consumption in households with children? A comparison between participants and non-participants in the Supplemental Nutrition Assistance Program (SNAP). *Appetite, 134*, 1–8. https://doi.org/10.1016/j.appet.2018.12.009.
77. Moubarac, J. C. (2015). Ultra-processed food and drink products in Latin America: Trends, impact on obesity, policy implications. *Pan American Health Organization World Health Organization*.
78. Moubarac, J. C., Batal, M., Louzada, M., Steele, E. M., & Monteiro, C. (2017). Consumption of ultra-processed foods predicts diet quality in Canada. *Appetite, 108*, 512–520. https://doi.org/10.1016/j.appet.2016.11.006.
79. Moubarac, J. C., Parra, D. C., Cannon, G., & Monteiro, C. A. (2014). Food classification systems based on food processing: Significance and implications for policies and actions: A systematic literature review and assessment. *Current Obesity Reports, 3*(2), 256–272. https://doi.org/10.1007/s13679-014-0092-0.
80. Movassagh, E. Z., Baxter-Jones, A. D., Kontulainen, S., Whiting, S. J., & Vatanparast, H. (2017). Tracking dietary patterns over 20 years from childhood through adolescence into young adulthood: The Saskatchewan Pediatric Bone Mineral Accrual Study. *Nutrients, 9*(9), 990. https://doi.org/10.3390/nu9090990.
81. Mullen, A. (2020). Ultra-processed food and chronic disease. *Nature Food, 1*(12), 771–772.
82. Nardocci, M., Leclerc, B. S., Louzada, M. L., Monteiro, C. A., Batal, M., & Moubarac, J. C. (2019). Consumption of ultra-processed foods and obesity in Canada. *Canadian Journal of Public Health, 110*(1), 4–14. https://doi.org/10.17269/s41997-018-0130-x.
83. Ng, S. W., Rivera, J. A., Popkin, B. M., & Colchero, M. A. (2019). Did high sugar-sweetened beverage purchasers respond differently to the excise tax on sugar-sweetened beverages in Mexico? *Public Health Nutrition, 22*(4), 750–756. https://doi.org/10.1017/S136898001800321X.
84. Oliveira, T., Ribeiro, I., Jurema-Santos, G., Nobre, I., Santos, R., Rodrigues, C., et al. (2020). Can the consumption of ultra-processed food be associated with anthropometric indicators of obesity and blood pressure in children 7 to 10 years old? *Foods, 9*(11), 1567. https://doi.org/10.3390/foods9111567.
85. Ostfeld, R. J., & Allen, K. E. (2021). Ultra-processed foods and

cardiovascular disease: Where do we go from here? *American College of Cardiology Foundation Washington DC.*

86. Pagliai, G., Dinu, M., Madarena, M., Bonaccio, M., Iacoviello, L., & Sofi, F. (2021). Consumption of ultra-processed foods and health status: A systematic review and meta-analysis. *British Journal of Nutrition, 125*(3), 308–318. https://doi.org/10.1017/S0007114520002688.
87. Peñalvo, J. L., Cudhea, F., Micha, R., Rehm, C. D., Afshin, A., Whitsel, L., et al. (2017). The potential impact of food taxes and subsidies on cardiovascular disease and diabetes burden and disparities in the United States. *BMC Medicine, 15*(1), 1–13. https://doi.org/10.1186/s12916-017-0971-9.
88. Pérez-Andrés, J. M., Charoux, C. M., Cullen, P., & Tiwari, B. K. (2018). Chemical modifications of lipids and proteins by nonthermal food processing technologies. *Journal of Agricultural and Food Chemistry, 66*(20), 5041–5054.
89. Popkin, B. M., Barquera, S., Corvalan, C., Hofman, K. J., Monteiro, C., Ng, S. W., et al. (2021). Towards unified and impactful policies to reduce ultra-processed food consumption and promote healthier eating. *The Lancet Diabetes & Endocrinology.*
90. ProCon.org: Global Obesity Levels. (2020). Retrieved from: https://obesity.procon.org/global-obesity-levels/ (accessed on 21 October 2024).
91. Rancière, F., Lyons, J. G., Loh, V. H., Botton, J., Galloway, T., Wang, T., et al. (2015). Bisphenol A and the risk of cardiometabolic disorders: A systematic review with meta-analysis of the epidemiological evidence. *Environmental Health, 14*(1), 1–23. https://doi.org/10.1186/s12940-015-0036-5.
92. Rauber, F., Campagnolo, P. D. B., Hoffman, D. J., & Vitolo, M. R. (2015). Consumption of ultra-processed food products and its effects on children's lipid profiles: A longitudinal study. *Nutrition, Metabolism & Cardiovascular Diseases, 25*(1), 116–122. https://doi.org/10.1016/j.numecd.2014.08.001.
93. Rauber, F., Chang, K., Vamos, E. P., da Costa Louzada, M. L., Monteiro, C. A., Millett, C., et al. (2021). Ultra-processed food consumption and risk of obesity: A prospective cohort study of UK Biobank. *European Journal of Nutrition, 60*(4), 2169–2180. https://doi.org/10.1007/s00394-020-02367-1.
94. Ruíz-Roso, M. B., de Carvalho, P. P., Matilla-Escalante, D. C., Brun, P., Ulloa, N., Acevedo-Correa, D., et al. (2020). Changes of physical activity and ultra-processed food consumption in adolescents from different

countries during Covid-19 pandemic: An observational study. *Nutrients, 12*(8), 2289. https://doi.org/10.3390/nu12082289.

95. Sadler, C. R., Grassby, T., Hart, K., Raats, M., Sokolović, M., & Timotijevic, L. (2021). Processed food classification: Conceptualisation and challenges. *Trends in Food Science & Technology*.
96. Santos, F., Dias, M., Mintem, G., Oliveira, I., & Gigante, D. (2020). Food processing and cardiometabolic risk factors: A systematic review. *Revista de Saúde Pública, 54*, 70. https://doi.org/10.11606/s1518-8787.2020054001704.
97. Schnabel, L., Kesse-Guyot, E., Allès, B., Touvier, M., Srour, B., Hercberg, S., et al. (2019). Association between ultra-processed food consumption and risk of mortality among middle-aged adults in France. *JAMA Internal Medicine, 179*(4), 490–498. https://doi.org/10.1001/jamainternmed.2018.7289.
98. Setyowati, D., Andarwulan, N., & Giriwono, P. E. (2018). Processed and ultraprocessed food consumption pattern in the Jakarta Individual Food Consumption Survey 2014. *Asia Pacific Journal of Clinical Nutrition, 27*(4), 840.
99. Silva Meneguelli, T., Viana Hinkelmann, J., Hermsdorff, H. H. M., Zulet, M. Á., Martínez, J. A., & Bressan, J. (2020). Food consumption by degree of processing and cardiometabolic risk: A systematic review. *International Journal of Food Science & Nutrition, 71*(6), 678–692. https://doi.org/10.1080/09637486.2020.1725961.
100. Simões, B. d. S., Barreto, S. M., Molina, M. d. C. B., Luft, V. C., Duncan, B. B., Schmidt, M. I., et al. (2018). Consumption of ultra-processed foods and socioeconomic position: A cross-sectional analysis of the Brazilian Longitudinal Study of Adult Health. *Cadernos de Saúde Pública, 34*, e00019717.
101. Srour, B., Fezeu, L. K., Kesse-Guyot, E., Allès, B., Debras, C., Druesne-Pecollo, N., et al. (2020). Ultraprocessed food consumption and risk of type 2 diabetes among participants of the NutriNet-Santé prospective cohort. *JAMA Internal Medicine, 180*(2), 283–291. https://doi.org/10.1001/jamainternmed.2019.5942.
102. Srour, B., Fezeu, L. K., Kesse-Guyot, E., Allès, B., Méjean, C., Andrianasolo, R. M., et al. (2019). Ultra-processed food intake and risk of cardiovascular disease: Prospective cohort study (NutriNet-Santé). *BMJ, 365*.
103. Steele, E. M., Baraldi, L. G., da Costa Louzada, M. L., Moubarac, J. C., Mozaffarian, D., & Monteiro, C. A. (2016). Ultra-processed foods and

added sugars in the US diet: Evidence from a nationally representative cross-sectional study. *BMJ Open, 6*(3).

104. Tavares, L. F., Fonseca, S. C., Rosa, M. L. G., & Yokoo, E. M. (2012). Relationship between ultra-processed foods and metabolic syndrome in adolescents from a Brazilian Family Doctor Program. *Public Health Nutrition, 15*(1), 82–87. https://doi.org/10.1017/S1368980011001571.
105. Turner, S., & Jones, C. (2018). The Milkybar 'Kid': Nestlé, childhood obesity and the regulation of advertising of high sugar and fat ultra-processed products. *British Dental Journal, 225*(8), 771–776. https://doi.org/10.1038/sj.bdj.2018.868.
106. van Boekel, M., Fogliano, V., Pellegrini, N., Stanton, C., Scholz, G., Lalljie, S., et al. (2010). A review on the beneficial aspects of food processing. *Molecular Nutrition & Food Research, 54*(9), 1215–1247. https://doi.org/10.1002/mnfr.200900608.
107. Vedovato, G. M., Vilela, S., Severo, M., Rodrigues, S., Lopes, C., & Oliveira, A. (2021). Ultra-processed food consumption, appetitive traits and BMI in children: A prospective study. *British Journal of Nutrition, 125*(12), 1427–1436. https://doi.org/10.1017/S0007114520003712.
108. Wang, L., Steele, E. M., Du, M., Pomeranz, J. L., O'Connor, L. E., Herrick, K. A., et al. (2021). Trends in consumption of ultraprocessed foods among US youths aged 2–19 years, 1999–2018. *JAMA, 326*(6), 519–530. https://doi.org/10.1001/jama.2021.10238.
109. Zinöcker, M. K., & Lindseth, I. A. (2018). The Western diet–microbiome-host interaction and its role in metabolic disease. *Nutrients, 10*(3), 365. https://doi.org/10.3390/nu10030365.

Chapter 4

Cardiovascular Health Implications

Contents

INTRODUCTION

Unlike processed foods, which are whole foods preserved by conventional methods like canning or pickling, ultra-processed foods are industrially produced ready-to-eat or ready-to-heat formulations with little to no whole foods and food additives. According to recent epidemiological research, consuming more ultra-processed food may raise your risk of cardiovascular disease (CVD). Nevertheless, biological plausibility standards must be used to support epidemiological data.

As the primary cause of death in both the US and the rest of the world, cardiovascular diseases (CVDs) account for about one-third of all fatalities. Over half of all US deaths from coronary heart disease (CHD) and stroke are thought to be related to poor diet, making it the primary risk factor for the country's total CVD burden. Without a doubt, nutrition is crucial for preventing both primary and secondary CVD. Rather than concentrating on specific nutrients, current research and practice are more concerned with overall dietary patterns and quality (Mozaffarian, 2016).

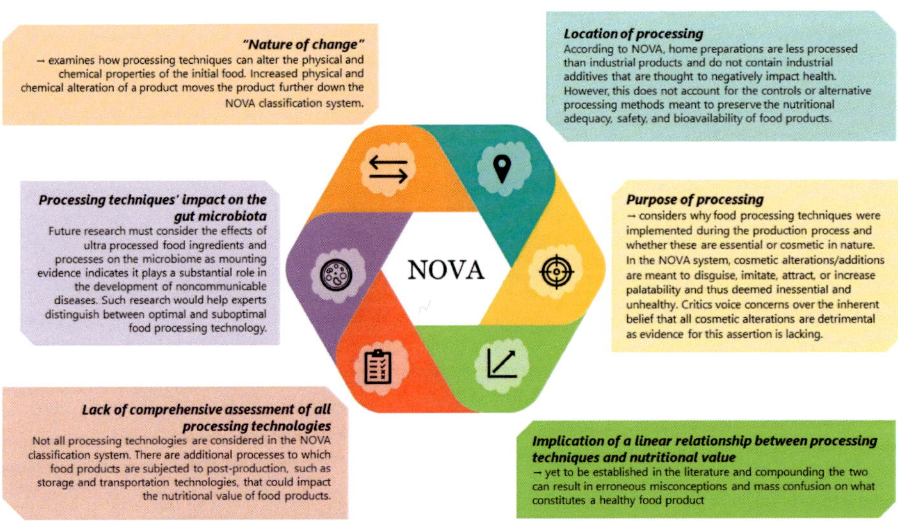

Figure 4.1. *Food-processing spectrum as defined by the NOVA framework.*

Source: Nees S, Lutsiv T, Thompson HJ. Ultra-Processed Foods—Dietary Foe or Potential Ally? Nutrients. 2024; 16(7):1013. https://doi.org/10.3390/nu16071013.

Food processing has a significant impact on the quality of a diet, but there is a wide range of processing, from minimal processing (e.g., frozen veggies, pasteurized milk, dried fruits free of added sugar or additives, and frozen vegetables) to ultra-processing (e.g., hot dogs, fast food, carbonated soft drinks, and industrially made breads) (Figure 4.1). The majority of energy in the US diet comes from ultra-processed foods, which are defined as industrially produced, ready-to-eat, or ready-to-heat formulations containing little whole foods. Not only do these foods include junk food, but they also include a lot of foods that are marketed and thought to be healthy, like yogurts with flavors, low-fat/low-calorie products, breakfast cereals, and goods that have been "enriched" with healthy nutrients.

Although data is still being gathered, recent epidemiological studies indicate that consuming more ultra-processed foods may raise one's risk of cardiovascular disease. After 18 years of follow-up, the Framingham Offspring Study found that a daily serving of ultra-processed foods increased the risk of incident CVD by 7%. Similarly, a 12% higher risk of CVD was linked to higher consumption of ultra-processed foods in the French NutriNet-Santé cohort study. Additionally, there is strong evidence from meta-analyses that certain ultra-processed foods (like processed meat and beverages with added sugar) and nutrients (like trans fats and sodium) found in large amounts in ultra-processed foods raise the risk of CVD. Epidemiological studies have also connected ultra-processed foods to several CVD risk factors, such as obesity, hypertension, and metabolic syndrome. The American College of Cardiology/American Heart Association Guideline on the Primary Prevention of Cardiovascular Disease recommends avoiding trans fats and minimizing certain ultra-processed foods (processed red meats, foods high in refined carbohydrates, such as refined grain products, and sweetened beverages) due to the strength and level of evidence of these associations that currently exists. However, physicians, and public health experts are still largely unaware of the potential health risks linked to a wide variety of ultra-processed foods, aside from the products already mentioned.

The lack of large randomized clinical trials (RCTs) with hard cardiovascular endpoints is a major obstacle to establishing evidence in cardiovascular nutrition. As a result, to prove causation, data from carefully planned prospective observational studies must be supported by biologically plausible standards. A fundamental aspect of preventive cardiology's theory and application is comprehending the mechanisms of action of different food ingredients and dietary patterns that comprise the human "food-ome." Thus, this chapter aims to provide an overview of the data that are currently available concerning the potential biological mechanisms that underlie the correlations between the consumption of ultra-processed foods and the risk of atherosclerotic and cardiometabolic CVDs.

This chapter tried to provide the most recent information on the state of the evidence regarding healthy dietary choices for CVD by going over the research that is currently available on the dietary components that may affect its progression and prevention. We summarize the findings from population-based trials, observational studies, and meta-analyses with an emphasis on the impacts of specific macronutrients, whole food products, and dietary patterns on the risks of CVD. Unhealthy eating practices, such as consuming large amounts of saturated fats, sugar-sweetened drinks, red, and processed meat, and salt, are linked to an increased risk of CVD. On the other hand, it has been demonstrated that consuming more plant-based foods like fruits, vegetables, nuts, and dietary fiber can effectively lower the risk factors associated with CVD. One of the dietary regimens that seems to have the most scientific support for preventing CVD is the Mediterranean diet. The question of whether taking vitamin and mineral supplements offers cardioprotective advantages is still hotly contested. This chapter offers fresh perspectives on the role of dietary components that can either prevent or worsen CVD and can be implemented to enhance cardiovascular health.

4.1. POTENTIAL MECHANISMS LINKING ULTRA-PROCESSED FOODS TO CVD RISK

Foods may become less healthful as a result of processing, which can also change a food's physical (food structure), chemical (presence of additives, artificial sweeteners, and neo-formed contaminants, glycemic index, and load), and nutritional (macro- and micronutrient content) qualities. Food reward systems, satiety signaling, and long-term dietary habits can all be impacted by food processing. Unknown complex mechanisms and synergies between a variety of ultra-processed food components and attributes may be involved in the biological pathways through which these foods affect cardiovascular health (Figure 4.2). A number of pathways are involved in the intricate physio-pathological relationships that underlie atherogenesis and the development of CVD. Numerous variables, including endothelial dysfunction, prothrombotic, proinflammatory, prooxidative, and metabolic dysfunction, interact, and intensify one another. There are a lot of subtleties. For example, different glucose metabolism levels trigger distinct patterns of inflammation, and immune factors have a bilateral relationship with gut microbiota. Furthermore, the majority of cardiovascular risk factors contribute to the maintenance of a prothrombotic, proinflammatory molecular milieu as well as the induction of endothelial dysfunction and injury. A complex network of molecular feedback loop mechanisms, in conjunction with these factors, cause atherogenesis to intensify and persist, ultimately leading to a variety of cardiovascular events.

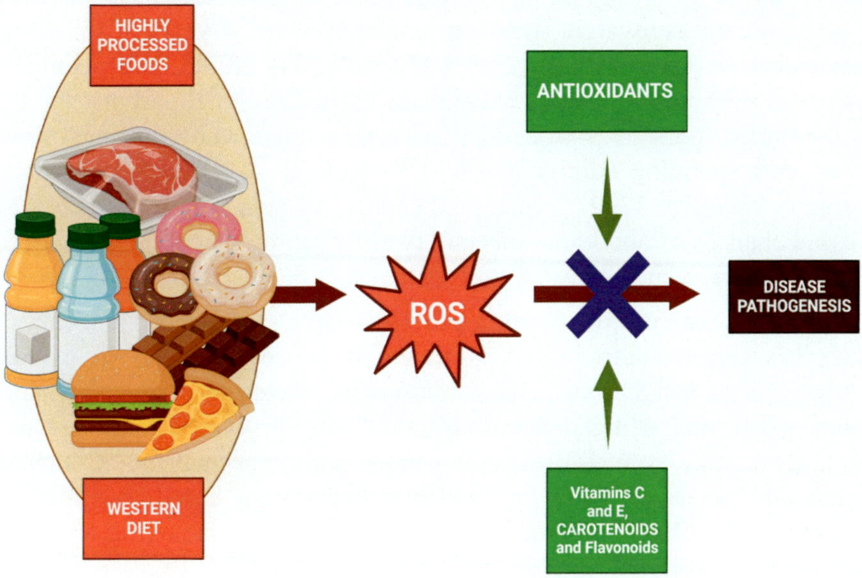

Figure 4.2. *Potential biological mechanisms underlying the association between ultra-processed foods and CVD. The diagram describes plausible biological pathways through which a diet high in ultra-processed foods may contribute to CVD. Key mechanisms include altered serum lipid concentrations, modified gut microbiota and host–microbiota interactions, obesity, inflammation, oxidative stress, dysglycemia, insulin resistance, hypertension, and hormonal imbalances.*

Source: Clemente-Suárez VJ, Beltrán-Velasco AI, Redondo-Flórez L, Martín-Rodríguez A, Tornero-Aguilera JF. Global Impacts of Western Diet and Its Effects on Metabolism and Health: A Narrative Review. Nutrients. *2023; 15(12):2749. https://doi.org/10.3390/nu15122749.*

4.1.1. Excess Weight and Body Adiposity

In both epidemiological and experimental research, a higher intake of ultra-processed foods has been linked to excess obesity. Interestingly, an RCT conducted by Hall and colleagues. showed that compared to a minimally processed diet, an ultra-processed diet increased ad libitum energy intake by approximately 500 kcal/d and resulted in weight gain. In the Brazilian Longitudinal Study of Adult Health and the prospective Seguimiento University of Navarra study, adult participants with ultra-processed food intakes in the highest quartile compared to the lowest quartile had a 26–29% increased risk of becoming overweight over approximately 9 and 4 years of follow-up, respectively [HR (95% CI): 1.26 (1.10, 1.45) and 1.29 (1.08, 1.53), respectively]. Participants in three cross-sectional studies, carried out in Brazil and the US, who consumed more ultra-processed foods than those who consumed fewer (top vs. corresponding quintile/quartile) showed 31–48% increased chances of being overweight, 47–97% increased chances of being obese, and 41–62% increased chances of abdominal obesity (waist circumference ≥88/102 cm for men and women, respectively).

The nutritional profile of ultra-processed foods and their tendency to replace low-calorie, nutrient-dense, minimally processed foods in the diet can both lead to weight gain. Due to the intense flavoring produced by high fat, salt, sugar, and artificial flavoring levels, ultra-processed foods are very appealing and may even override endogenous satiety mechanisms. Compared to minimally processed foods and culinary preparations based on minimally processed foods, ultra-processed foods are typically less satisfying and more hypercaloric. Higher energy density foods may encourage excessive energy intake because human satiety mechanisms respond better to volume than to calorie content (Rolls, 2017).

Ultra-processed foods have a higher energy density and orosensory qualities (i.e., easier-to-chew textures that are softer and less fibrous) may also make it possible to consume more energy in less time. According to experimental research, eating more frequently may lead to higher energy intakes, possibly as a result of delayed satiety signals. In fact, the ultra-processed diet had a significantly higher average eating rate (measured in grams/minute and kilocalories/minute) than the minimally processed diet in the Hall et al. RCT. Additionally, compared to the ultra-processed diet, higher concentrations of the hormone peptide YY, which suppresses appetite, were observed during the minimally processed diet (Hall et al., 2019). Lastly, ultra-processed food's accessibility, affordability, huge portion sizes, and persuasive marketing may encourage overindulgence in snacking and eating, which may lead to an increase in energy intake. According to experimental research, visual cues and pictures of appetizing foods, like those in food advertisements, can stimulate the parts of the brain linked to overeating and may even motivate eating when one is not actually hungry (Stanhope et al., 2018).

4.1.2. Glycemic Response and Related Insulin Dysregulation

The primary source of added and total sugar in the diet of Americans is ultra-processed food items. Consuming too much sugar is linked to a number of

CVD risk factors, such as insulin resistance, type 2 diabetes, abdominal, and total adiposity, hypertension, and dyslipidemia. This is especially the case with sugar-sweetened beverages. Diets high in glucose are linked to a higher risk of type 2 diabetes and coronary heart disease (CHD). Food matrix structure (e.g., liquid or solid), amount, and quality of carbohydrates (source and digestibility), fat, protein, and fiber content, and level of processing all influence a food's glycemic response. Foods with high processing levels are generally more hyperglycemic than those with low and moderate processing levels, according to experimental data. Due to its promotion of endothelial dysfunction, oxidative stress, inflammation, and weight gain, hyperglycemia raises the risk of cardiovascular disease (Blaak et al., 2012). Low-calorie sweeteners' physiological effects are not fully understood. Regular consumption may be linked to a higher risk of type 2 diabetes, according to observational data from prospective studies. Experimental research on humans and animals has revealed that some low-calorie sweeteners, like sucralose, may affect brain regulatory centers to change glucose metabolism and may impair gut-brain regulatory pathways to decrease insulin sensitivity (Dalenberg et al., 2020).

4.1.3. Hypertension

One of the main risk factors for CVD and stroke is hypertension, which is linked to an excessive dietary intake of sodium. The intricate and incompletely understood mechanisms that link a higher sodium intake to elevated blood pressure are still being studied. A number of pathways are involved, including extrarenal sodium handling, altered renal sodium homeostasis, direct effects on vascular walls, and local and systemic neuro-hormonal pathways in combination with phenotypes associated with salt sensitivity. Changes in metabolism, hemodynamics, and inflammation cause endothelial stiffness, volume enlargement, water retention, elevated peripheral resistance, and ultimately elevated blood pressure. Higher potassium intakes are linked to lower risks of CVD mortality, while a higher sodium-to-potassium ratio ($\geq$1.0) is associated with an increased risk. These findings are supported by epidemiological studies. The main dietary sources of potassium in the US are minimally processed foods like milk and vegetables, while the main sources of sodium are commercially processed foods. Therefore, a higher intake of ultra-processed foods and a lower intake of foods high in potassium but minimally processed foods may affect the risk of CVD by raising sodium intake and changing the diet's sodium-to-potassium ratio.

4.1.4. Gut Microbiome

With possible effects on host health, diet is a major modulator of the composition and activity of the gut microbiota. Excessive fat, insulin resistance, type 2 diabetes, and cardiovascular disease have all been connected to disruptions in the gut microbial ecosystem, which include variations in the relative abundance of particular microbial taxa or in the diversity of gut bacteria, as well as intestinal barrier dysfunction. The underlying mechanisms are thought to involve weight gain, endotoxemia-induced low-grade systemic inflammation, increased bacterial production of atherogenic metabolites like choline, trimethylamine N-oxide, and betaine, and modification of the host immune system. Additional mechanisms could include the host consuming more calories, changes to energy homeostasis, and the buildup of lipids in the liver.

The composition, function, and interactions between bacteria and hosts of the gut microbiota may be affected by changes made to the food matrix, the amount of fat and fiber in food, and the addition of specific food additives during processing. High-fiber diets are linked to increased microbial gene richness and microbial production of SCFA fermentation end-products, which have important roles in regulating host metabolism and the immune system and may attenuate inflammation. Dietary fibers, primarily soluble fiber, are abundant in many unrefined plant foods and provide a substrate for microbial fermentation. In contrast to diets high in minimally processed plant foods, Western dietary patterns—which are marked by low fiber intake and high consumption of sugar, fat, and animal protein—are linked to a unique and less varied microbiota. Significantly, low-fiber diets cause the intestinal mucus layer to deteriorate and make people more vulnerable to chronic inflammatory diseases by changing the metabolism of gut microbes toward the use of proteins and host mucins. However, high-fiber diets based on processed, extruded grains led to less enrichment in beneficial butyrate-producing bacteria and decreased bacterial diversity in animal models when compared to diets based on unprocessed whole grains. Furthermore, high-fat Western diets may exacerbate metabolic disorders and low-grade systemic inflammation by raising circulating concentrations of lipopolysaccharides (LPS) derived from intestinal bacteria, as evidenced by studies on humans and animals (metabolic endotoxemia). Diets high in glucose or fructose have also been shown in animal studies to decrease gut microbiota diversity and increase intestinal permeability. In a human experiment, endotoxemia rose following a meal consisting primarily of highly processed, high-fat, high-carbohydrate foods, but not following an isocaloric meal consisting of whole grains, fruits, and nuts with little to no processing.

Processing-related modifications to the food matrix's physical structure can affect the kinetics and bioaccessibility of nutrients, which can have an impact on the growth, metabolism, and composition of the gut microbiota. The nutrients in ultra-processed foods are mostly acellular, whereas plant-based, minimally processed foods have intact fibrous cell walls that guarantee a slow release of nutrients along the digestive tract and provide substrate for fiber-degrading bacteria in the colon. According to experimental research, a high nutrient availability in the small intestine is caused by the high percentage of acellular nutrients found in ultra-processed foods. This, in turn, fosters an inflammatory gut microbiota linked to cardiometabolic conditions.

Furthermore, a growing body of research from studies on humans and animals indicates that consuming low-calorie sweeteners may upset the balance and diversity of the gut microbiota, which in turn may lead to metabolic problems and insulin resistance. Studies on animals have demonstrated that emulsifiers, which are used in industrial food processing, increase the pro-inflammatory potential of the microbiome by boosting microbiotic virulence factors.

4.1.5. Plasma Lipid Profile

The amount and quality of dietary fats and carbohydrates, which are influenced by food processing, have an impact on plasma lipid concentrations. It is widely accepted in science that partially hydrogenated vegetable oils, which are found in industrially produced trans fatty acids, negatively impact the blood lipoprotein profile and raise the risk of coronary heart disease (CHD). Consequently, the FDA has moved to eliminate partially hydrogenated vegetable oils from the food supply in the United States. Saturated fat consumption should be limited due to long-standing recommendations based on evidence that it raises low-density lipoprotein cholesterol. Whole-fat dairy products and other minimally processed foods, as well as processed ingredients used in ultra-processed foods like palm oil and palm kernel oil, are major sources of saturated fats in the diet. Nonetheless, a growing body of research indicates that the health effects of consuming saturated fat in the diet vary depending on the food source and the interactions between contaminants from high-heat oil processing and naturally occurring food components. For instance, it has been demonstrated that rats' serum cholesterol concentrations are raised by processed coconut oil but not by virgin coconut oil. Furthermore, eating unprocessed meat and full-fat dairy does not appear to raise the risk of cardiometabolic illnesses. Because of this, researchers have suggested substituting food-based recommendations for saturated fat intake—which consider the entire food matrix and level of processing—for dietary saturated fat targets.

Moreover, studies indicate that the source and degree of processing of carbohydrates affects their impact on serum lipid concentrations. Consuming whole grains lowers total and LDL cholesterol as well as serum triglycerides, albeit to a lesser extent when combined with sugar. Therefore, by encouraging healthy plasma lipid concentrations, eating minimally processed whole grains like oatmeal rather than ultra-processed, high-sugar, refined grain products may lower the risk of CVD.

4.1.6. Endocrine Pathways

Because inorganic phosphate salts are widely used as additives in industrial food processing, concerns have been raised about the total amount of phosphorus consumed in the US diet. Compared to the 40–60% of organic phosphorus that naturally occurs in food, inorganic phosphate is absorbed to a greater extent (80–100%). Additional evidence for a connection with CVD comes from experimental research on excess dietary phosphorus and epidemiological studies measuring serum phosphate. These effects may be mediated by increased secretion of fibroblast growth hormone 23 and parathyroid hormone, two hormones that promote arterial calcification and disrupt the hormonal regulation of extracellular phosphate. Decreased endothelial function and increased oxidative stress of the endothelial cells are also associated with elevated tissue and serum phosphate concentrations.

Foods that are industrially produced are frequently sold in ornate packaging that may include chemicals that disrupt hormones, like bisphenol A (BPA). There is insufficient epidemiological data to support the hypothesis that higher BPA exposure is linked to a higher prevalence of key CVD risk factors, such as diabetes, total, and abdominal obesity, and hypertension. It has been demonstrated that BPA, which shares structural similarities with 17β-estradiol, increases insulin resistance, oxidative stress, inflammation, adipogenesis, and pancreatic B-cell dysfunction by attaching itself to estrogen-related receptors. There is still much to learn about the specific mechanisms of action. Though foods kept or reheated in BPA-lined containers are thought to be the main source of human exposure, BPA is widely present in the environment. Diets centered on fresh and minimally processed foods appear to be associated with lower urinary concentrations of BPA, despite the paucity of empirical evidence. This is supported by observational and intervention studies. Increased consumption of sugar-sweetened beverages, but not ultra-processed foods in general, was linked to higher urinary concentrations of BPA, according to an analysis using data from the nationally representative NHANES 2013–2014. Although, due to safety concerns, BPA is increasingly replaced by its structurally homologous compounds (e.g., bisphenol S and bisphenol F), as well as others, that may disrupt hormones in a similar way.

4.1.7. Other Mechanisms

Advanced glycation end-products (AGEs) are produced during processing and preparation that involve high heat treatment. AGEs have been connected to elevated oxidative stress and inflammation and may be involved in the pathogenesis of CVD. Animal foods high in fat and protein are especially high in AGE formation, which rises with cooking temperatures, cooking times, and the absence of moisture. Deep-frying and dry-heat processing of foods high in carbohydrates (e.g., foods like chips, crackers, cookies, and French fries) also quicken the formation of AGEs. It is still unclear how important dietary AGEs are for the development of CVD and whether there is a causal relationship between AGEs and CVD.

According to observational studies, the consumption of fruits and vegetables is negatively correlated with major CVD risk factors, such as type 2 diabetes, CHD, stroke, and CVD. There is evidence to support a protective effect against stroke and coronary heart disease. The health benefits of fruits and vegetables are ascribed to their high fiber, micronutrient, and phytochemical content.

These components may lower the risk of CVD through several mechanisms, including lowered blood pressure, improved insulin sensitivity, and serum lipid profiles. It is conceivable that increasing consumption of ultra-processed foods displaces less-processed, more nutrient-dense foods like fruit and vegetables from the diet, increasing the risk of CVD and its intermediate risk factors. On the other hand, consuming a lot of fruits and vegetables may have a protective effect in part because fewer unhealthy, highly processed foods are consumed.

4.1.8. Areas for Future Research

In addition to determining the relative harm associated with the nutritional makeup, food additives, physical structure, and other characteristics of ultra-processed foods, research is necessary to better understand the biological mechanisms through which these foods may influence the risk of CVD. To eliminate unhealthy processing methods and ingredients and determine "optimal" processing levels, it is necessary to comprehend how ultra-processing alters whole foods and the pathways through which these foods impact health. Given the growing body of research on the role of the gut microbiome in cardiometabolic health and the relationship between diet and disease, the effects of ultra-processed foods on the gut microbiota and microbiota–host interactions are of particular scientific interest.

The health effects of long-term exposure to the various food additives and substances found in ultra-processed foods and their packaging also need to be clarified, as current regulations limiting the maximum levels of individual food additives do not account for the potential "cocktail effects" of combinations of compounds from different food sources (Chen et al., 2023).

4.2. HEALTHY DIETARY CHOICES FOR CARDIOVASCULAR DISEASE

Globally, cardiovascular disease (CVD) continues to be the leading cause of death. Nutrition is essential to cardiovascular health and directly influences the onset of CVD. A multitude of human studies have demonstrated the connection between diet and CVD.

A class of disorders affecting the heart and blood vessels are known as cardiovascular diseases (CVDs), and they include peripheral artery disease, aortic atherosclerosis, coronary heart disease (CHD), and stroke. They are the primary cause of death globally and a significant public health concern. The American Heart Association (AHA) reports that over 800,000 people in the US pass away from cardiovascular diseases (CVDs) each year, and 261 million people have some type of CVD. Over USD 316 billion is spent on direct and indirect costs associated with CVD deaths each year. In the world, cardiovascular disease (CVD) accounts for 17.9 million deaths annually, or roughly 32% of all deaths. More than 80% of CVD-related deaths are thought to occur in low- and middle-income nations. Thus, the development of cost-effective and efficient methods for CVD prevention and treatment is essential.

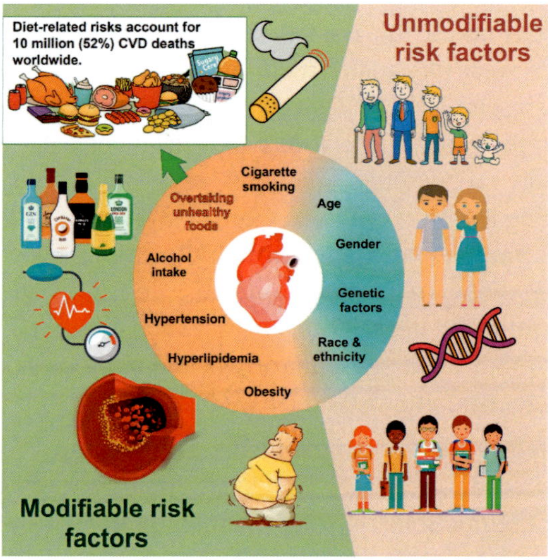

Figure 4.3. *Factors contributing to the development of CVD.*

Source: Chen W, Zhang S, Hu X, Chen F, Li D. A Review of Healthy Dietary Choices for Cardiovascular Disease: From Individual Nutrients and Foods to Dietary Patterns. Nutrients. 2023; 15(23):4898. https://doi.org/10.3390/nu15234898.

The development of CVD is influenced by unchangeable factors such as age, gender, and genetic predispositions. The influence of factors that can be changed has received a lot of attention lately. Inadequate eating patterns rank among the worst lifestyle choices for CVD morbidity and mortality (Figure 4.3), and they are responsible for roughly 10 million deaths globally. Over the past few decades, dietary patterns have replaced individual nutrients and particular foods as the focus of nutritional research in the context of CVD prevention. The best dietary ingredients to enhance cardiometabolic health are still unknown, though. Therefore, defining standards for a healthy diet and creating intervention plans to lower cardiovascular risk require a thorough and methodical assessment of the relationship between nutrition and CVD (Chen et al., 2023). In this section, we conducted a current review of the literature from observational studies, meta-analyses, and population-based randomized control trials (RCTs) to try to fully investigate the contribution of individual dietary components and patterns to CVD risk and health outcomes. For the management of CVD and general health, we offered an evidence-based summary of nutritious foods, dietary options, and recommendations.

4.2.1. Macronutrients

Carbohydrates, fat, and protein are called macronutrients. They are the nutrients you use in the largest amounts.

4.2.1.1. Fat

Triacylglycerols, which are primarily found in dietary fats, are composed of three distinct fatty acids bonded to a glycerol backbone via an ester bond. FAO and WHO recommend reducing dietary fat intake to prevent the development of CVD. There are four main categories of dietary fats found in food: monounsaturated, polyunsaturated, trans, and saturated fats.

Chains of carbon atoms joined by hydrogen bonds make up saturated fats. The present dietary guidelines advise reducing dietary intake of saturated fat because it is thought to improve cardiovascular health. The scientific justification for reducing dietary saturated fat is predicated on the theory that consuming too much of it may raise levels of low-density lipoprotein cholesterol, a significant risk factor for the advancement of cardiovascular disease. RCTs and prospective cohort studies have both reported a link between the consumption of saturated fat and the risk of CVD (Table 4.1). The risk of cardiovascular events appears to be decreased by consuming less saturated fat, though the benefits may vary depending on the food source. In particular, saturated fats from meat have a positive correlation with the risk of CVD, while saturated fats from dairy have an inverse correlation. Total fat and its subtypes, such as saturated and unsaturated fatty acids, are linked to a lower risk of stroke and overall mortality in the Prospective Urban Rural Epidemiology (PURE) study, but not with CVD mortality. Moreover, consumption of saturated fatty acids is inversely correlated with stroke mortality, according to the Japan Collaborative Cohort Study for Evaluation of Cancer Risk Study. Another meta-analysis shows that the relationships between fatty acid biomarkers and coronary risk are heterogeneous. These results demonstrate that the consumption of saturated fatty acids had negligible or no impact on cardiovascular or all-cause mortality.

Figure 4.4. *Suggested individual nutrient consumption according to the evidence on the association between dietary choices and cardiovascular disease management. (A) Fat; (B) Carbohydrate; (C) Protein; (D) Vitamins and minerals; (E) Dietary fiber.*

Source: Chen W, Zhang S, Hu X, Chen F, Li D. A Review of Healthy Dietary Choices for Cardiovascular Disease: From Individual Nutrients and Foods to Dietary Patterns. Nutrients. *2023; 15(23):4898. https://doi.org/10.3390/nu15234898.*

Similarly, there is insufficient data to support the notion that dietary saturated fat intake lowers the risk of cardiovascular disease. According to the Women's

Health Initiative RCTs, postmenopausal women's total mortality and CVD risk are not reduced by following a low-fat diet. Thus, the question of whether a diet high in saturated fatty acids can raise the risk of CVD is still open.

Consumption of saturated and trans fats is linked to an increased risk of cardiovascular disease (CVD), while consumption of monounsaturated and polyunsaturated fats is inversely correlated with CVD death, according to a prospective analysis of the PREvención con DIeta Mediterranean (PREDIMED) study, which included 7038 participants at high risk of CVD. Consistent with these results, a recent dose-response meta-analysis of cohort studies indicates that consumption of trans fatty acids in the diet is linked to an increased risk of cardiovascular disease (CVD), while consumption of polyunsaturated fatty acids is inversely associated with CVD risk. It has been shown that the Mediterranean diet, which is low in trans and saturated fats and high in polyunsaturated and monounsaturated fats, effectively lowers the risk of major cardiovascular events. More specifically, a lower risk of total CHD events and deaths is linked to higher consumption of linoleic acid, an n-6 polyunsaturated fat mainly found in nuts and vegetable oils. Thus, to lower the incidence of clinical CHD events, substituting monounsaturated and polyunsaturated fats for saturated and trans unsaturated fats may be advised (Figure 4.4A).

4.2.1.2. Carbohydrate

The primary energy source in the diet and a significant factor in health outcomes are carbohydrates. Since there may be a link between dietary fat consumption and the risk of CVD, replacing fats in the diet with carbohydrates is thought to help prevent CVD. The results, however, are mixed when it comes to the relationship between dietary carbohydrate intake and CVD events.

There is a correlation between lower mortality and lower carbohydrate intake, according to trial and cohort study results (Table 4.1). A low-carb, high-protein, high-fat diet, for example, has been shown to help obese people lose weight and improve their risk factors for coronary heart disease. In a meta-analysis of 11 RCTs with 1369 participants, individuals following a low-carb diet lost more weight but had higher cholesterol than those following a low-fat diet. A modest intake of carbohydrates is linked to a lower risk of heart disease and stroke, according to another prospective cohort study. Glycemic management is an essential strategy for preventing CVD. Research indicates that diets low in carbohydrates are beneficial for glycemic control. A non-calorie-restricted low-carbohydrate diet did not negatively impact cardiovascular risk factors in patients with type 2 diabetes, and it improved glycemic control and abdominal adiposity in a recent randomized controlled trial. These results indicate that cutting back on carbohydrates seems to be a good nutritional approach to lower cardiovascular risk factors. The findings of these studies, however, are unable to clarify whether the advantages of dietary intervention are related to reduced energy intake, high protein (animal or plant-based), or low carbohydrate intake.

An increased risk of total mortality was found to be associated with a high intake of carbohydrates in a large prospective cohort study comprising 135,335 individuals from 18 countries. A recent dose-response meta-analysis of 19 cohort studies with 15,663,111 participants revealed that while there is no correlation between higher carbohydrate intake and CVD risk in men, it is associated with a slight increase in risk in women. It's interesting to note that a different meta-analysis of prospective cohort studies revealed a possible U-shaped relationship between carbohydrate intake and mortality, showing that both high and low carbohydrate intake are associated with increased mortality. Different types of carbohydrates, such as sugar, starch, and fiber, have different associations with mortality and CVD risk, according to data from the UK Biobank cohort. This suggests that it is important to consider the sources and types of carbohydrates rather than the total amount of carbohydrates. The incremental area under the blood glucose response curve following consumption of a food is measured to determine the GI value, which is attributed to that particular food. The GI is influenced by both the amount and quality of carbohydrates, with high GI foods generally being associated with low-quality carbs. High GI and load diets are linked to CHD events in women but not in men, according to a prior meta-analysis of prospective cohort studies. Diets with a high GI are linked to a higher risk of CVD and death than diets with a low GI, according to a recent prospective study involving 137,851 participants from various nations and regions. The GI may therefore be utilized as a marker of the quality of carbohydrates concerning CVD or other chronic illnesses, according to these studies. For dietary intervention to prevent the negative consequences linked to CVD, improving the quality of carbohydrates by lowering the GI is crucial (Figure 4.4B).

4.2.1.3. Protein

A vital part of promoting health is dietary protein. It affects the metabolic targets related to energy metabolism, appetite, and satiety. According to early research, eating a high-protein diet may help prevent obesity and metabolic syndrome as well as help lower triglycerides, body weight, and fat mass (Table 4.1). Furthermore, the results of prospective cohort studies have demonstrated that switching to protein instead of carbs is linked to a lower risk of ischemic heart disease (IHD) and an improvement in cardiometabolic risk factors. Interestingly, a prospective U.S. Nurses' Health Study. s. women indicate that while increased intake of fish and poultry protein is significantly associated with lower risk of CHD, higher intake of red meat protein is significantly associated with elevated risk of the disease.

Table 4.1. *Effects of Macronutrients on Cardiometabolic Risk Factors, Cardiovascular Disease, and Mortality*

Macronutrients	Study	Research Object	MA (No. of Studies)	Exposure Measure	Follow-Up Period	Main Findings
Polyunsaturated fat and saturated fat	Meta-analysis of randomized controlled trials	Studies that randomized adults to increased total or n-6 PUFA consumption for at least 1 year and reported incidence of CHD.	YES (8)			Consuming PUFA in place of SFA can reduce the occurrence of clinical CHD events.
Saturated fat	Meta-analysis of prospective cohort studies	Studies of CHD, stroke, and cardiovascular disease of 347,747 participants.	YES (21)		5–23 years	No significant evidence supports associations between the intake of dietary saturated fat and the risk of CHD or CVD.
Saturated fat	The Japan Collaborative Cohort Study	58,453 Japanese men and women aged 40–79	No	FFQ	14.1 years	Saturated fatty acid intake is inversely associated with mortality from total stroke.
Saturated fat	The Multi-Ethnic Study	Participants who were 45–84 y old at baseline ($n = 5209$).	No	FFQ	10 years	Meat saturated fats are positively associated with CVD risk, whereas dairy saturated fats are inversely associated with CVD risk.

Linoleic acid	Meta-analysis of prospective cohort studies	Studies of linoleic acid and CHD with 310,602 participants.	YES (13)		5.3–30 years	Increased intake of dietary linoleic acid is associated with a low risk of CHD in a dose–response manner.
Fatty acids	Meta-analysis of observational studies	Studies of fatty acids, unsaturated fatty acids, and coronary disease of 659,298 participants.	YES (76)			There are no significant associations in prospective studies of coronary disease that involved assessment of dietary intake of long-chain w-3 and w-6 polyunsaturated fatty acids.
Dietary fat	The PREvención con DIeta MEDiterránea (PREDIMED) study	7038 participants with men (aged 55–80 y) and women (aged 60–80 y) at high CVD risk	NO	FFQ	6 years	Saturated fatty acid and trans-fat intake are associated with a high risk of CVD, whereas MUFA and PUFA intake are inversely associated with CVD death.
Low-fat dietary pattern	The Women's Health Initiative randomized controlled trial	Participants comprised 48,835 postmenopausal women aged 50–79 y.	NO	FFQ and performing laboratory analysis of blood specimens	8.3 years	Low-fat diets do not have beneficial effects on CVD risk and total mortality.

Fat	Prospective Urban Rural Epidemiology (PURE) study	A study of 135,335 individuals aged 35–70 years in 18 countries from five continents.	NO	FFQ	7.4 years	Total fats are associated with low risk of total mortality and stroke, but not with the risk of CVD, myocardial infarction, or cardiovascular disease mortality.
A Mediterranean diet supplemented with extra-Virgin olive oil or nuts	The PREDIMED study	A multicenter trial in Spain with 7447 participants (55 to 80 years of age, 57% women) who were at high cardiovascular risk.	NO	FFQ	4.8 years	The Mediterranean diet effectively prevents the risk of major cardiovascular events.
Dietary total fat	A dose–response meta-analysis of cohort studies	Studies of cohort studies reporting associations of dietary fat intake and risk of CVDs.	YES (43)			Total fat, SFA, MUFA, and PUFA intake are not associated with the risk of cardiovascular disease.
Low-carbohydrate diet	Randomized trial	A multicenter, controlled trial involving 63 obese men and women.	NO	participants met with registered dietitian to review dietary issues	1 years	In the first six months, the low-carbohydrate diet produced a greater weight loss compared with the conventional diet, but the differences are not significant at one year.

Low carbo-hydrate–high protein diet	A population-based prospective study	A cohort of 42,237 Swedish women (30–49 years old at baseline)	NO	FFQ	12 years	Low carbohydrate–high protein diet is associated with increased total and particularly cardiovascular mortality among women.
Low in carbohydrate diet	Randomized trial	Participants with 311 free-living, overweight non-diabetic, premeno-pausal women	NO	Received weekly instruc-tion for 2 months, then an additional 10-month follow-up	1 years	Low-Carbohydrate diet benefit to weight loss and metabolic effects outcomes.
Low-carbo-hydrate diets	A prospective cohort study	Nurses' Health Study and Health Professionals' Follow-up Study with 85,168 women (aged 34–59 years at baseline) and 44,548 men (aged 40–75 years at base-line) without heart disease, cancer, or diabetes.	NO	FFQ	20–26 years	A low-carbohydrate diet based on animal sources is associated with higher all-cause mortality, whereas a vegetable-based low-car-bohydrate diet is associated with lower cardiovascular disease mortality rates.

Diet with high glycemic index and load	Meta-analysis of prospective cohort studies	Studies showed associations of glycemic index and glycemic load with incidence of CHD including 240,936 participants.	YES (10)		6–25 years	Diet with high glycemic index and glycemic load diets are significantly associated with CHD events in women but not in men.
Low-carbohydrate diets	Meta-analysis of randomized controlled trials	Studies showed associations of low-carbohydrate diet, low-fat diet, weight loss, and cardiovascular disease with 1369 participants.	YES (11)		6–24 months	Low-carbohydrate diets have greater weight loss but increased LDL cholesterol.
Carbohydrate	Prospective Urban Rural Epidemiology (PURE) study	A large, epidemiological cohort study of 135,335 individuals aged 35–70 years	NO	FFQ	7.4 years	High intake of carbohydrate is associated with an increased risk of total mortality but is not associated with the risk of CVD or CVD mortality.
Carbohydrate	A prospective cohort study and meta-analysis	15,428 adults aged 45–64 years, in four US communities	YES (7)		25 years	Both high and low carbohydrate intake are associated with increased mortality.

Carbohydrate	Prospective population-based study of UK Biobank participants	The UK Biobank cohort of general population with 195,658 participants	NO	24-h recall	3 years	There were nonlinear associations between macronutrient intakes and health (mortality and CVD risk).
Diet with a high glycemic index	Prospective study on five continents	The study included 137,851 participants between the ages of 35 and 70 years living on five continents.	NO	FFQ	9.5 years	Diet with high glycemic index is associated with an increased risk of CVD and CVD mortality.
Carbohydrate and saturated fat	Prospective cohort study	9899 women (aged 50–55 years) were recruited into the Australian Longitudinal Study on Women's Health.	NO		15 years	A moderate carbohydrate intake is associated with reduced risk of heart disease and stroke.
Low-carbohydrate diet	An open-label randomized controlled trial in Denmark	Study included 73 patients older than 18 years with type 2 diabetes.	NO	visits and telephone call	6 months	A non-calorie-restricted low-carbohydrate diet high in fat is significantly beneficial for glycemic control and body composition, without adversely affecting CVD risk factors in patients with T2D.

Carbohydrate	A dose–response meta-analysis	Studies about the relationship between dietary carbohydrate and the incidence of cardiovascular events and mortality.	YES (19)			Higher carbohydrate intake is associated with a slight increase in CVD risk in women but no association is found in men.
Dietary protein	A prospective cohort study	In the Nurses' Health Study cohort of 80,082 women aged 34–59 y and without a previous diagnosis of ischemic heart disease, stroke, cancer, hypercholesterolemia, or diabetes.	NO	FFQ	14 years	Replacing carbohydrates with protein may be associated with a lower risk of ischemic heart disease.
Dietary protein	A prospective cohort study	84,136 women aged 30–55 years in the Nurses' Health Study with no known cancer, diabetes, angina, myocardial infarction, stroke, or other cardiovascular disease	NO	FFQ	26 years	Higher intake of red meat and high-fat dairy are significantly associated with elevated risk of CHD, and CHD risk may be reduced by replacing sources of protein.

Energy-restricted high-protein, low-fat diet	Meta-analysis of randomized controlled trials	Studies that compared energy-restricted, isocaloric, high-protein, low-fat (HP) diets with standard-protein, low-fat (SP) diets included 1063 individuals.	YES (24)			Compared with an energy-restricted standard-protein, low-fat diet, an isocalorically prescribed high-protein, low-fat diet provided more benefits in reducing body weight, fat mass, and triglycerides.
Animal and plant protein	Two prospective US cohort studies	85,013 women and 46,329 men from the Nurses' Health Study (1980–2012) and Health Professionals Follow-up Study (1986–2012)	NO	FFQ	32 years	Higher animal protein intake is positively, whereas plant protein is inversely, associated with all-cause mortality.
Plant and animal protein	The Adventist Health Study-2 cohort	81,337 men and women in the Adventist Health Study-2	NO	FFQ	9.4 years	Higher animal protein intake is associated with high CVD mortality, but no associations between plant protein intake and CVD mortality.

Animal and plant protein	A large prospective cohort study	Study included 70,696 participants in the Japan Public Health Center–based Prospective Cohort who were aged 45 to 74 years.	NO	FFQ	18 years	Higher plant protein intake is associated with lower total and CVD-related mortality, but animal protein intake is not associated with mortality outcomes.
Dietary protein	Meta-analysis of prospective cohort studies	Studies about associations of dietary protein from different sources with all-cause and cause-specific mortality with 350,452 participants.	YES (11)		12–28 years	Total protein intake is positively associated with all-cause mortality, driven mainly by a harmful association of animal protein with CVD mortality. Plant protein intake is inversely associated with all-cause and CVD mortality.
Plant and animal protein	A large prospective cohort study	Study included 416,104 men and women aged 50 to 71 in the US National Institutes of Health–AARP Diet and Health Study.	NO	FFQ	16 years	There are small but significant associations between high intake of plant protein and low overall and CVD mortality.

Total, animal, and plant proteins	Dose–response meta-analysis of prospective cohort studies	Studies of the dose–response relation between intake of total, animal, and plant protein and the risk of mortality from all causes, cardiovascular disease, and cancer.	YES (32)		3.5–32 years	Intake of plant protein is associated with low CVD mortality risk.
Dietary antioxidant vitamins	A prospective study	34,486 postmenopausal women aged 55 to 69 years with no cardiovascular disease	NO	FFQ and 24-h recall	7 years	The intake of vitamin E, but not vitamins A and C, from food is inversely associated with the risk of death from CHD.
Vitamin D	Framingham Offspring Study	1739 participants (mean age 59 years; 55% women; all white)	NO	FFQ		Deficiency of vitamin D is associated with incident cardiovascular disease.
Vitamin D	Cross-sectional study	The data from the National Health and Nutrition Examination Survey (NHANES) with 8351 participants	NO			Vitamin D deficiency is associated with increased risk of CVD.

Vitamin D	Cross-sectional analysis	The data from the Third National Health and Nutrition Examination Survey (1988–1994) with 16,603 men and women aged 18 years or older	NO			25-hydroxyvitamin D deficiency is found to be associated with high prevalence of angina, myocardial infarction, and heart failure.
Vitamin D	A randomized clinical trial	The study recruited participants mostly from family practices in Auckland, New Zealand, with 5110 participants aged 50 to 84 years.	NO	Questionnaire	3.3 years	Monthly high-dose vitamin D supplementation does not prevent CVD.
Supplemental vitamins and minerals	Meta-analyses of randomized controlled trials	Studies of dietary supplements and cardiovascular disease outcomes and all-cause mortality	YES (179)			Folic acid and B vitamins had preventive benefits for stroke.
Vitamin D	Randomized trials	Among men 50 years of age or older and women 55 years of age or older in the United States	NO	FFQ	5.3 years	Supplementation with vitamin D does not result in a lower incidence of invasive cancer or cardiovascular events compared with placebo.

Supplemental vitamins and minerals	Meta-analyses of randomized controlled trials	Studies of dietary supplements and cardiovascular disease outcomes and all-cause mortality	YES (35)			Niacin shows an increased risk of all-cause mortality. However, multivitamins, vitamins C and D, β-carotene, calcium, and selenium do not exhibit positive effects in reducing the risk of CVD.
Vitamin and mineral supplements	Pooled analyses of RCTs and observational cohort studies	RCTs of vitamin or mineral use among adults without cardiovascular disease or cancer and with no known vitamin or mineral deficiencies; observational cohort studies examining serious harms.	YES (84)			Supplementation with vitamins and minerals is associated with little or no benefit in preventing cancer, CVD, and death.
vitamin D	Non-linear Mendelian randomization analyses	The analysis was conducted in the UK Biobank with 44,519 CVD cases and 251,269 controls.	NO			Vitamin D deficiency can increase the risk of CVD.

Vitamin D	A randomized controlled trial	The randomized, placebo-controlled trial among 2495 male participants ≥60 years and postmenopausal female participants ≥65 years from a general Finnish population who were free of prior CVD or cancer.	NO	Annual study questionnaires and national registry data	5 years	Supplementation with vitamin D3 does not lower the incidences of major CVD events or invasive cancer.
Dietary fiber	A pooled analysis of cohort studies	Studies about the association between dietary fiber intake and the risk of coronary heart disease	YES (10)			There is an inverse relationship between intake of dietary fiber and the risk of CHD.
Dietary fiber	Meta-analysis	Studies about the association of dietary fiber and cardiovascular or coronary heart disease	YES (19)		≥3 years	A higher intake of dietary fiber is associated with a lower risk of both CVD and CHD.

Rice bran extract	Randomized controlled trial	Single-blind design study with 60 postmenopausal Vietnamese women (45–65 y old) with high LDL cholesterol levels (over 140 mg/dL)	NO	Questionnaires	6 months	Pre-germinated brown rice bran extract containing acylated steryl glucosides associated with the reduction in the risk of atherosclerosis.
Barley β-glucan	Randomized controlled trial	Crossover study with mild hypercholesterolemia participants ($n = 45$)	NO		5 weeks	Consumption of barley β-glucan is found to be effective in circulating cholesterol levels.
Quinoa	Randomized controlled trial	Crossover designed study with 37 healthy overweight men (35–70 years) completed a 4-week crossover intervention.	NO	FFQ	6 months	Daily consumption of 20 g quinoa can reduce CVD risk markers including blood cholesterol and blood glucose in overweight participants.
Dietary fiber	Meta-analyses	Studies about indicators of carbohydrate quality and noncommunicable disease incidence, mortality, and risk factors	YES (243)			High intake of dietary fiber is associated with lower risk of mortality and incidence of cardiometabolic events.

Dietary fiber	Meta-analyses	Studies about dietary fiber in hypertension and cardiovascular disease	YES (15)			Higher fiber intake is shown to be associated with an improvement in cardiometabolic risk factors.

Source: Chen W, Zhang S, Hu X, Chen F, Li D. A Review of Healthy Dietary Choices for Cardiovascular Disease: From Individual Nutrients and Foods to Dietary Patterns. Nutrients. 2023; 15(23):4898. https://doi.org/10.3390/nu15234898.

Proteins found in food can be categorized according to whether they come from plants or animals, and different protein sources can have different effects on health outcomes. Higher total protein intake was found to be associated with higher all-cause mortality in the Rotterdam Study, a population-based cohort in the Netherlands. This association is primarily driven by higher animal protein intake and CVD mortality. According to data from two prospective US cohort studies, eating more animal protein is linked to a higher risk of death from cardiovascular disease (CVD), but eating more plant protein is linked to a lower risk of death from all causes. The Adventist Health Study-2 cohort study's results consistently demonstrate that while plant protein intake is not linked to CVD mortality, higher animal protein intake is associated with a higher risk of CVD mortality. The relationship between intake of total, animal, and plant protein and CVD risk was investigated in a recent systematic review and dose-response meta-analysis of prospective cohort studies. The findings indicate that intake of plant protein is associated with low CVD risk and mortality, but not that of total or animal protein. A second prospective cohort study including 70,696 Japanese adults also demonstrated a correlation between reduced overall and cardiovascular disease mortality and higher plant protein intake. These results are supported by a large US prospective cohort that shows modest but significant associations between high plant protein intake and low overall and CVD mortality.

It should be remembered that dietary protein is ingested in conjunction with other foods, not separately. The bioactive ingredients and nonprotein nutrients in a diet matrix may also have an impact on CVD risk factors. It has been reported that substituting plant-derived protein or fat for animal-derived carbohydrates results in reduced mortality. Furthermore, there is a lower risk of death from CVD when plant protein is substituted for red or processed meat protein.

All of these results demonstrate that consuming a lot of plant-based protein improves long-term health outcomes and ought to be promoted in worldwide dietary guidelines. To fully understand the underlying mechanisms influencing the health effects of different sources of plant-based protein, more research is required. Furthermore, the current dietary guidelines ought to make clear how much of a contribution plant-based protein sources make (Figure 4.4C) (Chen et al., 2023).

4.2.1.4. Vitamins and Minerals

The B vitamins and the chemically organic vitamins A, C, D, E, and K are necessary for sustaining a regular metabolism within cells. Important inorganic substances that support human health's normal function are minerals, such as calcium, iron, and zinc. Due to their potential to promote health and prevent disease, vitamins, and minerals are frequently used as dietary supplements. It has been reported that the risk of CHD is inversely related to postmenopausal women's consumption of vitamin E, but not of A or C. Additionally documented are the prophylactic advantages of folic acid and B vitamins in the prevention of stroke. Multivitamins, vitamins C and D, β-carotene, calcium, and selenium, on the other hand, show no benefit in lowering the risk of CVD, and taking excessive amounts of vitamin supplements may even have negative side effects. In a systematic review conducted for the US Preventive Services Task Force, the advantages and disadvantages of vitamin and mineral supplementation in healthy adults were assessed. The findings indicate that vitamin and mineral supplementation has little to no benefit in preventing cardiovascular disease. Therefore, there is currently insufficient data to conclusively support the advantages of vitamins and minerals for the prevention of CVD.

As a steroid hormone, vitamin D is a fat-soluble vitamin. Vitamin D's primary role in bone mineralization is to improve calcium absorption. There is evidence from earlier cross-sectional studies that vitamin D deficiency is linked to a higher risk of cardiovascular disease. For example, the National Health and Nutrition Examination Survey data indicate that adults in the US with specific CVDs, such as heart failure and congestive heart failure, frequently have hypovitaminosis D. Heart failure and myocardial infarction have been found to be highly prevalent in people who are deficient in 25-hydroxyvitamin D. Cohort and RCT studies have assessed the link between vitamin D deficiency and CVD risk. According to the Framingham Offspring Study, people who have 25-hydroxyvitamin D. < 15 ng/mL have a hazard ratio of 1.62 for developing cardiovascular events when compared with those with 25-hydroxyvitamin D $>$ or $=$ 15 ng/mL, indicating a possible link between vitamin D insufficiency and incident CVD. A nonlinear Mendelian randomization analysis consistently demonstrates that vitamin D deficiency contributes to the risk of CVD. Recent RCT research, however, contradicts the notion that vitamin D supplementation lowers the risk of CVD. Monthly high-dose vitamin D supplementation did not prevent CVD in an RCT involving participants from New Zealand. Daily high-dose vitamin D3 (2000 IU/day) supplementation for five years was not able to lower the incidence of major cardiovascular events, such as myocardial infarction, stroke, and CVD mortality, in an RCT involving 25,871 healthy African Americans (The Vitamin D and OmegA-3 Trial). Similarly, vitamin D3 supplementation at doses of 1600 IU or 3200 IU per day for five years did not reduce the incidence of major cardiovascular disease events in another RCT involving elderly subjects from Finland. As a result, neither the substantial benefits nor harms of vitamin D on cardiovascular risk are supported by the data currently available (Figure 4.4D).

4.2.1.5. Dietary Fiber

One well-known component of a balanced diet is dietary fiber. It is described as plant edible portions or similar carbohydrate polymers that are not absorbed or digested in the small intestine. Dietary fiber can be divided into soluble and insoluble categories based on its water solubility characteristics. Insoluble fiber comprises cellulose, hemicellulose, lignin, and chitin, while soluble fiber comprises fructans, β-glucan, pectin, and mucilage. Dietary guidelines recommend eating fruits, vegetables, and cereals on a regular basis as they are significant sources of dietary fiber. According to the WHO, a healthy diet should include more than 25 g of dietary fiber daily. The majority of European countries advise daily intakes of 25–30 g of fiber. Several RCTs have looked into how foods high in fiber affect CVD risk factors. For example, in overweight individuals, daily consumption of 20 g of quinoa for four weeks has been shown to lower blood glucose and cholesterol, two CVD risk markers. A different RCT study found that eating barley β-glucan reduced blood cholesterol levels. A short-term, prospective, open-labeled, randomized controlled, parallel-group study discovered that giving hypercholesterolemic Indians oat-based soluble fiber daily improves lipid parameters. Atherosclerosis risk in postmenopausal Vietnamese women has been found to be lowered by pre-germinated brown rice bran extract containing acylated steryl glucosides.

The risk of coronary heart disease (CHD) is inversely correlated with dietary fiber intake, according to a prior meta-analysis of 10 cohort studies. Comparably, data from an additional meta-analysis of cohort studies indicate that a lower risk of CVD and CHD is linked to a higher intake of dietary fiber, particularly from vegetable or cereal sources and rich in the insoluble type. A recent meta-analysis of 185 prospective studies and 58 clinical trials yielded data that demonstrated a positive correlation between a high dietary fiber intake and a lower risk of death and the incidence of cardiometabolic events. Increased fiber consumption was linked to improved cardiometabolic risk factors in a different meta-analysis that looked at the relationship between fiber intake and CVD mortality. This study emphasizes the potential advantages of dietary fiber for patients with hypertension and CVD as an adjuvant therapy.

When combined with the results of observational studies and RCTs mentioned above, the evidence strongly suggests that dietary fiber intake is linked to reduced risk factors for CVD. It is necessary to conduct additional research to examine the impact of particular dietary fiber types on the prevention of CVD (Figure 4.4E).

4.2.2. Foods and Food Products

Food is non-processed meat, eggs, milk, cheese, yogurt, fresh or frozen fruits and vegetables, beans, nuts, and nut butters, seeds, and olive, canola, and coconut oils.

4.2.2.1. Sugar-Sweetened Beverages (SSBs)

The main source of added sugar in the diet in high-income nations is sugar-sweetened beverages (SSBs), which include fruit drinks, sports drinks with added caloric sweeteners, and carbonated and noncarbonated soft drinks. Health organizations like the World Health Organization and the Dietary Guidelines for Americans have recommended limiting the consumption of added sugars (SSBs) to less than 10% of total energy intake due to the growing correlation between SSBs and cardiometabolic risk factors. Consuming sugar-sweetened beverages (SSBs) may raise the risk of CVD in American populations, particularly in children.

Data from the Coronary Artery Risk Development in Young Adults (CARDIA) study indicate that many cardiometabolic outcomes, including elevated triglycerides, high low-density lipoprotein cholesterol, and hypertension, are linked to increased consumption of sugar-sweetened beverages (SSBs). Additionally, it has been reported that SSB consumption is linked to an increased risk of CHD as well as unfavorable changes in leptin, inflammatory factors, and blood pressure. In addition, results from prospective analyses of two sizable US cohorts (Health Professionals Follow-up Study and Nurses' Health Study) and cross-sectional studies have demonstrated that a habitual SSB intake is positively correlated with a higher risk of CVD mortality as well as negative levels of several cardiometabolic biomarkers.

It's interesting to note that urbanization and beverage marketing are to blame for the rise in SSB consumption in low- and middle-income nations. While a Chinese adult cohort in Singapore did not show a significant correlation between

SSB intake and CVD mortality, a Chinese younger adult cohort did show a positive correlation between SSB intake (≥2 servings/day) and CVD mortality. Reducing SSB consumption is therefore crucial to enhancing the overall quality of the diet and cardiometabolic health. Globally reduced consumption of sugar-sweetened beverages (SSBs) should remain a goal of public nutrition policies and regulatory approaches.

Since artificially sweetened beverages (ASBs) have fewer calories than sugar-sweetened beverages (SSBs) while maintaining a sweet taste, they have been recommended as a substitute. It may be advised to switch from SSBs to ASBs to enhance longevity and health, even though a high level of ASB intake is suggested to have a marginally positive correlation with CVD mortality. Using ASBs instead of SSBs may lead to lower body weight and lower levels of cardiometabolic risk factors like body mass index (BMI), body fat, and lipid percentage, according to a recent network meta-analysis of 17 RCTs. The advantages of switching from SSBs to ASBs as a replacement strategy are demonstrated by these results. Nonetheless, certain studies also report results that are controversial.

For example, it has been shown that consuming a lot of artificial sweeteners is directly linked to an increased risk of CVD, which implies that ASBs might not be a healthy replacement for SSBs. A prior meta-analysis comprising 308,420 participants from seven prospective cohort studies indicated a correlation between ASB consumption and cardiovascular risk. Results from the French NutriNet-Santé study, which included 104,760 participants, consistently indicate that a higher consumption of ASBs and sugar-filled beverages is linked to an increased risk of cardiovascular disease (CVD). Additionally, a second meta-analysis of prospective cohort studies shows that a high risk of CVD incidence and mortality is linked to regular consumption of both SSBs and low-calorie sweetened beverages.

As such, the question of whether ASBs can replace SSBs is still up for debate. To fully comprehend the underlying mechanisms of the possible metabolic impact of ASBs on health, more research is required. Policymakers must take into account both new research and existing literature when formulating recommendations regarding sugar-sweetened beverages (SSBs) and cardiometabolic health. Furthermore, there is growing interest in healthy substitutes for sugar-filled drinks like 100% fruit juice, water, tea, and coffee. Future research is necessary to determine the long-term health effects of these drinks as well as how they can prevent CVD (Figure 4.5A).

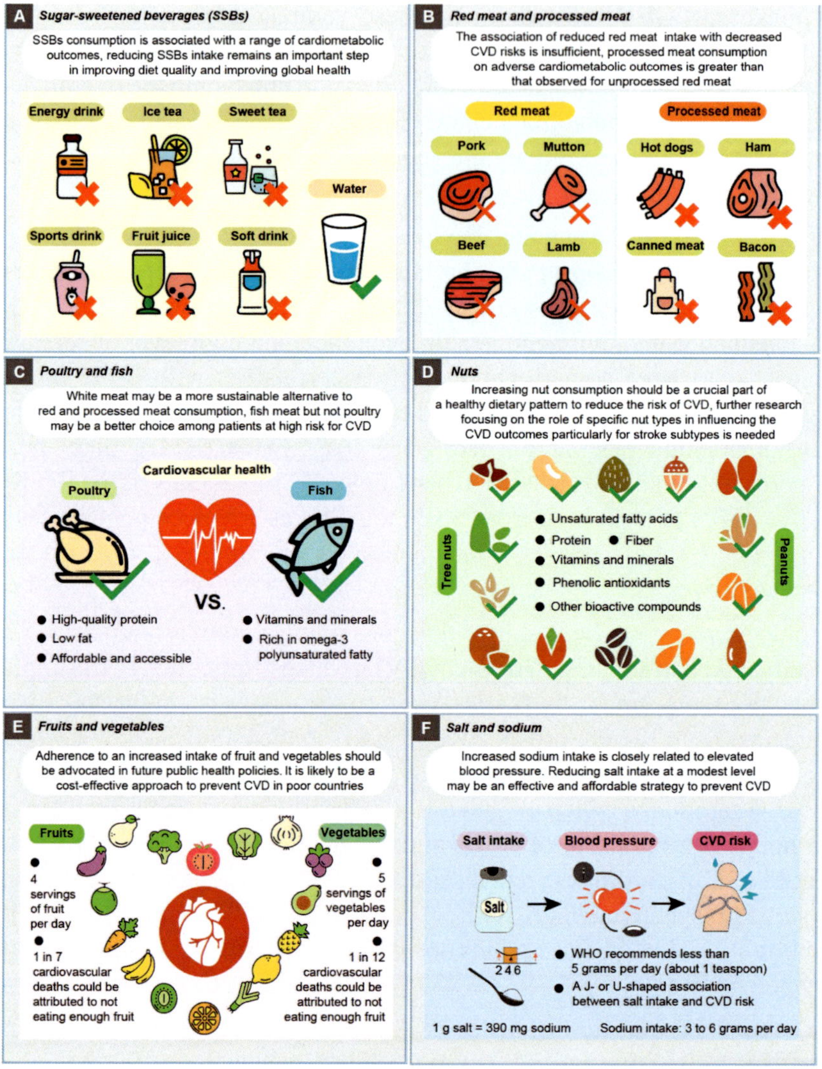

Figure 4.5. *Suggested specific food consumption according to the evidence on the association between dietary choices and cardiovascular disease management. (A) Sugar-sweetened beverages; (B) Red meat and processed meat; (C) Poultry and fish; (D) Nuts; (E) Fruits and vegetables; (F) Salt and sodium.*

Source: Chen W, Zhang S, Hu X, Chen F, Li D. A Review of Healthy Dietary Choices for Cardiovascular Disease: From Individual Nutrients and Foods to Dietary Patterns. Nutrients. 2023; 15(23):4898. https://doi.org/10.3390/nu15234898.

4.2.2.2. Red Meat and Processed Meat

Eating meat is thought to increase the risk of metabolic and cardiovascular disorders. Both processed and unprocessed red meat are staples of the American diet, and eating them has been connected to a higher risk of death from cardiovascular disease (CVD) among US citizens. The World Health Organization states that while eating red meat is "probably carcinogenic" to humans, eating processed meat is "carcinogenic" to humans. The US Dietary Guidelines for Americans and the World Health Organization both advise against consuming a lot of red or processed meat.

Consuming red and processed meat is positively correlated with the risk of developing cardiovascular disease (CVD), according to a growing body of evidence from observational studies. For example, a meta-analysis of 17 prospective cohort studies demonstrates that consuming large amounts of processed and red meat is linked to a higher risk of death from cardiovascular disease (CVD). Red meat and red plus processed meat consumption are positively correlated with CVD mortality but negatively correlated with stroke risk, according to data from the Isfahan Cohort Study. A larger intake of total, processed, and unprocessed red meat is linked to an increased risk of congestive heart failure (CHD) in a sizable prospective cohort of the Health Professionals Follow-up Study. Notably, a lower risk of coronary heart disease (CHD) is linked to replacing red meat with high-quality plant-based protein foods.

There is insufficient data to support the idea that consuming less processed and red meat lowers the risk of CVD. A meta-analysis involving 6,035,051 participants revealed very-low certainty evidence in favor of dietary patterns that reduce consumption of processed and red meat and may lower the risk of CVD. Reduced consumption of processed and unprocessed red meat was linked to a slightly lower risk of CVD mortality, stroke, and myocardial infarction, according to a different meta-analysis of prospective cohort studies. A systematic review of randomized trials consistently demonstrated that diets low in red meat may have negligible or no impact on nonfatal CVD and all-cause mortality.

It's also important to note that processed and red meat may have different effects on the risk of CVD. For instance, consumption of processed meat was found to be associated with a high incidence of CHD in a meta-analysis involving 17 prospective cohorts and 3 case-control studies with 1,218,380 individuals from 10 countries. However, there was no significant correlation found between the risk of CHD and the consumption of unprocessed red meat and total meat. A recent large-scale multinational prospective study found no significant correlation between major CVD and unprocessed red meat intake and mortality, but there is a correlation between processed meat intake and major

CVD. According to these results, eating processed meat may have a slightly bigger negative impact on cardiometabolic outcomes than eating unprocessed red meat. It is critically necessary to conduct additional research to identify the mechanisms by which the consumption of processed and unprocessed red meat affects the risk of cardiometabolic diseases in order to develop dietary and policy recommendations (Figure 4.5B) (Chen et al., 2023).

4.2.2.3. Poultry and Fish

While it is known that eating processed or unprocessed red meat reduces the risk of cardiovascular disease (CVD), it is still unclear how eating white meat affects the incidence of CVD-related events. Consuming white meat is thought to have health benefits over red meat, whether it is processed or not. A lower risk of all-cause mortality has been found when substituting white meat, especially unprocessed white meat, according to the National Institutes of Health (NIH)-AARP Diet and Health Study. According to a meta-analysis of prospective cohort studies, eating a lot of white meat lowers your risk of having a stroke and dying from one. A further meta-analysis of 13 cohort studies with 1,674,272 participants reveals no meaningful correlation between consumption of white meat and death from CVD. Moreover, a recent meta-analysis of prospective cohort studies found a strong and inverse relationship between the consumption of white meat and all-cause mortality. All things considered, these results point to the possibility that eating white meat instead of red and processed meat is more sustainable and healthful.

Fish and poultry make up the majority of white meat consumption. Due to its high protein content and tendency to have less fat than other animal meats, poultry meat is typically regarded as a health food. In addition, chicken meat is widely available, reasonably priced, and consumed at a high rate worldwide. People who live near the sea have traditionally relied heavily on fish in their diets. High-quality protein, vitamins, essential fatty acids, and micronutrients are all provided by the diversity of the fish family. Many observational studies have been carried out to investigate the relationships between the consumption of fish and poultry and the incidence of CVD, despite the paucity of evidence from RCTs. Intake of poultry but not fish is significantly associated with the incidence of CHD, stroke, heart failure, and CVD, according to a prospective cohort study of 29,682 US adults. However, consumption of poultry or fish was found to be not significantly associated with all-cause mortality. Fish consumption is linked to a lower risk of negative cardiovascular outcomes, such as stroke, myocardial infarction, and heart failure, according to data from another UK Biobank prospective study. This is in contrast to poultry consumption. According to these results, patients at high risk for CVD might do better by substituting fish, but not poultry, for red and processed meats. Robust evidence demonstrating a dose-dependent inverse relationship between fish consumption and heart failure incidence, cerebrovascular risk, IHD risk, and CHD death has been presented by prospective cohort studies and randomized controlled trials. A borderline significant inverse association is observed in the substitution analyses of the European Prospective Investigation Into Cancer and Nutrition, a large prospective study including 409,885 participants from nine European countries. The study reports that there is no clear association between the risk of IHD and the consumption of white or fatty fish. A weekly fish intake of 175 g (two servings) is linked to a lower risk of major CVD events and overall mortality among high-risk individuals or patients with pre-existing vascular disease, according to a pooled analysis of four prospective cohort studies involving 191,558 people from 58 countries. The American Heart Association and European Society of Cardiology (ESC) have recommended two servings of fish per week, which is consistent with these findings. Omega-3 polyunsaturated fatty acids, such as eicosapentaenoic and docosahexaenoic acids, are primarily responsible for the protective effect of fish consumption on cardiovascular incidents. To ascertain whether additional nutrient components in fish may also enhance its cardioprotective effect on cardiovascular risk, more research is necessary. Further details must be taken into account in upcoming clinical trials (Figure 4.5C).

4.2.2.4. Nuts

Nuts are special plant foods that are high in unsaturated fatty acids, protein, fiber, vitamins, minerals, and other bioactive substances like phytosterols and phenolic antioxidants. Nuts include peanuts and tree nuts. Owing to their distinct nutritional makeup, nuts are believed to be advantageous for enhancing well-being.

Nut consumption provides protection against CVD, according to strong data. According to the results of the early prospective cohort studies of the Physicians' Health Study and the Adventist Health Study, eating nuts lowers the risk of nonfatal myocardial infarction, sudden cardiac death, and fatal coronary heart attack. Soon after, findings from other sizable prospective cohort studies, including the Nurses' Health Study and the Health Professionals Follow-up Study, indicated the inverse relationship between nut intake and death from heart disease, including IHD and CVD. Nut consumption is associated with a lower risk of CVD mortality in subjects with low socioeconomic status, according to a prospective evaluation from three large cohorts, including 134,265 Chinese participants and 71,764 US residents of African and European descent. This suggests that nut intake is a cost-effective way to improve cardiovascular health.

According to studies, compared to not eating nuts, a one-serving (28 g) daily increase in nut intake is linked to a reduction in the relative risk of CHD and CVD of 29% and 21%, respectively. Furthermore, regular consumption of nuts, such as walnuts, peanuts, and tree nuts, has been linked to a lower incidence and mortality rate from CVD. Notably, there is currently conflicting and little evidence linking nut consumption to stroke. Research from prospective cohort studies has demonstrated an inverse relationship between the risk of stroke and a high dietary nut intake. However, the prospective cohorts of 26,285 participants in the European Prospective Investigation into the Cancer and Nutrition Potsdam Study and 21,078 participants in the Physicians' Health Study show no significant correlation between nut consumption and the risk of ischemic or total stroke.

Overall, these studies point to the importance of promoting nut consumption as an integral component of a healthy diet in order to lower the risk of CVD. It is necessary to conduct more research on the impact of particular nut varieties on the outcomes of CVD, especially for stroke subtypes (Figure 4.5D).

4.2.2.5. Fruits and Vegetables

Elevated intake of fruits and vegetables is recognized as the fundamental element of a nutritious diet to avert cardiovascular disease. Consumption of fruits and vegetables may lower the risk of CVD and may have positive effects on the overall mortality and incidence of CHD, according to a number of epidemiological studies conducted over the past few decades. These studies include the Atherosclerosis Risk in Communities study and the first National Health and Nutrition Examination Survey Epidemiologic Follow-up Study. Still, there is some discrepancy between individuals from various ethnic backgrounds. For example, fruit, and vegetable consumption is inversely correlated with the risk of coronary heart disease (CHD) in Western populations, but not in Asian populations. A lower risk of cardiovascular disease (CVD), myocardial infarction, cardiovascular mortality, non-cardiovascular mortality, and total mortality in non-Western countries was found to be associated with higher fruit and vegetable consumption in a prospective cohort study (Prospective Urban Rural Epidemiology) that included 135,335 participants from 18 low-income, middle-income, and high-income countries. Furthermore, in Chinese cohort studies, a lower risk of CVD mortality has been linked to a higher consumption of fruits rather than vegetables. In conclusion, these studies largely corroborate the positive relationship between a diet rich in fruits and vegetables and a lower risk of cardiovascular disease.

Research has been done on the dose-response relationship between fruit and vegetable consumption and CVD. Numerous dietary guidelines suggest consuming at least 400 grams of fruits and vegetables per day. Consuming more than five servings of fruits and vegetables each day has been linked to a decreased risk of coronary heart disease (CHD). Reduced risk of CVD can be achieved by consuming 800 g/day of combined apples/pears, citrus fruits, green leafy vegetables/salads, and cruciferous vegetables. A meta-analysis of 26 prospective cohort studies and the findings from two prospective cohort studies (the Health Professionals Follow-up Study and the Nurses' Health Study) indicate that eating roughly five servings of fruits and vegetables a day is linked to the lowest mortality rate; however, higher intake does not lower the additional risk. In low- and lower-middle-income countries, however, a daily intake of three to four servings (375–500 g) of fruits and vegetables is found to be just as beneficial in lowering overall mortality as higher amounts of intake. Consuming fruits and vegetables is therefore probably a cheap way to prevent CVD in low-income nations.

In developed nations, stroke ranks as the third most common cause of death and disability. There is evidence to support the inverse relationship between fruit and vegetable intake and stroke from early cohort studies. Consuming cruciferous and green leafy vegetables, as well as citrus fruit and juice, has been shown to have protective effects on lowering the risk of ischemic stroke. Likewise, a prospective cohort study involving 54,506 Danish individuals demonstrated that consuming more fruit lowers the risk of ischemic stroke. Consuming green-yellow fruits and vegetables on a daily basis is linked to a decreased risk of cerebral infarction mortality, intracerebral hemorrhage, and total stroke, according to a prospective cohort study involving 40,349 Japanese individuals. When compared to people who consume fewer than three servings of fruits and vegetables daily, those who consume more than five servings are linked to a decreased risk of stroke. A recent large cohort comprising 418,329 participants from nine European countries provided data indicating an inverse relationship between total and ischemic stroke risk and higher fruit and vegetable consumption. Therefore, it is advised to stick to a higher fruit and vegetable intake in order to prevent stroke (Figure 4.5E).

4.2.2.6. Salt and Sodium

A healthy diet should include a sufficient amount of salt because the human body requires it for cellular homeostasis. The results of a global dietary survey and a systematic analysis of 24-hour urine sodium excretion show that sodium intake has exceeded recommended levels in nearly every country. An estimated 1.65 million deaths in 2010 were attributed to cardiovascular causes, with over 2.0 g of sodium consumed daily being the primary cause. As a result, American, and European guidelines recommend reducing sodium intake to prevent hypertension and the onset of CVD. The World Health Organization has established a global goal to lower adult salt consumption to <5 g per day.

Maintaining intravascular volume is sodium's function in cardiovascular health. A growing body of research has demonstrated the strong correlation between high blood pressure—a crucial risk factor for cardiovascular disease—and higher sodium consumption. The INTERSALT study found a direct correlation between blood pressure and salt intake in 10,079 adults from 32 different countries. It has been reported that cutting back on salt helps to effectively control blood pressure. According to a study looking at how much sodium is in food; blood pressure can be lowered by cutting back on sodium consumption to levels below the current recommendation of 2.3 grams per day. A longer-term, modest reduction in salt intake from 9–12 to 5–6 g/day consistently lowers blood pressure and may prevent cardiovascular disease (CVD), according to a systematic review and meta-analysis of randomized trials. A further meta-analysis comprising five RCTs and 14 cohort studies demonstrates unequivocally the beneficial effects of consuming less sodium on lowering blood pressure.

A growing body of research indicates that eating too much salt may raise your risk of cardiovascular disease. For example, an early prospective study involving 1173 individuals in Finland found that elevated urinary sodium excretion is a significant risk factor for CVD and can predict mortality and CHD risk. According to a meta-analysis of 13 prospective studies involving 170,000 participants, consuming a lot of salt is linked to an increased risk of cardiovascular events and stroke. An additional meta-analysis of prospective studies revealed a correlation between increased sodium intake and a higher death rate from CVD. According to the findings of two completed lifestyle intervention trials with long-term follow-up, participants who reduced their sodium intake had a lower risk of cardiovascular events 10 to 15 years after the trial ended. Increased risk of CVD events and death in people with hypertension is linked to high sodium intake (greater than 6 g/day), according to a pooled analysis of data from four international prospective studies involving 133,118 participants from 49 countries. Significantly, low sodium intake (less than 3 g/day) is also linked, in people with or without hypertension, to an increased risk of cardiovascular outcomes, suggesting a J- or U-shaped relationship between salt consumption and the risk of CVD. Consistent with these conclusions, data from observational studies and clinical trials have demonstrated that lowering daily sodium intake to less than 5 g is beneficial in lowering blood pressure and cardiovascular disease.

A few recent cohort studies, however, present mixed findings. For example, a prospective cohort study finds no strong correlation between dietary salt intake and increased risk of cardiovascular disease (CVD), despite a linear relationship between blood pressure and sodium intake. A recent prospective cohort from the UK biobank of 176,570 participants found that a lower frequency of salt addition to food was linked to a lower risk of CVD, especially heart failure and IHD. All things considered, lowering salt intake to a moderate degree may be a practical and cost-effective way to prevent CVD, even though both high and low salt intakes are linked to an increased risk of the disease. To verify the efficacy and safety of salt reduction for CVD events and mortality, more extensive and prolonged trials are required (Figure 4.5F) (Chen et al., 2023).

4.2.2.7. Dairy Products

Dairy products, such as cheese, yogurt, butter, milk, and milk products, are consumed all over the world and are a significant source of dietary saturated fats. As previously mentioned, high consumption of saturated fats has been linked to a higher risk of cardiovascular disease. This may suggest that cutting back on dairy products can lower plasma levels of low-density lipoprotein cholesterol and improve cardiovascular health. Dairy fat intake can be objectively measured thanks to circulating biomarkers of dairy fat. There are no discernible effects of circulating dairy fat biomarkers on overall CVD, CHD, or stroke, according to a systematic review and meta-analysis of prospective studies. This suggests that a higher exposure to dairy fat does not translate into a higher risk of CVD. A different systematic review and meta-analysis revealed a correlation between a lower risk of CVD and higher levels of both odd-chain dairy fat biomarkers. Furthermore, data from a sizable, international prospective cohort study with participants from 21 nations across five continents showed a correlation between lower mortality and CVD, especially stroke, risks, and higher dairy consumption. Dairy foods are actually a diverse range of goods with varying nutritional value and biochemical characteristics. This implies that depending on the particular food type, dairy products may have different effects on cardiovascular health. In fact, there is ongoing debate regarding the available data regarding the link between the consumption of dairy products and health outcomes related to CVD.

According to a prior meta-analysis of prospective cohort studies, there is a negative correlation between the total risk of CVD and dairy consumption. Dairy consumption was also discovered to have an inverse relationship with stroke, CHD, and CVD, supporting these findings. Dairy products and cardiovascular mortality are not related, according to a different meta-analysis of 29 prospective cohort studies involving 938,465 participants. This study found no correlation between the risk of CVD and the consumption of yogurt, but it did show an inverse relationship between the intake of all fermented dairy products, including cheese and yogurt. A systematic review and meta-analysis of cohort studies investigating the relationship between the consumption of dairy products and the risk of major atherosclerotic cardiovascular diseases (CVDs) in the general adult population also noted the heterogeneity of individual dairy products. The findings indicate that cheese has an inverse relationship with the risk of CHD and that high-fat milk has a positive association. Consuming fermented dairy products is consistently linked to a lower risk of CVD. High consumption of fermented dairy products was found to be inversely correlated with the risk of congestive heart failure (CHD) in the Kuopio Ischemic Heart Disease Risk Factor Study, while high consumption of non-fermented dairy products was linked to an increased risk of CHD. Furthermore, consuming non-fermented milk has been linked to a higher chance of developing cardiometabolic illnesses. Yogurt consumption was found to be inversely associated with the risk of all-cause and CVD mortality in a recent meta-analysis of cohort studies involving 896,871 participants, but not significantly associated with the risk of cancer mortality.

Overall, these results imply that the risk of CVD may be inversely correlated with both fermented and non-fermented dairy products. Consuming dairy products may have different effects on cardiovascular health depending on the type of food consumed rather than the amount of fat in the products. To clarify the processes by which fermented dairy products enhance cardiovascular health, more research is necessary.

4.2.3. Dietary Patterns

4.2.3.1. Mediterranean Diet

One of the diets that has been studied the most for cardiovascular health is the Mediterranean diet. This is a traditional diet found in the Mediterranean region, with a high percentage of plant foods like fruits, vegetables, nuts, and cereals; moderate to low consumption of poultry and fish; olive oil as the primary fat source; low consumption of dairy products, red meat, processed meats, and sweets; and occasional wine consumption. It is important to note that there are variations in the Mediterranean diet based on the nutritional composition. The absence of a common definition for the Mediterranean diet could result in inconsistent findings from various research projects. Despite this drawback, both the American and European societies advise the Mediterranean diet for lowering cardiovascular risk.

Numerous interventional studies have accumulated data suggesting that the Mediterranean diet can significantly decrease the likelihood of cardiovascular disease (CVD) complications and fatalities. The Lyon Diet Heart Study and the

Seven-Countries Study have demonstrated that a stronger commitment to the Mediterranean diet can lessen the chance of myocardial infarction recurrence and lower the dangers of initial coronary heart disease (CHD) and stroke. In comparison to low-fat diets, Mediterranean diets have been found to be more efficient in enhancing crucial cardiovascular risk factors and inflammatory indicators. In the PREDIMED trial, which involved individuals with a high risk of cardiovascular complications, a Mediterranean diet enriched with extra-virgin olive oil or nuts was proven to be effective in reducing the occurrence of major cardiovascular events, outperforming a reduced-fat diet. Moreover, a recent extensive, long-term clinical trial evaluating dietary pattern effectiveness concluded that the Mediterranean diet is more successful than a low-fat diet in preventing further CVD complications. Collectively, these findings underscore the positive impact of the Mediterranean diet on both primary and secondary CVD prevention.

The Mediterranean diet is effective in lowering the incidence and mortality of cardiovascular disease (CVD), as evidenced by many observational studies. Following the Mediterranean diet more closely is linked to a lower incidence and mortality of cardiovascular disease, according to an umbrella review of meta-analyses on prospective cohort studies. The Mediterranean Diet Score showed no inverse relationship with stroke incidence or mortality, but it did show an association with cardiovascular incidence and mortality in a Danish cohort study. Similarly, despite being linked to a lower risk of CVD, the Mediterranean diet only provided protection against ischemic stroke—not hemorrhagic stroke. Furthermore, lowering the incidence of CVD and myocardial infarction in diabetics is another benefit of the Mediterranean diet.

Overall, the Mediterranean diet is strongly supported as the best eating strategy for cardiovascular health by the evidence that is currently available from clinical trials, observational studies, and meta-analyses. The benefits of the Mediterranean diet may come from a combination of the diet's many advantageous elements rather than from any one of them acting alone. This implies that the current effective dietary strategy for the management of CVD may be represented by overall food patterns.

4.2.3.2. Vegetarian and Vegan Diets

A vegetarian diet is characterized by abstaining from meat products like beef, chicken, pork, and seafood and is generally defined as a plant-based dietary profile. A vegetarian diet may contain eggs and dairy products. The suggested dietary pattern that complies with the AHA recommendations is vegetarianism. Furthermore, the EAT-Lancet Commission has suggested a plant-based diet as a healthy diet for the management and prevention of CVD.

Studies from meta-analyses and systematic reviews have demonstrated the link between a vegetarian diet and a lower risk of CVD. For example, nonvegetarians are reported to have a higher mortality from IHD than vegetarians in a meta-analysis assessing the effect of a vegetarian diet on CVD mortality. Again, a systematic review and meta-analysis of 98 cross-sectional studies and 10 cohort prospective studies with 130,000 vegetarian subjects shows that a vegetarian diet reduces the incidence and mortality from IHD. Further evidence for the association between a vegetarian diet and a moderate cardiovascular benefit comes from a meta-analysis of eight studies involving 183,321 participants. However, this association was primarily driven by a particular cohort, the Seventh-Day Adventists. Furthermore, it was shown by a recent meta-analysis of 13 prospective cohort studies involving 844,175 participants that eating a vegetarian diet lowers the risk of CVD and IHD but not stroke.

Research has demonstrated that while not all plant foods are healthful, the quality of plant-based diets is a significant factor linked to the risk of CVD. Nuts and legumes, fruits, vegetables, whole grains, oils, tea, and coffee are examples of nutritious plant foods. Sweets, refined grains, potatoes, and fries, juices, and sweetened beverages are the main categories of unhealthy plant foods. High intakes of plant foods that are healthy are linked to lower risks of CHD, while high intakes of plant foods that are unhealthy are linked to higher risks of the disease. Higher intakes of plant foods that are high in nutrients are linked to lower mortality rates from CVD, while higher intakes of plant foods that are high in nutrients are linked to increased risk of cardiometabolic disease. For example, potatoes are classified as plant foods, but their glycemic load and index are high. In an Iranian population cohort, the consumption of potatoes is significantly correlated with CVD risk factors. There isn't any evidence, though, to support the link with cardiovascular death. Furthermore, evidence for the significance of maintaining a nutritious plant-based diet in reducing the risk of both ischemic and total stroke can be found in a recent study investigating the relationship between the quality of a plant-based diet and the risk of stroke. A vegetarian diet was linked to a lower risk of hemorrhagic and ischemic strokes in a Taiwanese study involving two prospective cohorts. To improve cardiometabolic health outcomes in the future, nutrition, and public health policies should therefore emphasize and recommend adherence to a diet rich in healthful plant foods.

A vegan diet is a dietary pattern distinguished from a vegetarian diet by the complete avoidance of all animal-origin products, such as dairy and eggs. Veganism was linked to a lower risk of recurrent cardiovascular disease events, according to a systematic review of research assessing the diet's health effects on the risks of primary, intermediate, and recurrent cardiovascular disease. Vegans are notable for having a higher risk of ischemic stroke, which may be caused by their low intake of certain nutrients that are essential for preventing disease. A vegan diet was linked in another study to an 18% lower relative risk of IHD; however, no significant correlation was found between the vegan diet and CVD or stroke. It was discovered in a cross-sectional study that children who were vegans had a better profile of cardiometabolic risk factors than children who were omnivores. The restriction of animal-based foods in vegan children may lead to nutritional deficiencies that cause abnormal development. It's interesting to note that children who follow a vegan diet have low levels of height and bone mineral content. Accordingly, there may be a connection between a vegan diet and micronutrient deficiencies, and further research should focus on particular nutritional supplements. Overall, these results emphasize the significance of maintaining a balanced diet to lower the risk of CVD while also lowering the possibility of nutritional deficiencies.

4.2.3.3. Ultra-Processed Foods

The NOVA classification system divides foods and food products into four categories based on the degree and intent of industrial processing: (1) minimally or fully processed foods; (2) processed culinary ingredients; (3) processed foods; and (4) ultra-processed foods. Ultra-processed meals don't qualify as "real food." The idea of ultra-processed food was first put forth by São Paulo University researchers. "Formulations of ingredients, mostly of exclusive industrial use, that result from a series of industrial processes" is how they defined ultra-processed foods. Soft drinks, packaged snacks (sweet or savory), reconstituted meat products, and frozen dishes that have already been prepared are a few examples of this type of food.

The availability of ultra-processed foods has grown significantly over the past few decades, raising concerns about their consumption and the health of the cardiometabolic system on a global scale. Increased consumption of ultra-processed foods may be linked to a higher risk of cardiovascular disease (CVD), according to recent epidemiological studies. A higher overall risk of cardiovascular, coronary heart, and cerebrovascular diseases was found to be associated with higher consumption of ultra-processed foods in a large prospective cohort study called NutriNet-Santé. Eating foods high in processing was linked to a higher risk of cardiovascular events and death in the Framingham Offspring Cohort study, which was prospective. Consuming ultra-processed food is linked to an increased risk of death from CVD-related causes, according to a second dose-response meta-analysis of seven cohort studies. Higher intake of ultra-processed foods is consistently linked to CVD and all-cause mortality, according to the UK Biobank Cohort prospective study. According to the Moli-Sani Study, people who have experienced previous cardiovascular events are more likely to die from CVD and to have a diet high in ultra-processed foods. A prospective analysis conducted as part of the Third National Health and Nutrition Examination Study (NHANES III) found a significant correlation between the consumption of ultra-processed foods and mortality. Consuming a high level of processed food may also be a modifiable risk factor for cardiovascular problems in kids and teenagers.

The correlation between the consumption of ultra-processed food and the risk of CVD may be explained by the nutritional makeup of the products, food additives, and neo formed contaminants present in ultra-processed food. To fully understand the causal role of ultra-processed food in CVD, more research is necessary to elucidate the precise contributions and biological mechanisms of these factors. Furthermore, official nutritional guidelines and public health authorities should advise against consuming a large amount of ultra-processed food and to consume a greater amount of unprocessed or minimally processed food in the diet.

4.2.3.4. Ketogenic Diet

A diet that lowers the proportion of carbohydrates and raises the proportion of proteins and fats is known as a ketogenic diet. It has a proven track record of success in treating neurodegenerative illnesses and epilepsy. Moreover, it has been extensively employed in weight loss regimens for metabolic disorders. Studies on the possible application of ketogenic diets for CVD treatment and prevention have recently been conducted. Significant reductions in BMI, abdominal circumference, systolic blood pressure, fasting plasma glucose, and glycated hemoglobin have been linked to the ketogenic diet. Ketogenic diets have the potential to improve cardiovascular risk factors like body weight, diastolic blood pressure, triglycerides, and HDL cholesterol over the long term when compared to traditional low-fat diets. Furthermore, the anthropometric, hemodynamic, and metabolic parameters linked to CVD risk can be improved with a short-term ketogenic diet under the supervision of a general practitioner. Comparing the ketogenic diet to isoenergetic diets, however, appears to have little to no effect on changes in cardiovascular risk factors, according to another meta-analysis that included 19 RCTs and 3209 participants.

It should be mentioned that because a ketogenic diet involves consuming more dietary fats, there may be some unfavorable side effects and an elevated risk of CVD. For example, a three-week low-carbohydrate, high-fat diet results in metabolic inflexibility by raising levels of low-density lipoproteins, total cholesterol, and free fatty acids—all of which are linked to an increased risk of cardiovascular disease (CVD). Giving healthy women a ketogenic diet that is low in carbohydrates and high in fat for four weeks resulted in a negative blood lipid profile, which suggests a higher risk of cardiovascular disease in a randomized controlled feeding trial. Because of this, there is inconsistent information about how ketogenic diets affect cardiovascular health, even though there have been some improvements in cardiovascular risk factors. Additional research is required to confirm the long-term effectiveness and safety of ketogenic diets as well as to expand our understanding of potential mechanisms.

4.2.3.5. Intermittent Fasting

One definition of intermittent fasting is a type of time-restricted eating. Both humans and animals have shown the potential health benefits of intermittent fasting for aging and lifespan. Numerous indicators of cardiovascular health have benefited from intermittent fasting, as demonstrated by recent studies. For example, it has been reported that an intermittent fasting regimen is a useful dietary therapy for reducing visceral fat and adipokines, two CHD risk factors. Sustained intermittent fasting improves blood pressure and lipid profile, two metabolic and cardiovascular risk factors, in healthy, non-obese individuals. The positive impact of intermittent fasting on cardiometabolic outcomes associated with obesity was recently documented by a review of 11 meta-analyses that included 130 RCTs.

Time-restricted eating, a particular type of intermittent fasting protocol that involves 16 hours of fasting and 8 hours of unlimited feeding, has been demonstrated to help lower cardiovascular risk factors in healthy males. Time-restricted eating did not significantly alter blood biomarkers linked to metabolic and cardiovascular risk, but it did aid in weight loss in an obese woman's non-randomized controlled trial. Furthermore, eating within an 8-hour window, from 7:00 to 15:00, has been proven to be more effective than eating over a 12-hour window at 14 weeks for improving diastolic blood pressure and aiding in weight loss. However, time-restricted eating does not alter any significant metabolic markers, according to data from the TREAT Randomized Clinical Trial.

"Two-day" intermittent fasting is defined as "an eating pattern consisting of an ad libitum diet for five days and two nonconsecutive days of calorie restriction per week." Two days of intermittent fasting were linked to improvements in gut microbiota homeostasis and reduction of cardiometabolic risk factors in adults with metabolic syndrome, according to a randomized controlled trial. An effective strategy with cardioprotective effects in non-obese subjects was demonstrated by another RCT study: alternate day fasting. This involves strict 36-hour periods without calorie intake followed by 12-hour intervals with ad libitum food consumption. Within four weeks, alternate-day fasting improves cardiovascular parameters and the risk of CVD in adults who are not obese, according to a prospective cohort study with an embedded RCT. Alternate-day fasting significantly lowers total cholesterol, according to a systematic review and meta-analysis. Another kind of alternate-day fasting, which involves a 24-hour fast and a 150% energy intake on alternate days, was found to have no particular effects on systemic markers of cardiovascular health in an RCT involving lean, healthy adults. All things considered, data from clinical and observational studies support the idea that intermittent fasting may be beneficial for cardiovascular disease (Arnett et al., 2019).

REFERENCES

1. Arunima, S., & Rajamohan, T. (2014). Influence of virgin coconut oil-enriched diet on the transcriptional regulation of fatty acid synthesis and oxidation in rats—A comparative study. *British Journal of Nutrition, 111*(10), 1782–1790.
2. Astrup, A., Magkos, F., Bier, D. M., Brenna, J. T., de Oliveira Otto, M. C., Hill, J. O., King, J. C., Mente, A., Ordovas, J. M., & Volek, J. S., et al. (2020). Saturated fats and health: A reassessment and proposal for food-based recommendations: JACC state-of-the-art review. *Journal of the American College of Cardiology, 76*(7), 844–857.
3. Azad, M. B., Abou-Setta, A. M., Chauhan, B. F., Rabbani, R., Lys, J., Copstein, L., Mann, A., Jeyaraman, M. M., Reid, A. E., & Fiander, M., et al. (2017). Nonnutritive sweeteners and cardiometabolic health: A systematic review and meta-analysis of randomized controlled trials and prospective cohort studies. *Canadian Medical Association Journal, 189*(28), E929–939.
4. Benjamin, E. J., Virani, S. S., Callaway, C. W., Chamberlain, A. M., Chang, A. R., Cheng, S., Chiuve, S. E., Cushman, M., Delling, F. N., Deo, R., et al. (2018). Heart Disease and Stroke Statistics-2018 Update: A Report from the American Heart Association. *Circulation, 137*, e67–e492.
5. Bischoff, S. C., Barbara, G., Buurman, W., Ockhuizen, T., Schulzke, J. D., Serino, M., Tilg, H., Watson, A., & Wells, J. M. (2014). Intestinal permeability—A new target for disease prevention and therapy. *BMC Gastroenterology, 14*, 189.
6. Blaak, E. E., Antoine, J. M., Benton, D., Bjorck, I., Bozzetto, L., Brouns, F., Diamant, M., Dye, L., Hulshof, T., Holst, J. J., et al. (2012). Impact of postprandial glycaemia on health and prevention of disease. *Obesity Reviews, 13*(10), 923–984.
7. Bonaccio, M., Costanzo, S., Di Castelnuovo, A., Persichillo, M., Magnacca, S., De Curtis, A., Cerletti, C., Donati, M. B., de Gaetano, G., & Iacoviello, L. (2022). Ultra-processed food intake and all-cause and cause-specific mortality in individuals with cardiovascular disease: The Moli-sani Study. *European Heart Journal, 43*, 213–224.
8. Boulange, C. L., Neves, A. L., Chilloux, J., Nicholson, J. K., & Dumas, M. E. (2016). Impact of the gut microbiota on inflammation, obesity, and metabolic disease. *Genome Medicine, 8*(1), 42.
9. Brenna, J. T., & Kothapalli, K. S. (2014). Commentary on "Influence of virgin coconut oil-enriched diet on the transcriptional regulation of fatty acid synthesis and oxidation in rats—A comparative study" by Sakunthala

Arunima and Thankappan Rajamohan. *British Journal of Nutrition, 112*(9), 1425–1426.

10. Buckley, J. P., Kim, H., Wong, E., & Rebholz, C. M. (2019). Ultra-processed food consumption and exposure to phthalates and bisphenols in the US National Health and Nutrition Examination Survey, 2013–2014. *Environmental International, 131*, 105057.
11. Calvo, M. S., & Uribarri, J. (2013). Public health impact of dietary phosphorus excess on bone and cardiovascular health in the general population. *American Journal of Clinical Nutrition, 98*(1), 6–15.
12. Calvo, M. S., Moshfegh, A. J., & Tucker, K. L. (2014). Assessing the health impact of phosphorus in the food supply: Issues and considerations. *Advances in Nutrition, 5*(1), 104–113.
13. Canhada, S. L., Luft, V. C., Giatti, L., Duncan, B. B., Chor, D., Fonseca, M., Matos, S. M. A., Molina, M., Barreto, S. M., & Levy, R. B., et al. (2020). Ultra-processed foods, incident overweight and obesity, and longitudinal changes in weight and waist circumference: The Brazilian Longitudinal Study of Adult Health (ELSA-Brasil). *Public Health Nutrition, 23*(6), 1076–1086.
14. Chan, C. B., Hashemi, Z., & Subhan, F. B. (2017). The impact of low and no-caloric sweeteners on glucose absorption, incretin secretion, and glucose tolerance. *Applied Physiology, Nutrition, and Metabolism, 42*(8), 793–801.
15. Chassaing, B., Van de Wiele, T., De Bodt, J., Marzorati, M., & Gewirtz, A. T. (2017). Dietary emulsifiers directly alter human microbiota composition and gene expression *ex vivo* potentiating intestinal inflammation. *Gut, 66*(8), 1414–1427.
16. Chaudhuri, J., Bains, Y., Guha, S., Kahn, A., Hall, D., Bose, N., Gugliucci, A., & Kapahi, P. (2018). The role of advanced glycation end products in aging and metabolic diseases: Bridging association and causality. *Cell Metabolism, 28*(3), 337–352.
17. Dalenberg, J. R., Patel, B. P., Denis, R., Veldhuizen, M. G., Nakamura, Y., Vinke, P. C., Luquet, S., & Small, D. M. (2020). Short-term consumption of sucralose with, but not without, carbohydrate impairs neural and metabolic sensitivity to sugar in humans. *Cell Metabolism, 31*(3), 493–502, e7.
18. Delzenne, N. M., Olivares, M., Neyrinck, A. M., Beaumont, M., Kjolbaek, L., Larsen, T. M., Benitez-Paez, A., Romani-Perez, M., Garcia-Campayo, V., & Bosscher, D., et al. (2020). Nutritional interest of dietary fiber and

prebiotics in obesity: Lessons from the MyNewGut consortium. *Clinical Nutrition, 39*(2), 414–424.

19. Do, M. H., Lee, E., Oh, M. J., Kim, Y., & Park, H. Y. (2018). High-glucose or -fructose diet cause changes of the gut microbiota and metabolic disorders in mice without body weight change. *Nutrients, 10*(6), 761.
20. Elizabeth, L., Machado, P., Zinocker, M., Baker, P., & Lawrence, M. (2020). Ultra-processed foods and health outcomes: A narrative review. *Nutrients, 12*(7), 1955.
21. Fardet, A. (2016). Minimally processed foods are more satiating and less hyperglycemic than ultra-processed foods: A preliminary study with 98 ready-to-eat foods. *Food & Function, 7*(5), 2338–2346.
22. Freeman, A. M., Morris, P. B., Barnard, N., Esselstyn, C. B., Ros, E., Agatston, A., Devries, S., O'Keefe, J., Miller, M., & Ornish, D., et al. (2017). Trending cardiovascular nutrition controversies. *Journal of the American College of Cardiology, 69*(9), 1172–1187.
23. Ghanim, H., Abuaysheh, S., Sia, C. L., Korzeniewski, K., Chaudhuri, A., Fernandez-Real, J. M., & Dandona, P. (2009). Increase in plasma endotoxin concentrations and the expression of Toll-like receptors and suppressor of cytokine signaling-3 in mononuclear cells after a high-fat, high-carbohydrate meal: Implications for insulin resistance. *Diabetes Care, 32*(12), 2281–2287.
24. Grundy, M. M., Lapsley, K., & Ellis, P. R. (2016). A review of the impact of processing on nutrient bioaccessibility and digestion of almonds. *International Journal of Food Science & Technology, 51*(9), 1937–1946.
25. Guerville, M., Leroy, A., Sinquin, A., Laugerette, F., Michalski, M. C., & Boudry, G. (2017). Western-diet consumption induces alteration of barrier function mechanisms in the ileum that correlates with metabolic endotoxemia in rats. *American Journal of Physiology-Endocrinology and Metabolism, 313*(2), E107–E120.
26. Hall, K. D., Ayuketah, A., Brychta, R., Cai, H., Cassimatis, T., Chen, K. Y., Chung, S. T., Costa, E., Courville, A., Darcey, V., et al. (2019). Ultra-processed diets cause excess calorie intake and weight gain: An inpatient randomized controlled trial of ad libitum food intake. *Cell Metabolism, 30*(1), 226–237.
27. He, F. J., Tan, M., & MacGregor, G. A. (2020). Salt reduction to prevent hypertension and cardiovascular disease: JACC state-of-the-art review. *Journal of the American College of Cardiology, 75*(6), 632–647.
28. Hollaender, P. L., Ross, A. B., & Kristensen, M. (2015). Whole-grain and blood lipid changes in apparently healthy adults: A systematic review

and meta-analysis of randomized controlled studies. *American Journal of Clinical Nutrition, 102*(3), 556–572.

29. Honicky, M., Cardoso, S. M., Kunradi Vieira, F. G., Hinnig, P. F., Back, I. C., & Moreno, Y. M. F. (2022). Ultra-processed food intake is associated with children and adolescents with congenital heart disease clustered by high cardiovascular risk factors. *British Journal of Nutrition, 129*, 1163–1171.
30. Juul, F., Martinez-Steele, E., Parekh, N., Monteiro, C. A., & Chang, V. W. (2018). Ultra-processed food consumption and excess weight among US adults. *British Journal of Nutrition, 120*(1), 90–100.
31. Juul, F., Vaidean, G., Lin, Y., Deierlein, A., & Parekh, N. (2021). Ultra-processed foods and incident cardiovascular disease in the Framingham Offspring Study. *Journal of the American College of Cardiology, 77*(12), 1520–1531.
32. Keller, A., Heitmann, B. L., & Olsen, N. (2015). Sugar-sweetened beverages, vascular risk factors and events: A systematic literature review. *Public Health Nutrition, 18*(7), 1145–1154.
33. Kim, H., Hu, E. A., & Rebholz, C. M. (2019). Ultra-processed food intake and mortality in the USA: Results from the Third National Health and Nutrition Examination Survey (NHANES III, 1988–1994). *Public Health Nutrition, 22*, 1777–1785.
34. Lopez-Moreno, J., Garcia-Carpintero, S., Jimenez-Lucena, R., Haro, C., Rangel-Zuniga, O. A., Blanco-Rojo, R., Yubero-Serrano, E. M., Tinahones, F. J., Delgado-Lista, J., & Perez-Martinez, P., et al. (2017). Effect of dietary lipids on endotoxemia influences postprandial inflammatory response. *Journal of Agricultural and Food Chemistry, 65*(35), 7756–7763.
35. Louzada, M. L., Baraldi, L. G., Steele, E. M., Martins, A. P., Canella, D. S., Moubarac, J. C., Levy, R. B., Cannon, G., Afshin, A., & Imamura, F., et al. (2015). Consumption of ultra-processed foods and obesity in Brazilian adolescents and adults. *Preventive Medicine, 81*, 9–15.
36. Ludwig, D. S. (2011). Technology, diet, and the burden of chronic disease. *JAMA, 305*(13), 1352–1353.
37. Makki, K., Deehan, E. C., Walter, J., & Backhed, F. (2018). The impact of dietary fiber on gut microbiota in host health and disease. *Cell Host & Microbe, 23*(6), 705–715.
38. Martinez Steele, E., Baraldi, L. G., Louzada, M. L., Moubarac, J. C., Mozaffarian, D., & Monteiro, C. A. (2016). Ultra-processed foods and added sugars in the US diet: Evidence from a nationally representative cross-sectional study. *BMJ Open, 6*(3), e009892.

39. Mendonca, R. D., Pimenta, A. M., Gea, A., de la Fuente-Arrillaga, C., Martinez-Gonzalez, M. A., Lopes, A. C., & Bes-Rastrollo, M. (2016). Ultraprocessed food consumption and risk of overweight and obesity: The University of Navarra Follow-Up (SUN) cohort study. *American Journal of Clinical Nutrition, 104*(5), 1433–1440.

40. Menini, S., Iacobini, C., Vitale, M., & Pugliese, G. (2020). The inflammasome in chronic complications of diabetes and related metabolic disorders. *Cells, 9*(8), 1812.

41. Micha, R., Penalvo, J. L., Cudhea, F., Imamura, F., Rehm, C. D., & Mozaffarian, D. (2017). Association between dietary factors and mortality from heart disease, stroke, and type 2 diabetes in the United States. *JAMA, 317*(9), 912–924.

42. Micha, R., Shulkin, M. L., Penalvo, J. L., Khatibzadeh, S., Singh, G. M., Rao, M., Fahimi, S., Powles, J., & Mozaffarian, D. (2017). Etiologic effects and optimal intakes of foods and nutrients for risk of cardiovascular diseases and diabetes: Systematic reviews and meta-analyses from the Nutrition and Chronic Diseases Expert Group (NutriCoDE). *PLoS One, 12*(4), e0175149.

43. Moen, B., Berget, I., Rud, I., Hole, A. S., Kjos, N. P., & Sahlstrom, S. (2016). Extrusion of barley and oat influences the fecal microbiota and SCFA profile of growing pigs. *Food & Function, 7*(2), 1024–1032.

44. Monteiro, C. A. (2009). Nutrition and health: The issue is not food, nor nutrients, so much as processing. *Public Health Nutrition, 12*(5), 729–731.

45. Monteiro, C. A., Cannon, G., Levy, R. B., Moubarac, J. C., Louzada, M. L., Rauber, F., Khandpur, N., Cediel, G., Neri, D., & Martinez-Steele, E., et al. (2019). Ultra-processed foods: What they are and how to identify them. *Public Health Nutrition, 22*(5), 936–941.

46. Monteiro, C. A., Cannon, G., Moubarac, J. C., Levy, R. B., Louzada, M. L. C., & Jaime, P. C. (2018). The UN Decade of Nutrition, the NOVA food classification and the trouble with ultra-processing. *Public Health Nutrition, 21*(5), 5–17.

47. Moubarac, J. C., Martins, A. P., Claro, R. M., Levy, R. B., Cannon, G., & Monteiro, C. A. (2013). Consumption of ultra-processed foods and likely impact on human health: Evidence from Canada. *Public Health Nutrition, 16*(12), 2240–2248.

48. Mozaffarian, D. (2016). Dietary and policy priorities for cardiovascular disease, diabetes, and obesity: A comprehensive review. *Circulation, 133*(2), 187–225.

49. Mundi, S., Massaro, M., Scoditti, E., Carluccio, M. A., van Hinsbergh,

V. W. M., Iruela-Arispe, M. L., & De Caterina, R. (2018). Endothelial permeability, LDL deposition, and cardiovascular risk factors: A review. *Cardiovascular Research, 114*(1), 35–52.

50. Nettleton, J. E., Reimer, R. A., & Shearer, J. (2016). Reshaping the gut microbiota: Impact of low-calorie sweeteners and the link to insulin resistance? *Physiology & Behavior, 164*(Pt B), 488–493.
51. O'Donnell, M., Mente, A., Rangarajan, S., McQueen, M. J., Wang, X., Liu, L., Yan, H., Lee, S. F., Mony, P., & Devanath, A., et al. (2014). Urinary sodium and potassium excretion, mortality, and cardiovascular events. *New England Journal of Medicine, 371*(7), 612–623.
52. Peng, C. Y., Tsai, E. M., Kao, T. H., Lai, T. C., Liang, S. S., Chiu, C. C., & Wang, T. N. (2019). Canned food intake and urinary bisphenol A concentrations: A randomized crossover intervention study. *Environmental Science and Pollution Research, 26*(27), 27999–28009.
53. Penttinen, R., Kinnula, H., Lipponen, A., Bamford, J. K., & Sundberg, L. R. (2016). High nutrient concentration can induce virulence factor expression and cause higher virulence in an environmentally transmitted pathogen. *Microbial Ecology, 72*(4), 955–964.
54. Pilic, L., Pedlar, C. R., & Mavrommatis, Y. (2016). Salt-sensitive hypertension: Mechanisms and effects of dietary and other lifestyle factors. *Nutrition Reviews, 74*(10), 645–658.
55. Poston, R. N. (2019). Atherosclerosis: Integration of its pathogenesis as a self-perpetuating propagating inflammation: A review. *Cardiovascular Endocrinology & Metabolism, 8*(2), 51–61.
56. Poti, J. M., Braga, B., & Qin, B. (2017). Ultra-processed food intake and obesity: What really matters for health—processing or nutrient content? *Current Obesity Reports, 6*(4), 420–431.
57. Poulsen, M. W., Hedegaard, R. V., Andersen, J. M., de Courten, B., Bugel, S., Nielsen, J., Skibsted, L. H., & Dragsted, L. O. (2013). Advanced glycation end products in food and their effects on health. *Food and Chemical Toxicology, 60*, 10–37.
58. Rancière, F., Lyons, J. G., Loh, V. H., Botton, J., Galloway, T., Wang, T., Shaw, J. E., & Magliano, D. J. (2015). Bisphenol A and the risk of cardiometabolic disorders: A systematic review with meta-analysis of the epidemiological evidence. *Environmental Health, 14*, 46.
59. Robinson, E., Almiron-Roig, E., Rutters, F., de Graaf, C., Forde, C. G., Tudur Smith, C., Nolan, S. J., & Jebb, S. A. (2014). A systematic review and meta-analysis examining the effect of eating rate on energy intake and hunger. *American Journal of Clinical Nutrition, 100*(1), 123–151.

60. Rolls, B. J. (2017). Dietary energy density: Applying behavioral science to weight management. *Nutrition Bulletin, 42*(3), 246–253.
61. Roth, G. A., Johnson, C., Abajobir, A., Abd-Allah, F., Abera, S. F., Abyu, G., Ahmed, M., Aksut, B., Alam, T., & Alam, K., et al. (2017). Global, regional, and national burden of cardiovascular diseases for 10 causes, 1990 to 2015. *Journal of the American College of Cardiology, 70*(1), 1–25.
62. Roth, G. A., Mensah, G. A., Johnson, C. O., Addolorato, G., Ammirati, E., Baddour, L. M., Barengo, N. C., Beaton, A. Z., Benjamin, E. J., & Benziger, C. P., et al. (2020). Global burden of cardiovascular diseases and risk factors, 1990–2019: Update from the GBD 2019 study. *Journal of the American College of Cardiology, 76*(25), 2982–3021.
63. Sacks, F. M., Lichtenstein, A. H., Wu, J. H. Y., Appel, L. J., Creager, M. A., Kris-Etherton, P. M., Miller, M., Rimm, E. B., Rudel, L. L., & Robinson, J. G., et al. (2017). Dietary fats and cardiovascular disease: A presidential advisory from the American Heart Association. *Circulation, 136*(3), e1–e23.
64. Silva, F. M., Giatti, L., de Figueiredo, R. C., Molina, M., de Oliveira Cardoso, L., Duncan, B. B., & Barreto, S. M. (2018). Consumption of ultra-processed food and obesity: Cross-sectional results from the Brazilian Longitudinal Study of Adult Health (ELSA-Brasil) cohort (2008–2010). *Public Health Nutrition, 21*(12), 2271–2279.
65. Sonnenburg, E. D., & Sonnenburg, J. L. (2014). Starving our microbial self: The deleterious consequences of a diet deficient in microbiota-accessible carbohydrates. *Cell Metabolism, 20*(5), 779–786.
66. Spreadbury, I. (2012). Comparison with ancestral diets suggests dense acellular carbohydrates promote an inflammatory microbiota, and may be the primary dietary cause of leptin resistance and obesity. *Diabetes, Metabolic Syndrome and Obesity, 5*, 175–189.
67. Srour, B., Fezeu, L. K., Kesse-Guyot, E., Allès, B., Méjean, C., Andrianasolo, R. M., Chazelas, E., Deschasaux, M., Hercberg, S., & Galan, P., et al. (2019). Ultra-processed food intake and risk of cardiovascular disease: Prospective cohort study (NutriNet-Santé). *BMJ, 365*, l1451.
68. Stanhope, K. L., Goran, M. I., Bosy-Westphal, A., King, J. C., Schmidt, L. A., Schwarz, J. M., Stice, E., Sylvetsky, A. C., Turnbaugh, P. J., & Bray, G. A., et al. (2018). Pathways and mechanisms linking dietary components to cardiometabolic disease: Thinking beyond calories. *Obesity Reviews, 19*(9), 1205–1235.
69. Suksatan, W., Moradi, S., Naeini, F., Bagheri, R., Mohammadi, H., Talebi, S., Mehrabani, S., Hojjati Kermani, M. A., & Suzuki, K. (2021). Ultra-

Processed Food Consumption and Adult Mortality Risk: A Systematic Review and Dose-Response Meta-Analysis of 207,291 Participants. *Nutrients, 14*(1), 174.

70. Te Morenga, L. A., Howatson, A. J., Jones, R. M., & Mann, J. (2014). Dietary sugars and cardiometabolic risk: Systematic review and meta-analyses of randomized controlled trials of the effects on blood pressure and lipids. *American Journal of Clinical Nutrition, 100*(1), 65–79.
71. Virani, S. S., Alonso, A., Aparicio, H. J., Benjamin, E. J., Bittencourt, M. S., Callaway, C. W., Carson, A. P., Chamberlain, A. M., Cheng, S., & Delling, F., et al. (2021). Heart disease and stroke statistics—2021 update: A report from the American Heart Association. *Circulation, 143*(8), e254–e743.
72. Woodruff, R. C., Zhao, L., Ahuja, J. K. C., Gillespie, C., Goldman, J., Harris, D. M., Jackson, S. L., Moshfegh, A., Rhodes, D., & Sebastian, R. S., et al. (2020). Top food category contributors to sodium and potassium intake—United States, 2015–2016. *MMWR Morbidity and Mortality Weekly Report, 69*(32), 1064–1069.
73. Ye, X., Wong, L. Y., Kramer, J., Zhou, X., Jia, T., & Calafat, A. M. (2015). Urinary concentrations of bisphenol A and three other bisphenols in convenience samples of U.S. adults during 2000–2014. *Environmental Science & Technology, 49*(19), 11834–11839.
74. Yeh, C. F., Chen, Y. H., Liu, S. F., Kao, H. L., Wu, M. S., Yang, K. C., & Wu, W. K. (2020). Mutual interplay of host immune system and gut microbiota in the immunopathology of atherosclerosis. *International Journal of Molecular Sciences, 21*(22), 8729.
75. Yusuf, S., Rangarajan, S., Teo, K., Islam, S., Li, W., Liu, L., Bo, J., Lou, Q., Lu, F., & Liu, T., et al. (2014). Cardiovascular risk and events in 17 low-, middle-, and high-income countries. *New England Journal of Medicine, 371*(9), 818–827.
76. Zinocker, M. K., & Lindseth, I. A. (2018). The Western diet-microbiome-host interaction and its role in metabolic disease. *Nutrients, 10*(3), 365.

Chapter 5

Mental Health and Cognitive Function

Contents

INTRODUCTION

Ultra-processed foods (UPFs) have been connected to several detrimental effects on cognitive function and mental health. UPFs are linked to higher risks of anxiety, depression, and cognitive decline because of their high levels of unhealthy fats, artificial additives, and added sugars. These foods frequently lack important nutrients that are critical to the health of the brain, such as vitamins, minerals, and antioxidants. Because UPFs have a high glycemic load, blood sugar levels can spike and fall quickly, which may have an impact on mood and energy levels. In addition, consuming UPFs has been linked to inflammation, which is a factor in the emergence of neurodegenerative diseases and mental health issues. Research indicates that a diet rich in UPFs may cause disruptions in gut microbiota, which are important for brain health, and consequently impair memory, learning, and executive function. Furthermore, some UPFs have an addictive quality that can result in overconsumption and unhealthy eating habits, which exacerbates the detrimental effects on mental and cognitive health. The complex relationships between nutrition and brain function are still being studied, but it is becoming more and more obvious that maintaining optimal mental and cognitive health requires limiting the consumption of UPFs and encouraging a diet high in whole, nutrient-dense foods.

5.1. ULTRA-PROCESSED FOODS AND MENTAL HEALTH

The strongest evidence for a connection between UPFs and poor mental health outcomes comes from treating patients with depression and depressive symptoms. Although some research has linked UPFs to anxiety, the majority of the data points to a dose-response risk for anxiety in contrast to depression. Extensive prospective research has shown correlations between higher UPF intake and higher incident depression, even after adjusting for dietary patterns associated with UPFs (e.g., exercise, sleep schedule, and TV consumption) and additional possible confounders (e.g., status of marriage and employment). Adolescence is when these dose-response findings frequently surface, even after controlling for variables like bullying victimization and self-perceived body image. Increased internalizing symptoms like boredom, crying, fear, worry, loneliness, unhappiness, restless nights, and melancholy are also correlated with higher UPF consumption. Although consumption of UPF has been linked to ADHD in children, more research is required to investigate this relationship in adults.

Higher consumption of UPF is correlated with characteristics like being single, smoking, drinking too much alcohol, having a poor opinion of one's health, and feeling a lot of stress (n = 1270). A recent meta-analysis found that adults were 11% more likely to experience depression for every 10% increase in UPF consumption over daily calorie intake. Seven randomized controlled trials have demonstrated to date that counseling interventions that promote MPF intake while lowering UPFs result in a significant reduction in depression levels. Because body weight positively correlates with depressive symptoms, the systematic review excluded studies created specifically to induce weight loss. The Mediterranean diet has shown the strongest correlation with a lower risk of depression to date.

A higher daily percentage of UPFs was linked to cognitive decline in an ethnically diverse sample of adults (n = 10,775), aged 35–74. High UPF consumption has been associated with dementia in many long-term studies. Dementia risk has been linked to substituting MPFs for UPFs. It is evident that UPFs affect cognitive health throughout life. To provide one more example, the McCarthy Scales of Children's Abilities revealed that children of pregnant women who ingested the greatest UPFs during the third trimester also had children with lower verbal scores.

Additional depression-specific mechanisms can augment the list above:

1. Nutrient deficiencies (B vitamins, vitamin D, omega-3, magnesium, zinc, selenium, iron).
2. Oxidative stress (which can lead to inflammation).
3. Blood–brain-barrier permeability (associated with intestinal permeability).
4. Hypothalamic–pituitary–adrenal (HPA) axis (responsible for stress response via cortisol).
5. Neuroendocrine hormones (e.g., leptin, and insulin resistance).
6. Endocrine disruption (from neo-formed compounds, food additives, and contaminants).
7. Mitochondrial dysfunction (responsible for energy production).
8. Brain function and volume (functional connectivity related to eating modulation).
9. Brain integrity (distorted fatty acid composition of brain membrane phospholipids).
10. Brain-derived neurotrophic factors (BDNF) (related to synaptic plasticity).
11. Serotonin neurotransmission (tryptophan diverted to kynurenine rather than serotonin).
12. Dopamine overactivation (related to addiction-like eating).

A large number of the above-mentioned mechanisms represent changes along multiple levels of the gut–brain axis. A daily multimineral supplement won't help with the list. The gut microbiota is becoming more widely acknowledged as a factor in the transition of food intake mechanisms from homeostatic to hedonic. The research on food addiction that examines hedonic eating is referred to as ultra-processed food addiction (UPFA) in the following section. It adds the specificity that most, but not all, UPFs are specifically related to this idea.

5.1.1. Modern World Mental Disorders

A combination of abnormal thoughts, feelings, behavior, and interpersonal relationships are the hallmarks of mental disorders. Some of these may manifest in the early stages of adolescence and have an impact on people's quality of life. In addition to an individual's capacity for managing emotional difficulties, other factors that influence this include social, political, environmental, occupational, and community support. The World Health Organization further affirms that environmental hazards, stress, nutrition, genetics, and perinatal infections are risk factors for mental disorders.

"Roughly 1 in 7 people globally (11–18%) have one or more mental or substance use disorders," according to WHO sources. This indicates that, on a global scale, one billion people had one in 2017. "Mental disease is a broad definition that includes health disorders (such as depression, anxiety, bipolar disorder, eating disorders, and schizophrenia) and substance use disorders (such as alcohol and drug use disorders), as well as neurodevelopmental disorders like autism, attention-deficit hyperactivity disorder (ADHD), and developmental disability, according to the WHO's International Classification of Diseases (ICD-10) situation. The decline of the Western diet has been correlated with a rise in the prevalence of mental health disorders in developed nations. "People in the modern era are becoming more and more overfed, undernourished, sedentary, sleep deprived, and socially isolated. These lifestyle modifications have an impact on depression incidence and treatment, as well as poor physical health (Hidaka, 2012). It is still unclear whether primary micro- and macronutrient shortages or nutritional deficiencies as a symptom of depression, anxiety, or sadness lead to mental health disorders that, in both cases, only exacerbate the symptoms of these illnesses (Rao et al., 2008).

5.1.2. The Relationship Between Food and Mood

Not only should we pay attention to how food affects our mood, but also to how our mental health affects the foods we choose. Emotions have an impact on eating. Food affects mood like a vicious cycle; mood in turn influences food choices, which ultimately improve or worsen disposition once more. This cycle must be recognized in order to better manage mental health.

5.1.3. How does the food affect our mood?

Standard definitions describe food as any natural or processed product or substance that provides nourishment for vital processes, growth, and life maintenance. Food comprises macronutrients like proteins, fats, and carbohydrates as well as micronutrients like vitamins and minerals (Mahan et al., 2008). Any imbalance that persists for a long enough period of time can lead to both psychological and physical disorders.

The same nutrients that are found in food nourish the brain as they do any other organ. As a result, food's nutritional qualities have an effect on brain processes linked to emotion and mood. Food can affect mood through slower changes in brain chemistry as well as sensory stimulation from taste, savor, and smell, as well as hunger alleviation (Shepherd and Raats, 2006). Widespread opioid release in the brain occurs after eating, most likely indicating pleasure and satiety. According to a study, eating pizza causes the entire brain to release a significant amount of endorphins, and surprisingly, drinking the tasteless nutritional drink causes even more endorphin releases. The amount of opioid release was unrelated to the enjoyment of food (Tuulari et al., 2017).

The first connection between food and mood that typically springs to mind is that of coffee and chocolate. It's common knowledge that coffee energizes and heightens alertness, while a small piece of chocolate can elevate our mood and provide a happy boost. It has also been demonstrated that chocolate if consumed in sufficient quantities on an empty stomach, may promote the synthesis of serotonin. These effects are due to the excitants theobromine and caffeine, as well as the sweet taste and certain psychological mechanisms. Tryptophan (TRP), the precursor essential amino acid, is converted to 5-hydroxytryptamine (also known as 5-HT) by the enzyme tryptophan hydroxylase. This process produces the neurotransmitter serotonin. 5-HT has long been linked to affective disorders like anxiety and depression as well as sleep. Tryptophan and serotonin are recognized to enhance wellbeing (Rao et al., 2008).

5.1.4. Carbohydrates

Carbohydrate-rich foods stimulate the production of these brain chemicals (5-HT and TRP). Mood and behavior have been found to be affected by carbohydrates. These are the macronutrients that cause the body to release insulin, a hormone that permits cells to convert glucose into energy and aids in tryptophan's entry into the brain. While sugars, particularly sweets, can have an instantaneous but transient impact on mood, it is instead advised to eat foods high in complex carbohydrates, like whole grains and pasta, and low on the glycemic index (GI), which have a moderate but longer-lasting effect on mood, energy level, and brain chemistry (Rao et al., 2008).

5.1.5. Proteins

Consuming protein has an impact on mental health and brain function because it contains amino acids. Amino acids are the building blocks of the neurotransmitters that affect mood: tryptophan produces serotonin, and tyrosine produces dopamine. While an excess of these amino acids may cause mental retardation and brain damage, a shortage of these amino acids results in poor neurotransmitter synthesis and low mood. An increased proportion of protein was linked to an increased risk of depression and arousal, while an increased proportion of carbohydrates was linked to an increased likelihood of calmness and less depression. Furthermore, it appears that lowering TRP has a bigger effect on mood disturbance than raising carbohydrate intake. Tryptophan-containing food raises serotonin levels in the brain and modifies neural processing in neurocircuits that control mood. Tryptophan must, however, compete with other large-neutral amino acids (LNAA) for transport across the blood–brain barrier. This competition can be overcome by raising the ratio of tryptophan to LNAA. Tyrosine, threonine, methionine, valine, isoleucine, leucine, histidine, and phenylalanine are among the constituents of the LNAA. According to the findings of an experiment, increasing the ratio in a personalized drink can improve mood by influencing the neurocircuits that control mood.

5.1.6. Omega-3 Fatty Acids

The lipidic brain membrane, which covers the brain, is rich in fat and contains phospholipids, sphingolipids, and cholesterol. Polyunsaturated (PUFA) fatty acids, of which 33% are omega-3 family members, are thought to make up 50% of the grey matter in the brain. A deficiency in dietary n-3 polyunsaturated fatty acids (PUFA) may increase a person's susceptibility to depression and other psychiatric disorders, according to clinical and epidemiologic studies (McNamara, 2009). By adjusting the intensity and duration of the inflammatory response, N-3 PUFAs, specifically DHA and EPA, reduce inflammation. Accordingly, n-3 PUFAs may use anti-inflammatory mechanisms to support antidepressant effects and/or resistance to depression (McNamara, 2009).

The most prevalent polyunsaturated fatty acid (PUFA) in the brain is docosahexaenoic acid (DHA, 22:6n-3), which makes up about 15% of all the fatty acids in that tissue. Levant (2013) states that α-linolenic acid (18:3n-3) and eicosapentaenoic acid (EPA 20n-5:3) are the endogenous sources of these long-chain PUFAs. However, mammals are not able to synthesize DHA or arachidonic acid (AA)-(omega 6), so these nutrients must be obtained through diet. Human depression does not necessarily result from dietary or tissue n-3 PUFAs; however, when other risk factors (such as particular genotypes and stressors) are also present, they probably increase susceptibility to depression (Levant, 2013). It has been demonstrated that giving depressed patients 1.5 to 2 g of EPA daily can enhance mood elevation. However, omega-3 doses greater than 3 g are not more effective than placebos and might not be appropriate for certain patients, such as those on anticoagulant medications (Lakhan and Vieira, 2008). Omega-3 fatty acids, such as the 18-carbon omega-3 fatty acid alpha-linolenic acid, which is mostly found in plant oils, and DHA, which is mostly found in fish, are scarce in Western diets (Innis, 2008).

5.1.7. Micronutrients

The most prevalent sign of folate deficiency is depression. Blood folate (B9) levels are 25% lower in depressed patients than in healthy participants. Certain neurotransmitters are directly synthesized with the help of vitamins B6 and B12, among others. Cobalamin (B12) supplements enhance cerebral and cognitive processes and protect the integrity of the nervous fibers' myelin sheath (Rao et al., 2008). A study found that giving subjects an excess of nine vitamins for a year improved their mood in every subject. Furthermore, despite the fact that the blood levels peaked after three months, these alterations in mood were noted after a year. Vitamins B1, B2, and B6 were primarily associated with improvements in mood (Benton et al., 1995) (Table 5.1).

Table 5.1. Effect of Food Choice on Psychological Reactions and Behavioral Responses

Objective	**Key Sections**
Food and Mood	Stress and depression may cause emotional eating
	Unhealthy food choices reduce satisfaction and psychosomatic benefits
	Consuming fruits and vegetables reduces stress levels
	Food's antioxidants boost mood, reduce tension, and protect against anxiety
Food and Depression	Poor eating, smoking, and an inactive lifestyle raise the risk of depression
	Exclusion of certain foods increases depression risk
	Vitamin B complex, D, C, zinc, omega-3 fatty acids, and antioxidant deficiencies can negatively impact mood and potentially lead to depression. Nutrients like fruits, vegetables, and dairy are essential

Objective	Key Sections
	Consuming pro-inflammatory foods increases depression risk; low-inflammatory diets reduce symptoms
	Hormones like adrenaline, cortisol, and glucagon affect anxiety, hunger, and hypoglycemia
Food and Happiness	Research supports the idea that good dietary choices, like fruit and vegetable consumption, benefit physical and mental health
Romantic Relationships and Food	Couples eat over half of their meals together, strengthening, and preserving relationships. Examining food preferences in romantic relationships can reveal unhealthy eating habits and potential obesity. Male eating habits impact females when relationship creation goals are active, while female eating patterns impact males when relationship maintenance goals are active.
Emotion suppression in romantic relationships and food consumption	Overweight women consume more on emotion-suppression days than non-obese women
	Thematic templates reveal three themes: social environment, social economics, and food nostalgia, focusing on cultural, familial, and economic factors impacting mood
Food Insecurity effect on Violent Crime	Food insecurity increases violence and civil strife in developing countries
	Fish, micronutrient-rich diet, and good nutrition reduce violent behavior rates
Gut–brain axis communications and behavioral responses	Herbal extracts regulate gut microbiome structure, enhance depression-related behaviors
	Dietary flavonoids influence gut microbiota composition and affect the vagus nerve, which controls neurocognitive functions and adaptive behavioral responses
Halal and Tayyib foods and mental health	Halal and Tayyib meals improve physical, spiritual, and mental health

Source: Heidari M, Khodadadi Jokar Y, Madani S, Shahi S, Shahi MS, Goli M. Influence of Food Type on Human Psychological–Behavioral Responses and Crime Reduction. Nutrients. 2023; 15(17):3715. https://doi.org/10.3390/nu15173715.

5.1.8. Alcohol

The obvious relationship between alcohol and mood is another. There is no denying alcohol's impact on mood. It has long been believed that alcohol can momentarily reduce negative feelings and momentarily heighten enjoyment and relaxation. The Global Drug Survey (GDS), a large study that involved interviewing 30,000 individuals aged 18 to 34 from 21 different countries, looked at the relationship between the types of alcohol (beer, red, and white wine, spirits) and the reactions that consumers experience when they consume them. These reactions include feeling energized, relaxed, sexy, confident, tired, aggressive, ill, restless, and tearful. This study highlights the complex effects that drug consumption can have on an individual's temperament (Ashton et al., 2017). Ethanol may be helpful in small doses, as it acts as a kind of anesthetic by depressing brain function and promoting feelings of happiness and relaxation. Drinking alcohol modifies the reward, arousal, and addiction circuits by influencing the dopaminergic, serotoninergic, glutamate, and γ-amino butyric acid (GABA) pathways in the brain (Banerjee, 2014).

5.1.9. Hunger

However, it has been demonstrated that hunger can cause affective changes in mood, perception, and response. This primarily occurs because elevated blood glucose levels set off a chain reaction that releases hormones linked to specific emotional states. The term "hangry," which originates from the informal meaning of "bad tempered or irritable as a result of hunger" according to the Oxford Dictionary, suggests that hunger can influence emotional experiences and behaviors. The word "hungry," a pun derived from the words "hungry" and "angry," describes how someone's temper can become agitated and irritated when there isn't any food present or when they feel naturally hungry.

A scientific explanation relates blood sugar levels to mood swings. Without enough blood glucose, people cannot control their emotions, and when they are low on glucose, they are more likely to act impulsively, harshly, and aggressively (Anderberg, 2016). The metabolic hormone that signals hunger, ghrelin, is released when blood sugar falls. This release of hormones sets off a chain reaction that affects the sympathetic nervous system and causes unpleasant and intensely arousing physical changes. To bring blood sugar levels back to normal, the hormones cortisol and adrenaline are released when blood sugar levels drop. Additionally, the hormone neuropeptide Y contributes to the sensation of hunger when the body requires more food. These hormones have been connected to aggression and cause irritability (Naftulin, 2018). While some psychologists contend that hunger is only ever felt in negative contexts and can be perceived as a negative, high-arousal state, or "hanger," another theory contends that altered moods are the outcome of metabolic processes (MacCormack and Lindquist, 2019). Fasting can cause tryptophan levels to drop, and as a result, people with lower tryptophan levels exhibit more negative behaviors (Altman et al., 2010).

5.1.10. Our Microbes the Microbiota

The impact of gut microbiota on cognitive function has been demonstrated to be influenced by diet. The relationship between the gut, the brain, and disposition has come into sharper focus as evidence has grown. The enteric and central nervous systems are connected by the gut-brain axis, a bidirectional communication network that also includes humoral, metabolic, immunological, and endocrine pathways. The gut can affect mood, cognition, and mental health, and the brain can influence intestinal activities thanks to the gut-brain connection (Appleton, 2018).

The term microbiota, also known as microflora, refers to all of the microorganisms present in a given habitat. According to Wang and Kasper (2014) and Hooper, Littman, & Macpherson (2012), the term "microbiome" refers to the collective genomes of all the microorganisms within a colony. A complex community of microorganisms known as the human gut microbiota is made up of 100 trillion bacteria, quadrillion viruses, fungi, parasites, and archaea (Hooper et al., 2012). Functional Gastro-Intestinal tract (GI) disruptions are now known to be associated with anxiety, depression, and autism spectrum disorders; on the other hand, GI disease, such as irritable bowel syndrome, is frequently associated with psychological comorbidities (van den Elsen et al., 2017). Dysbiosis activates the so-called inflammasome pathway, which is why depression is increasingly understood to have an inflammatory component (van den Elsen et al., 2017). Eating an anti-inflammatory diet is essential to combating depression.

According to Appleton (2018), patients with major depression exhibit higher levels of antibody production in the human body when they have leaky gut syndrome, suggesting a possible correlation between gut and mental health. In particular, species with Bifidobacterium (B) and Lactobacillus (L) genes are known to be antidepressants. These involve mechanisms such as the inhibition of pro-inflammatory cytokines, control over the metabolism of tryptophan, and modulation of neurotransmitters in the central nervous system.

While consumption of *L. helveticus* and *B. longum* reduced somatization, depression, and anger-hostility, hospital anxiety and depression (Anderberg, 2016), gut-associated pathogens such as infection with *C. jejuni* can exacerbate anxiety (Wang & Kasper, 2014).

Altered gut microbiota composition, in particular a less diversified one, has been associated with stress as well. A solution might be balancing the gut bacteria, through diversifying the diet. Dietary formulas with higher fiber contents can improve microbiome richness. Moreover, the anxiolytic strains of Lactobacillus and Bifidobacterium genera that have anti-anxiety effects often display anti-stress effects as well (Wang & Kasper, 2014).

5.1.11. How Does the Mood Affect the Food?

According to research, people react through food to both positive and negative emotions. Emotional eaters eat a lot more when they're depressed than when they're joyful. Additionally, people who eat emotionally prefer sweet food over salty food (van Strien et al., 2013). In contrast to a negative mood, which motivates immediate, significant goals like mood control and results in a stronger preference for indulgent foods over healthy foods, a positive mood tends to project long-term health objectives (Gardner et al., 2014).

Comfort foods help people feel good by transporting them to a time in the past when they were cared for or pampered, giving them a momentary sense of well-being. Researchers have discovered that people's attraction to a particular food is influenced by their physiological and psychological needs.

Certain foods alter or modify emotional states or feelings and are linked to the alleviation of distress. This demonstrates that mood has a big influence on what people eat, and comfort foods are preferred in some situations, like when people are sick (Locher et al., 2005). When people are depressed or lonely, they tend to reach for comfort food (Spence, 2017). According to Wanink, Cheney, and Chan (2003), comfort foods related to warm, hearty meals, like steak, casseroles, and soup, are preferred by men, while chocolate and ice cream are preferred by women. Sugary, high-calorie foods like ice cream, cookies, and chocolate, for instance, improve mood because they produce opiates and serotonin (Stein, 2008).

5.1.12. Treatments for Affective Disorders

The nutritional approach has emerged as a viable and more healthful substitute for costly antidepressant or anxiety medication, which typically causes toxicity, undesirable side effects, and a high rate of noncompliance in patients with mental illnesses (Lakhan and Vieira, 2008). The most prevalent mood disorders, major depression, bipolar disorder, schizophrenia, anxiety disorders, eating disorders, attention deficit disorder/attention deficit hyperactivity disorder (ADD/ADHD), addiction, and autism, may be controlled with nutritional supplement treatment, according to new research and emerging evidence (Lakhan & Vieira, 2008).

Since neurotransmitter imbalances are thought to be directly linked to depression, anxiety, and other mood disorders, drugs that either stimulate or inhibit these chemicals' actions are frequently used in the treatment of these conditions. A focus on the nutritional deficits linked to mental health conditions may demonstrate the potential use of dietary supplements in the management of various conditions.

Lakhan and Vieira (2008) have reported that depression is caused by a deficiency of serotonin, dopamine/noradrenaline, γ-aminobutyric acid (GABA), omega 3 fatty acids, folate, and other B vitamins, magnesium, and SAM (S-adenosylmethionine), based on human pilot clinical trials, double blind and case studies, and placebo-controlled studies. Thus, tryptophan, which is a precursor to serotonin, tyrosine for dopamine and noradrenaline, GABA, omega 3, vitamins B complex, magnesium, and SAM should all be supplemented in effective treatments (Table 5.2).

Bipolar disorder can also be brought on by a deficiency of omega-3 fatty acids; excess acetylcholine receptors can be treated with taurine and lithium; deficiencies in vitamin B complex and/or tryptophan can be treated with corresponding supplements; and choline deficiency can be treated with lecithin? Omega 3 deficiency, decreased glycine synthesis, and/or impaired serotonin synthesis (tryptophan) are associated with schizophrenia. St. John's wort supplements have shown efficacy in treating obsessive-compulsive disorder (Holford, 2007; Lakhan & Vieira, 2008; Bolton & Bilbo, 2014) (Table 5.3).

Table 5.2. *Statistically Significant Results About the Beneficial Effect of Nutrient Supplementation on Mental Health and Well-Being*

Tested molecule(s)	Disorder	Study Design	Number of participants	Main results	Reference
Magnesium oxide, folic acid, multivitamin supplementation	Stress	8-week randomized controlled trial on women aged 25–45 and experiencing psychological distress. Post-intervention data collected 8 weeks after the start of the supplement intake	60 subjects randomized in an active group or in a placebo group	Positive impact of nutrient supplementation on pro-inflammatory cytokine profiles in the active group; no effect on psychological state	(19)

Magnesium, vitamin B6	Stress	8-week, Phase IV, randomized, controlled, investigator-blinded, parallel-group trial on subjects aged 18 to 50, with moderate to extremely severe stress and hypomagnesemia at screening	264 subjects randomized 1:1 to treatment with either the magnesium–vitamin B6 combination or magnesium alone	Significant reduction of perceived stress *vs.* baseline; no significant difference between the two treatment groups	(20)
DHA	Stress	6-week double blind randomized placebo-controlled clinical trial on academic personnel, aged 18–60, experiencing stress	93 subjects randomized in an active group treated with fish oil ($n = 16$), a control group treated with olive oil ($n = 14$), or in a control group with no treatment ($n = 63$)	Significant reduction of perceived stress in the active group *vs.* control groups	(21)
DHA, EPA	Burnout	8-week double-blind and placebo-controlled intervention on subjects aged 18–65, with self-reported signs of workplace-related exhaustion	43 subjects randomized in an active group treated with DHA and EPA ($n = 22$), or in a placebo control group ($n = 21$)	Significant improvement of burnout scores and decrease of saliva morning cortisol secretion	(22)
Omega-3 PUFAs	Psychotic disorders	6-month multicenter, double-blind, placebo-controlled trial on subjects aged 13–40, at ultra-high risk for psychotic disorders	304 subjects randomized in an active group treated with n-3 PUFA together with cognitive behavioral case management (CBCM) ($n = 66$) or in placebo group treated with CBCM only (n=62)	Increases of the erythrocyte n-3 index and DHA levels are predictive of better symptomatic and functional outcomes at month 6 in the active group	(23)

DHA, EPA	Depression and anxiety	16-week placebo-controlled trial on subjects aged 15–40, with recent onset schizophrenia-spectrum (n = 46) or bipolar (n = 4) disorders and current psychotic symptoms, treated with risperidone	50 subjects receiving risperidone randomized 1:1 in an active group treated with adjuvant n-3 PUFAs (n = 25) or in a placebo control group (n = 25)	Substantial decrease in depression-anxiety symptoms in the n-3 PUFA group *vs.* the placebo group	(24)
Omega-3 PUFAs	Major depressive disorder	12-week double-blind, randomized and placebo controlled clinical trial on subjects aged 18–65, with major depressive disorder, treated with sertraline	50 subjects receiving setraline randomized 1:1 in an active group treated with adjuvant n-3 PUFAs (n = 25) or in a placebo control group (n = 25)	Substantial improvement of symptoms of depression, and of dimensions of anxiety and sleep symptoms in the n-3 PUFA group *vs.* the placebo group	(25)
DHA	Sleep disorder	14–18-week longitudinal, randomized, double-blinded, placebo-controlled trial on pregnant women aged 18–35	48 pregnant women randomized in an active group receiving DHA-supplemented functional food (n = 27) or in a placebo group (n = 21)	Significant improvement of infant sleep organization in the first 48 h in the DHA group *vs.* the placebo group	(26)
Omega-3+Omega-6 PUFAs, antioxidant vitamins (alpha-tocopherol, gamma-tocopherol, vitamin A)	Mild cognitive impairment	6-month randomized, double-blind, placebo-controlled trial on subjects aged ≥65 years, with probable mild cognitive impairment	46 subjects randomized 1:1 in an active group treated with PUFAs (n = 23) or in a placebo group (n = 23)	Favorable improvement of functional capacity and cognitive function for the participants receiving the supplementation *vs.* the placebo group	(27)

DHA, EPA	Major depressive disorder	8-week randomized, double-blind, placebo controlled trial on subjects aged 18–80, with major depressive disorder and high plasma levels of inflammation biomarkers	196 subjects randomized in an active group treated with DHA (*n* = 66), an active group treated with EPA (*n* = 65) or in a control group treated with placebo (n=65)	Significant improvement of depressive symptoms in EPA-treated patients with high plasma levels of inflammation biomarkers	(28)
DHA, EPA	Antenatal depressive disorder	12-week randomized trial on pregnant women aged ≥20 years with antenatal depressive disorder	100 subjects randomized in an active group treated with omega-3 PUFAs (*n* = 49) or in a control group treated with placebo (*n* = 51)	Significant association between increased EPA and estradiol and decrease of depressive symptoms in the active group	(29)

Table 5.3. Prevalence of Stress, Depressive Symptoms, Suicidal Ideation, and Sugar-Sweetened Beverage and Fast-Food Consumption (N = 24,006)

Variables	**Categories**	***n* (%)**
Mental health		
Stress	With	19,806 (82.5)
	Without	4,200 (17.5)
Depressive symptoms	With	6,579 (27.4)
	Without	17,427 (72.6)
Suicidal ideation	With	2,852 (11.9)
	Without	21,154 (88.1)
Independent consumption of sugar-sweetened beverages and fast-foods		
Sugar-sweetened beverage consumption	Q1 (low)	8,917 (37.1)
	Q2 (medium)	9,855 (41.1)
	Q3 (high)	5,234 (21.8)
Fast-food consumption	Q1 (low)	17,301 (71.8)
	Q2 (medium)	5,321 (22.3)
	Q3 (high)	1,384 (5.9)
Combining consumption of sugar-sweetened beverages and fast-foods		
Sugar-sweetened beverage consumption	Fast-food consumption	

Q1 (low)	Q1 (low)	7,553 (31.5)
	Q2 (medium)	1,225 (5.1)
	Q3 (high)	139 (0.6)
Q2 (medium)	Q1 (low)	6,993 (29.1)
	Q2 (medium)	2,420 (10.1)
	Q3 (high)	442 (1.8)
Q3 (high)	Q1 (low)	2,755 (11.5)
	Q2 (medium)	1,676 (7.0)
	Q3 (high)	803 (3.3)
Covariates		
Biological factors		
Sex	Boys	12,441 (51.9)
	Girls	11,565 (48.1)
Sleep satisfaction	Dissatisfied	12,137 (50.6)
	Satisfied	11,869 (49.4)
Social factors		
Grade	1st	8,175 (31.8)
	2nd	8,348 (33.8)
	3rd	7,483 (34.4)
Academic achievement	Low	8,915 (36.9)
	Middle	7,839 (32.8)
	High	7,252 (30.3)
Living with family members	Yes	1,753 (6.4)
	No	22,253 (93.6)
Family's socioeconomic status	Low	3,430 (13.8)
	Middle	12,411 (51.3)
	High	8,165 (34.9)
Type of school	Specialized high school	4,255 (16.5)
	General high school	19,751 (83.5)
Area of residential location	Metropolis	12,110 (50.5)
	Middle sized city	10,430 (45.1)

	Rural area	1,466 (4.4)
Psychological factors		
Perceived health status	Unhealthy	2,563 (10.7)
	Fair	6,372 (26.5)
	Healthy	15,071 (62.8)
Perceived body shape	Being fat	9,661 (39.8)
	In average	8,575 (35.8)
	Skinny	5,770 (24.4)
Skipping breakfast	Yes	13,280 (55.4)
	No	10,726 (44.6)
Screen-based sedentary time (a day)	≥2 h	19,530 (81.5)
	<2 h	4,476 (18.5)
Moderate and vigorous physical activity (a week)	≥3 days	8,396 (34.5)
	<3 days	15,610 (65.5)
Current smoking consumption	Yes	1,985 (8.1)
	No	22,021 (91.9)
Current alcohol consumption	Yes	3,864 (15.9)
	No	20,142 (84.1)
Experience of sexual intercourse	Yes	2,026 (8.4)
	No	21,980 (91.6)
Experience of substance use	Yes	195 (0.8)
	No	23,811 (99.2)

n = unweighted, % = weighted.

Q1, first quantile; Q2, second quantile; Q3, third quantile.

Source: Ra JS (2022) Consumption of sugar-sweetened beverages and fast foods deteriorates adolescents' mental health. Front. Nutr. 9:1058190. doi: 10.3389/fnut.2022.1058190.

5.2. INDUSTRY INVOLVEMENT IN THE DEBATE ON ULTRA-PROCESSED FOODS AND HEALTH

Overseas billion-dollar corporations that make money from the sale of products known to accelerate poor health, such as alcohol, tobacco, and nicotine, as well as ultra-processed foods (UPFs, such as soda, chips, and snack cakes), pose a significant challenge to public health. Regulating public policy and market intervention, regrettably, usually depend on industry self-regulation (e.g., reformulating goods to merely satisfy school lunch requirements (e.g., making them 51% whole grain), a strategy that has proven futile in averting the unfavorable health consequences connected to these commodities.

To better understand the context surrounding the debate about the role of UPFs in mental health treatment, it is prudent to consider the range of "corporate scientific activities" that influence the public discourse around nutrition. A few examples of these activities are: (1) conducting nutrition research on their products internally (with publication bias favoring findings favorable to their products); (2) funding expert meetings and scientific seminars to develop consensus policies in their favor; (3) participating in policy committees to develop scientific standards; (4) publishing papers in scholarly journals (often at hefty article processing fees); (5) funding scientific front groups (like the International Food Information Council); and (6) offering nutrition education programs (granting continuing education credits). Industry-sponsored messaging seems to be effective in creating a sense of doubt about the reliability of the data pertaining to the risks associated with their products, which reduces the likelihood that UPFs will occur by raising public skepticism.

It is especially problematic to promote UPFs (such as sugar- or artificially-sweetened-beverages) to children because these individuals are deemed vulnerable, and governments ought to enact laws to safeguard their well-being, especially in times of mental health emergencies. It is commonly known that food companies frequently add fiber isolates, vitamins, and minerals to UPFs to make health claims on the packaging, which may mislead parents and consumers about the nutritional value of the product. The majority of dietary recommendations are based on nutrient-based theories that do not take into account the current knowledge of many of the problems with UPFs. However, an increasing number of nutritionists believe that worries about ingesting UPFs are the result of "diet culture" and fear mongering, which perpetuates "myths" about food and eating. With the crucial objective of ED prevention, attempts to "debunk" theories regarding the relationship between UPFs and homeostatic hunger may have good intentions. However, these messages have the potential to perpetuate shame and the internalization of weight stigma by invalidating people who struggle with hedonic eating and marginalizing them as a forgotten group within the mainstream nutrition narrative.

Additionally, in recent decades, UPFs have had a detrimental effect on environmental health. Negative exposure to the environment (e.g., socially disadvantaged groups are disproportionately affected by air pollution, hazardous wastes, and other toxins. Compared to their counterparts in more socially advantaged areas, residents in low socioeconomic status areas are exposed to more UPF advertisements. According to internal industry documents, food companies adapted the tobacco company's marketing playbook to specifically target racial and ethnic minority groups. It should come as no surprise that UPF consumption has been rising over time among younger, less educated, and lower-income groups, according to US data. This adds layers of complexity to this sociological inquisition because the discussion surrounding UPFs can legitimately be framed as a social justice issue rather than a question of personal responsibility (Holford, 2007; Lakhan & Vieira, 2008; Bolton & Bilbo, 2014).

5.2.1. Ultra-Processed Foods and Their Nutritional Characteristics

5.2.1.1. NOVA Classification for Food Processing

NOVA classifies food into four categories:

1. Unprocessed or minimally processed foods (e.g., washed/trimmed corn on the cob),
2. Processed culinary ingredients (e.g., oils, sugars, salts, flavors),
3. Processed foods (a combination of NOVA-1 and NOVA-2 foods, e.g., canned corn), and
4. Ultra-processed foods (containing little, if any, intact NOVA-1 foods, e.g., corn chips).

Processed food can be beneficial in many ways. Processing, for instance, improves the shelf life and palatability of NOVA-1 minimally processed foods (MPFs). Contrarily, UPFs include ingredients like flavors, colors, sweeteners, emulsifiers, stabilizers, and preservatives that aren't used in food preparation. These serve to mimic the sensory attributes of MPFs, mask the unfavorable aspects of the finished product, and enhance their lucrative attributes to optimize revenue. Large-scale industrial processes are usually used to create UPFs, which are impossible to fully replicate at home. NOVA is still the most popular classification scheme in use globally, despite the development of other food processing-based systems. One criticism is that food classification is a somewhat subjective process that leads to differences in ratings among raters and hinders scientifically sound assignment. Another argument is that UPFs acquire negative connotations, which causes the public to form unwarranted and counterproductive negative perceptions. This may be a crucial factor to take into account for people with restrictive EDs who already experience increased anxiety and discomfort when it comes to food. According to NOVA research,

UPFs are becoming more prevalent in middle-class countries and currently control the food supply chain in high-income countries. Higher volumes and a greater variety of UPFs are sold and consumed as nations become wealthier. Ultra-processed goods are designed to be incredibly tasty, long-lasting, and portable so they can be consumed at any time or place. Because UPFs are so profitable, multinational food companies can expand into new markets, keep investing billions in new technologies, and effectively advocate for their packaged branded products. The literature has a message that these foods are essential to feed the world, primarily found in publications where one or more authors disclosed their relationships with the food industry. This message may have some merit given that smaller farms are increasingly finding it difficult to compete with Big Ag. However, processing that lengthens shelf life or adds necessary nutrients (e.g., fortified rice) is not classified as UPFs in lower- and middle-income nations. UPFs are not made with public health goals in mind, but rather for taste and financial gain.

Another common nutrition narrative holds that UPFs—also known as "all foods fit"—are necessary for recovery for individuals suffering from any kind of disordered eating. Given the growing body of evidence regarding the addictive qualities of UPFs and the clinical manifestation of UPFA, this messaging frequently causes confusion among patients, physicians, researchers, and the general public. There is disagreement over the idea of dietary restriction; some believe that avoiding UPFs causes or results in EDs, while others believe that consuming UPFs is the main cause or effect.

5.2.1.2. Nutritional Characteristics of Ultra-Processed Foods

According to available data, almost two-thirds of US foods fit the NOVA-4 classification of UPFs. UPFs makeup 90% of added refined sugars, have higher levels of total fat from added saturated fats and have a negative correlation with the consumption of protein and fiber from food. Since fiber is the main source of energy for intestinal microbes and has a positive correlation with improved mental health, it is especially significant.

Less than half of the micronutrient content found in MPFs was present in UPFs. Those who consume higher amounts of UPFs tend to consume insufficient amounts of most essential vitamins and minerals, including copper, iron, phosphorous, potassium, magnesium, calcium, selenium, and zinc, as well as vitamins A, Bs (niacin, pyridoxine, folate, B12), C, D, and E. Some B vitamins (eg., riboflavin) are included in ultra-processed diets because they are added as enrichments to ultra-processed grain products. Antioxidant polyphenols, like flavonoids, are present in diets that promote health, but they are frequently disregarded. Ultra-processing readily destroys these compounds. Consuming

UPF protects against a total and five out of six flavonoid class intakes, which is not surprising. Polyphenols and the abundant fibers found in NOVA-1 plant foods have a synergistic interaction that is influenced by gut microbiota. Gut microbiota has been proposed as a potential gap in our knowledge of the etiology of ED.

Therefore, elevated levels of UPF-containing compounds combined with low intakes of fiber and polyphenols may result in elevated inflammatory markers. For instance, several studies have demonstrated a correlation between increased C-reactive protein (CRP), interleukins (e.g., TNF-α, IL-6, IL-8), and other related factors. Elevated inflammatory markers in the context of mental health (ie, depression) are important and will be covered in the section below. One mechanism is the invasion of microbes along the gastrointestinal wall, which causes intestinal permeability and sets off a series of inflammatory reactions around the periphery.

Although UPFs are typically not consumed in isolation, a meta-analysis indicates that consuming more UPFs may be linked to a lower-quality diet (e.g., less produce (fruits and veggies). Since there is probably a threshold (e.g., higher intake of fiber, polyphenols, omega-3 fatty acids, and protein from whole food sources), more research is required to evaluate the nutritional impact of UPFs among those who consume relatively higher amounts of these nutrients proportion of UPFs that pose a health risk when taken as a whole in the diet. Certain UPFs, like protein bars and plant-based meat substitutes, are promoted as completing nutritional gaps for people who must follow a restricted diet due to health issues (e.g., allergies) or alternative tastes. This argument appears to be part of the "planting doubt" playbook, which aims to exonerate manufacturers from other public health concerns about other UPFs (e.g., chips, and pastries) are more strongly linked to binge-type EDs and eating disorders resembling addictions.

UPFs negatively impact nutritional status through the following:

1. Displacement (consuming UPFs in place of MPFs).
2. Poor nutritional profile (low micronutrient/phytonutrient density of foods).
3. Neo-formed compounds from high-heat processes (e.g., acrylamide, advanced glycation end products, industrial trans-fats, furans).
4. Inflammatory responses to industrial food additives (e.g., emulsifiers, preservatives, thickeners, artificial sweeteners, flavoring agents, coloring agents).
5. Contaminants from packaging materials (e.g., bisphenols, phthalates).

6. Higher glycemic load reducing satiety signaling (e.g., refined grains and sugars).
7. Reduced fiber volume (leading to faster eating and decreased satiety).
8. Gut microbiota dysbiosis (associated with the loss of beneficial and overgrowth of opportunistic microbes, reducing short-chain fatty acid production in the colon).
9. Increased intestinal permeability (leading to peripheral inflammation).
10. Overconsumption (addiction-like eating) (Holford, 2007; Lakhan & Vieira, 2008; Bolton & Bilbo, 2014).

5.2.2. Nutrition for Mental Health

The promotion of mental health is significantly aided by nutrition. All-important nutrient-rich diets will guard against deficiencies, which can exacerbate mental health issues. A healthy diet that includes enough fiber intake supports the function of the gastrointestinal system (GI). The foods we eat can affect our mood and even the foods we eat in the future. Nutrition can impact mental health conditions such as, but not limited to:

1. Depression.
2. Anxiety disorder.
3. Schizophrenia.
4. Obsessive-compulsive disorder.

5.2.2.1. Where Food Meets Mental Health

Our eating habits, emotions, and mental well-being are closely related. Our dietary decisions are influenced by our mood. Our mood and subsequent decisions can be influenced by the foods we eat. This affects not just our mental health but also raises the risk of developing other chronic illnesses. We raise our risk of nutrition-related disorders like type 2 diabetes and hypertension when we habitually choose low-nutrient snack foods. Changes in appetite can be brought on by certain mental health issues. What's worse is that missing meals can result in nutrient deficiencies. There is a connection between mental health issues and cravings. Low serotonin levels are associated with cravings for carbohydrates. A balanced, healthful diet can help promote healthy serotonin levels, which will reduce cravings.

The GI tract, sometimes referred to as the gut, is impacted by the foods we eat. The kinds and quantity of bacteria in our stomachs can be influenced by the food we eat. Bacteria are critical to our well-being. Everybody has some pathogenic, or "bad," bacteria in their stomach, but everyone also has some symbiotic, or "good," bacteria. Good bacteria aid in shielding us from harmful bacteria. They also promote general GI health and a robust immune system. You might be wondering how gut health relates to mental health, and the answer is that it does! When we consume high-fiber foods, the fibers are not broken down and absorbed by our bodies. They travel to the large intestine, where they are fermented by bacteria. The gut bacteria's favorite food source can be conceptualized as fiber. Good bacteria can grow if they are fed. Consequently, this raises the amount of beneficial products the bacteria produce for us. Nutrients like biotin, vitamin K, short-chain fatty acids, and neurotransmitters are included in these products. The main energy source for the cells in our gastrointestinal tract is short-chain fatty acids. They aid in the efficient digestion and absorption of the nutrients required for optimal health. Diverse neurotransmitters can also be produced by the bacteria. These messengers facilitate neural communication within our brains. Serotonin is one of the neurotransmitters that is generated.

If you're unfamiliar with serotonin, it's commonly known as the "mood" neurotransmitter. 90% of our serotonin is thought to be produced in the gut. Serotonin controls mood, but it also affects sleep, appetite, and other bodily processes. These can all have an effect on mental health. For example, the hormone that induces sleep, melatonin, is produced by our body when serotonin levels are high enough. Getting enough sleep is crucial for maintaining mental wellness. Reduced serotonin can exacerbate depression, increase cravings for carbohydrates, and disrupt sleep. Serotonin synthesis requires a variety of nutrients. You might develop food cravings if your body isn't receiving enough nutrients. Short-term cravings can be met by consuming refined carbohydrates and sugar-filled foods. However, they will cause a blood sugar spike and fall, which will increase cravings. Your mental well-being may suffer as a result of this pattern.

5.2.2.2. Helpful Diet Patterns

Certain dietary habits have been shown to lower the risk of depression, mental health issues, and cognitive health issues. A diet pattern akin to the Mediterranean is one of these. There is a 30% lower risk of depression when following a Mediterranean diet. This diet pattern has benefits for both adults and adolescents. The pattern is rich in:

a. Fruits.

b. Vegetables.

c. Whole grains.

d. Fish.

e. Lean proteins.

f. Nuts and seeds.

It has more fiber, nutrient-dense foods, and healthy fats than the Western diet, which consists primarily of processed foods and added sugar. It is believed that these nutrients offer protection against disorders related to mental health. Making minor adjustments is simpler to stick to than trying to completely change your eating habits. Promoting mental health is best started with a goal like cutting back on processed foods and added sugar.

5.2.2.3. Other Considerations

1. **Genetics:** Mental health problems may be linked to specific genes. It's critical to keep in mind that having a particular gene does not automatically determine your fate. Our surroundings and way of life have an impact on how our genes are expressed. This is a different discussion entirely and will differ from person to person. The link between depression and the MTHFR gene is one illustration of how genes can impact diet and mental health. This gene may result in decreased MTHFR enzyme activity, which may impact the body's supply of the nutrient folate. Depression risk may rise as a result of this.
2. **Medications:** Many people who treat mental health conditions will get prescription drugs. It is crucial to take these drugs' side effects into account. Noting any interactions between food and drugs is also crucial. Certain drugs may make you feel more or less hungry. Your general consumption of vital nutrients will be impacted by a decrease in appetite. Conversely, an increase in appetite may result in overindulging in calories and weight gain. Extra weight can exacerbate physical activity limitations and worsen mental health effects (Holford, 2007; Lakhan & Vieira, 2008; Bolton & Bilbo, 2014).

5.3. COGNITIVE FUNCTION OF ULTRA-PROCESSED FOODS

The consumption of UPFs, which are industrially formulated products containing little to no whole foods and are often high in additives, sugars, unhealthy fats, and preservatives, has raised concerns regarding its impact on cognitive function. UPFs are typically characterized by their high energy density, low nutritional quality, and the presence of artificial ingredients designed to enhance flavor, color, and shelf life. Numerous mechanisms may be involved in the detrimental effects of a diet high in UPFs on cognitive health, according to research. First of all, these foods' high sugar and unhealthy fat content can cause oxidative stress and systemic inflammation, both of which are known to deteriorate brain function and raise the risk of neurodegenerative illnesses like Alzheimer's. Furthermore, deficiencies in cognitive processes, such as memory, attention, and executive functions, may be caused by the low concentrations of vital nutrients found in UPFs, such as vitamins, minerals, and omega-3 fatty acids. For synaptic plasticity, neurotransmitter function, and neuronal health to be maintained, these nutrients are essential. Furthermore, the high glycemic index of a lot of UPFs can cause blood sugar levels to spike quickly and then crash, which can affect mood and energy levels and possibly impair cognitive function.

Furthermore, because of the gut-brain axis, some additives and preservatives found in UPFs, such as artificial sweeteners and emulsifiers, have been linked to detrimental effects on the composition of the gut microbiota. Immune responses and neurotransmitter synthesis can be impacted by disruptions in this microbiota, which in turn can impact cognitive function. High UPF consumption has been associated in observational studies with worse cognitive function and an increased risk of mental health issues like anxiety and depression, which are frequently co-occurring with cognitive decline. Further investigation is necessary to establish causal relationships; however, existing data highlights the significance of dietary quality in preserving cognitive health. Lowering UPF consumption and encouraging a diet high in whole foods—such as fruits, vegetables, whole grains, lean meats, and healthy fats—may help maintain cognitive function and guard against cognitive decline brought on by aging. Understanding UPFs' effects on brain health is crucial for public health campaigns that aim to prevent cognitive impairments and improve dietary practices as the prevalence of UPFs rises globally.

5.3.1. Perinatal Diet Disrupts Cognitive Function Long-Term: A Role for Microglia

A poor diet during pregnancy and the first few years after delivery can lead to long-term modifications in many areas of central and metabolic functioning, including cognitive decline and accelerated aging of the brain. Offspring metabolic complications, such as diabetes and obesity, can result from maternal gestational diabetes as well as a junk food diet in non-diabetic individuals. Additionally, it may alter how the brain processes rewards, leading to the development of a preference for foods high in sugar and fat in the offspring. In a similar vein, premature solid food introduction and excessive consumption of fatty foods and sugar-sweetened beverages during childhood can hasten weight gain and cause metabolic issues over time that may be linked to impaired executive function. However, some dietary supplements have been shown to have a positive impact on cognition. For example, adding long-chain omega-3 PUFA to baby formula has been shown to improve infants' cognitive function. These randomized control trials (RCTs) found that giving 9-month-old infants an omega-3 PUFA-enriched formula starting soon after birth or breastfeeding for six weeks significantly improved the infant performance on problem-solving tasks.

It is evident from animal models that early-life nutrition has a wide range of effects. Even obesity in rat sires who do not participate in raising the young causes pancreatic beta cell dysfunction in the female progeny, a condition that can be inherited by future generations. In addition to having a detrimental effect on cognition, obesity, and high-fat diet feeding during pregnancy and lactation in rat and mouse dams causes impairments in many mood tests, including those that replicate depressive and anxious behaviors. Similar behaviors can be impacted by diet during the post-partum to weaning period.

In addition to the effects of a prenatal diet, male, and female rats that consume excessive amounts of their mother's milk during the first three weeks of their lives develop permanent obesity. Additionally, this overfeeding of newborns impairs cognitive function. For instance, when compared to control rats, neonatal overfed rats do poorly in the novel object recognition test and the delayed spatial win-shift radial arm maze as adults. These results bear comparison to the effects of poor diet in adults, wherein cognitive dysfunction requires a higher fat diet for a longer period of time (about 20 weeks in the rat) and/or a higher fat diet in combination with a pre-diabetic phenotype. These rats exhibit impaired microglial response to the learning task, even though there are no differences in post-learning synaptogenesis (synaptophysin) or apoptosis (caspase-3) to account for the effects observed in the neonatally overfed group.

One of the main immune cell populations in the brain is made up of microglia. They are crucial for synaptic pruning during development, but in an adult animal, their main function is to initiate an inflammatory immune response and phagocytose pathogens and damaged brain tissue. An appropriate microglial response may also be necessary for effective learning. Hyper-activated microglia can cause cognitive dysfunction through reduced production of molecules related to plasticity, such as brain-derived neurotrophic factor and insulin-like growth factor-1, excess pro-inflammatory cytokine production that impairs long-term potentiation-induction, and reduced synaptic plasticity. At postnatal day 14, the CA1 region of the hippocampus in neonatal overfed rats has a higher density of microglia while they are experiencing rapid weight gain and still have access to excess maternal milk. In addition, the retracted processes and larger some of these microglia suggest a more activated phenotype. When these rats grow to adulthood, the dentate gyrus's immunolabelled area with the microglial marker Iba1 continues to grow. The microglial response to a learning task is not as strong in neonatally overfed animals as it is in controls. This effect is linked to a suppression of cell proliferation in control animals compared to neonatal overfed animals, which may help to minimize novel inputs and maintain preexisting neural networks during the learning process. It's interesting to note that improving performance in the Barnes maze can result from global inducible depletion of monocytes and microglia, indicating that stopping microglial activity during particular learning stages is crucial for learning. These results imply that normal microglia must be able to robustly respond to learning tasks and that neonatal overfeeding impairs this ability. They also implicate microglia in the long-term effects of early-life overfeeding on cognition.

Neonatal diet can undoubtedly influence neuroinflammatory processes, including the function of microglia, and this is at least one mechanism by which cognitive function is impacted. Other early life events can also affect neuroinflammation and microglia, which are important in the way that stress during development modifies long-term physiology.

5.3.2. Early-life Stress (ES) Programs Vulnerability to Cognitive Disorders

Numerous preclinical and clinical studies have demonstrated that ES permanently modifies the structure and function of the brain, increasing the risk of developing emotional and cognitive disorders. The precise underlying processes of this kind of programming are still unknown. Numerous groundbreaking studies have demonstrated the critical importance of maternal sensory cues and neuroendocrine variables (e.g., stress hormones) in this programming, but new research indicates that these variables may work in concert with nutritional and metabolic components. Indeed, ES is linked to a higher risk of developing metabolic diseases like obesity, which typically coexist with cognitive deficits. Additionally, both ES and a poor early nutritional environment cause cognitive impairments that are remarkably similar in later life, indicating that nutritional components and metabolic factors may mediate some of the effects of ES on brain structure and function.

During this formative stage, the brain has an extremely high demand for nutrients, and nutritional imbalances disrupt normal neurodevelopment, leading to long-term cognitive deficits. It is essential to comprehend the function of particular nutrients and metabolic variables in this setting to create efficient peripheral (e.g., dietary) intervention techniques. It has been demonstrated that an aberrant maternal care regimen during the first postnatal week in a mouse model of chronic ES results in a decline in cognitive function in the offspring of the ES. These children who were exposed to ES have irreversibly changed hippocampal structure and function; a region of the brain essential for cognitive processes. Since the hippocampus continues to develop throughout the postnatal phase, it is in fact extremely sensitive to the early-life environment. A special kind of plasticity called adult neurogenesis (AN) occurs in the hippocampus and is characterized by the growth of neuronal progenitor cells that differentiate into fully functional neurons and then integrate into the hippocampal circuitry. The hippocampal-dependent processes of learning and memory are mediated by these recently developed neurons in different ways. ES has a lasting effect on AN, or more accurately, ES exposure first increases neurogenesis (i.e., the survival of the newly born cells is decreased at later time points (postnatal day 150), such as during the proliferation and differentiation of newborn cells) at postnatal day 9. Furthermore, ES has long-lasting effects on the neuroinflammatory profile; one such effect is an increase in CD68 (phagocytic microglia expression) in adulthood.

Crucially, peripheral adipose tissue metabolism is also persistently impacted by ES. Offspring exposed to ethylene straw (ES) exhibit a persistent reduction in white adipose mass (WAT), leptin mRNA expression in WAT, and plasma leptin, the adipokine released from the WAT. Furthermore, compared to controls, ES mice that are fed an unhealthy Western-style diet exhibit a greater increase in obesity. These results imply that, in a moderately obesogenic environment, exposure to ES causes metabolic dysregulation and increases susceptibility to obesity development. More research is necessary to determine whether these metabolic changes have a role in the cognitive deficits brought on by ES.

Apart from modifications in peripheral metabolism, the nutritional makeup of the dam's milk caused by ES and/or the pup's intake/absorption of nutrients may have long-term effects on the structure and function of the brain. In fact, postnatal day 9 offspring's plasma and hippocampus show decreased levels of the vital micronutrient methionine following ES exposure. Methionine is a crucial part of the one-carbon (1-C) metabolism and is needed for methylation as well as the synthesis of proteins, phospholipids, and neurotransmitters. Crucially, adding essential micronutrients linked to 1-C metabolism to the mother's diet for a brief period only during ES exposure not only raises methionine levels both centrally and peripherally, but it also mitigates the effects of ES on adult hippocampal cognitive measures and stops the hyperactivity of the hypothalamic-pituitary-adrenal axis that ES causes on postnatal day 9–25.

These findings emphasize how crucial it is to investigate nutritional and metabolic variables in relation to the effects of ES on the brain. To create completely new avenues for targeted nutrition for vulnerable populations, it will be crucial in the near future to better understand the precise mechanisms mediating the effects of nutrients and metabolic factors as well as the windows of opportunity for interventions on brain function. Diet at other stages of life is also significant in determining mood and cognition, even though the early life period is a window of particular vulnerability to the programming effects of diet and other environmental influences.

5.3.3. Adult Consumption of a High-Fat Diet: A Vulnerability Factor for Hippocampal-Dependent Memory

Adults in developed nations are consuming diets that are higher than ever in refined sugars and/or saturated fats. In fact, according to recent reports, Americans consume significantly more energy (12%) from saturated fats and 13% from added sugars each day than the US Department of Agriculture and the Department of Health and Human Services recommend (5–10%). Unsurprisingly, these eating patterns have led to an increase in adult obesity prevalence, which is currently about 37% in the US—a significant increase from the 13% prevalence rate in 1960.

These figures are concerning because obesity is now linked to mild cognitive impairments and dementia in addition to its well-known provocation of cardiovascular disease, metabolic syndrome, and type 2 diabetes. There is mounting evidence that suggests obesity-related cognitive deficits could be caused by neuroinflammation. Recent research has shown that consuming an unhealthy diet for just one to seven days can have negative effects high levels of sugar and/or saturated fat) start neuroinflammatory processes, indicating that obesity alone might not be enough to impair cognitive function. Because of its close proximity to the third ventricle, circumventricular organs, and medio basal eminence—areas where inflammatory signals from the periphery have easier entry into the brain—the hypothalamus has garnered the majority of attention regarding obesity-induced neuroinflammatory responses and functional declines over the last 10 to 15 years. In fact, it has been demonstrated that long-chain saturated fatty acids enter the hypothalamus directly and activate toll-like receptor 4 signaling, which in turn causes an inflammatory response. The hippocampus, a crucial part of the brain that mediates learning and memory, has not shown this active passage of saturated fatty acids, according to observations. However, it has been shown that both in humans and rodents, a high-fat diet impairs hippocampal-dependent memory function. As an illustration, rodents fed a high-fat and/or high-sugar diet showed marked impairments in different forms of memory (e.g., spatial, and contextual), as demonstrated by lower scores in the Morris water maze, Y-maze, radial arm maze, novel object recognition task, and contextual fear conditioning. In addition, compared to adult humans who followed a standard diet, those who consumed a high-fat diet for five days showed noticeably less focused attention and a slower rate of information retrieval from working and episodic memory.

Numerous studies, as well as other ones, have demonstrated that increased neuroinflammatory markers or responses in the hippocampus coincide with cognitive declines brought on by high-fat diets. It is unclear how exactly these neuroinflammatory processes signal to and/or impact the hippocampus, though. A growing body of research suggests that diets high in fat might damage hippocampal formation by making the immune cells (presumably microglia) in this area of the brain more sensitive, which would then trigger an inflammatory response in response to further stressful stimuli. In one study, for instance, it was shown that adult rats fed a high-fat diet for five months had a sensitized hippocampus. As a result, following a learning session, the rat neuroinflammatory response to a relatively mild stressor (a single, two-second, 1.5-milliampere foot shock) was amplified compared to the response of the rats fed regular chow. This response resulted in deficits in long-term contextual memory. Another study revealed that adult rats' hippocampal sensitivity could be achieved with just three days of a high-fat diet. Here, rats given a low-dose peripheral immune challenge (lipopolysaccharide, or LPS) experienced deficits in contextual memory in addition to an increased neuroinflammatory response in their hippocampal regions when compared to rats fed regular chow.

It has been demonstrated that markedly increased pro-inflammatory cytokines in the hippocampal regions impair several processes that support synaptic plasticity (including long-term potentiation) and, consequently, long-term memory. Sobesky and associates showed that eating a high-fat diet primes the hippocampal cells by increasing the glucocorticoid steroid hormone corticosterone in this area. More research is showing that corticosterone can prime hippocampal microglia and enhance the inflammatory response to a subsequent challenge, despite its traditional function as an immunosuppressant. Consider Frank et al. cleverly demonstrated that the inflammatory response in the hippocampus following a peripheral immune challenge (LPS) was amplified when corticosterone levels were raised beforehand. On the other hand, the neuroinflammatory response was inhibited when corticosterone levels rose following the immune challenge.

These results imply that whether an anti- or pro-inflammatory response occurs depends on the temporal relationship between the immune challenge and the corticosterone increase. Sobesky et al, discovered that rats fed a high-fat diet for three days showed noticeably higher hippocampal levels of corticosterone than rats fed either standard chow or a novel diet that was macronutrient-matched in control. Increases in the endogenous danger-associated molecular pattern high-mobility group box 1 (HMGB1), the interleukin (IL)-1 inflammasome-associated protein NLRP3, and the microglial activation marker cd11b were observed in conjunction with this rise in corticosterone caused by a high-fat diet. However, pro-inflammatory cytokine IL-1β was not elevated by a high-fat diet on its own unless rats were then challenged with a low dose of lipopolysaccharide (LPS). In contrast to the response to LPS in rats fed chow, the LPS challenge increased the pro-inflammatory response in the hippocampus of rats fed a high-fat diet. Sobesky et al. assessed the part corticosterone signaling plays in neuroinflammatory priming brought on by eating a high-fat diet given mifepristone, a glucocorticoid receptor antagonist, before consuming a high-fat diet.

As a result, the hippocampal IL-1β response to low-dose LPS was restored to normal. Moreover, mifepristone dramatically decreased the expression of HMGB1, IκBα, and NLRP3 that was induced by a high-fat diet plus LPS. Furthermore, the administration of mifepristone treatment successfully avoided the contextual memory impairments brought on by a high-fat diet in conjunction with an LPS challenge. These findings offer compelling evidence for the following theories: (a) eating a high-fat diet raises corticosterone levels in the hippocampus; (b) corticosterone plays a major role in sensitizing the hippocampus's immune cells, or microglia; and (c) sensitized microglia intensify their neuroinflammatory response to subsequent immune or stressful challenges, which results in cognitive deficits. Interestingly, however, while a high-fat diet in general can have a negative and substantial impact on cognitive processes, certain dietary elements may be able to counteract these effects; omega-3 polyunsaturated fats are one such potentially helpful element.

5.3.4. Dietary Omega-3 PUFA Regulate Neuroinflammation and eCBs: Role in Mood and Cognitive Disorders

The functions of PUFA in the brain have received a lot of attention since their discovery in the early 20th century. Being essential fatty acids, omega-3 and omega-6 polyunsaturated fatty acids must come from the diet. The Western diet is overly high in omega-6 polyunsaturated fatty acids (PUFA) relative to omega-3 PUFA, which results in an imbalanced ratio between these two fatty acids and has implications for brain and cardiovascular health. Though they come from different sources, essential omega-3 and omega-6 fatty acids can be found in green vegetables, seeds, and nuts. Linolenic acid (LA, 18:2 omega-6) is found in most plants, including coconut and palm, and α-linolenic acid (ALA,

18:3 omega-3) is found in green leafy vegetables, flax, and walnuts. Following ingestion, LA, and ALA are converted into docosahexaenoic acid (DHA, 22:6 omega-3), and arachidonic acid (AA, 20:4 omega-6).

The primary omega-3 and omega-6 long-chain polyunsaturated fatty acids (PUFA) in the brain are DHA and AA. Due to their ability to regulate basic neurobiological processes, particularly those related to mood and cognition, both long-chain polyunsaturated fatty acids (PUFA) play crucial roles in brain physiology. With a total brain phospholipid proportion of about 10% for AA and 20% for DHA, AA, and DHA are esterified to the phospholipid of neuronal and glial cell membranes. Owing to the brain's restricted ability to synthesize long-chain PUFA, oily fish can supply preformed DHA through diet. Therefore, when DHA-rich products are consumed in greater quantities, DHA partially replaces AA in brain cell membranes. On the other hand, a lower consumption of omega-3 PUFAs results in higher levels of AA and lower levels of DHA in the brain. Women report higher levels of DHA and AA than men do, indicating a gender difference in PUFA levels. These variations may be related to sex hormones because they have different effects on PUFA metabolism. Specifically, testosterone inhibits the conversion of omega-6 and omega-3 precursors into their corresponding long-chain metabolites, while estrogen stimulates this process. It has not, however, been adequately questioned whether these variations in PUFA are relevant to particular brain diseases that have a gender component, and more research is necessary.

DHA increases in the blood following direct ingestion and/or hepatic metabolization. As a non-esterified fatty acid, it is likely to enter the brain freely. It has been discovered more recently that Mfsd2a (major facilitator superfamily domain-containing protein 2a), which is expressed by adiponectin receptor 1 in the retina and brain endothelial cells, is crucial for DHA uptake and retention. Numerous randomized controlled trials (RCTs) have been conducted to assess the impact of long-chain omega-3 PUFA dietary supplementation on mood and cognitive disorders. This is because abnormal omega-3 PUFA levels have been extensively documented in both the peripheral tissues and the brain of patients with mood disorders or cognitive decline. All things considered, the results are inconsistent because different approaches were taken to assess the depressive and/or cognitive symptoms, the omega-3 PUFA supplementation was administered differently, for a longer period of time, and in different forms. Additionally, the nutritional intake and PUFA metabolism were not assessed before the supplementation began, and genotype-associated risk factors were not examined. Nevertheless, it is crucial to remember that several RCTs conducted in individuals with depressive disorders demonstrated an extra benefit of long-chain omega-3 PUFA supplementation to antidepressant treatments, even in spite of the inconsistent results. Notably, a recent study finds that long-chain omega-3 PUFA supplementation increases the responsiveness of depressed patients who exhibit elevated levels of inflammatory markers. Given that these PUFA are strong inflammatory regulators and that inflammation plays a critical role in mood disorders, this observation is extremely pertinent. Regarding cognitive decline, randomized controlled trials (RCTs) utilizing DHA supplementation in subjects carrying the apolipoprotein E ε4 (APOE4) allele, a risk factor for AD, show an improvement of pre-dementia despite the poor positive results of PUFA dietary supplementation in AD patients. Overall, disparities between clinical research strongly suggest that preclinical research is necessary to better target populations at risk of mood and cognitive disorders by illuminating the mechanisms of omega-3 PUFA on brain dysfunctions. Furthermore, the food industry faces a challenge in ensuring that the levels of omega-3 polyunsaturated fatty acids (PUFA) in food meet the body's need for these PUFA for optimal brain function.

DHA and AA influence neurotransmission and neuroinflammation, two important processes in mood and cognition, either directly or indirectly. As phospholipase A2 (PLA2) is activated, unesterified long-chain polyunsaturated fatty acids (PUFA) are released from cell membranes to commence their actions. Once released, cyclooxygenase (COX), lipoxygenases (LOX), and cytochrome P450 catalyze the metabolism of AA and DHA into bioactive mediators. Progression of inflammation, including brain inflammation, depends on the conversion of AA into several prostanoids, such as prostaglandins (PG), leukotrienes (LT), thromboxanes (TX), and lipoxins (LX). Additionally, DHA is metabolized via the COX/LOX pathways, producing metabolites that have pro-resolutive and anti-inflammatory characteristics. The brain contains neuroprotectin D1 (NPD1), resolvin D5 (RvD5), maresin 1 (MaR1), and specialized proresolving mediators (SPMs) derived from LOX. Some of these SPMs directly affect microglia, which allows them to potently modulate neuroinflammation both *in vivo* and *in vitro*. Patients with AD have compromised DHA and SPMs in the brain's periphery. Interestingly, synaptic loss rather than amyloid beta (Aβ) deposition is correlated with decreased DHA distribution in the brains of AD patients. Furthermore, DHA or SPMs influence the number and activation of microglia *in vivo* and facilitate the phagocytosis of Aβ42 by microglia. It is yet unknown if SPMs contribute to the protective action of long-chain omega-3 PUFA in mood and cognitive disorders linked to neuroinflammation.

Another important class of PUFA-derived lipid mediators in the brain is eCB. The fatty acid ethanolamides 2-arachidonoylglycerol (2-AG) and anandamide (AEA) are the primary brain AA-derived eCBs, whereas docosahexaenoylethanolamide (DHEA, also known as synaptamide) is an eCB-like derived from DHA. Certain catabolizing enzymes, such as fatty acid amide hydrolase for AEA and DHEA and monoacylglycerol lipase for 2-AG, control the half-life of ECPs in the brain. In terms of neuroinflammatory processes, COX, and LOX oxidize AA-derived eCBs into bioactive PG, which in turn promotes inflammation. At least two cannabinoid receptors—type 1 (CB1) and type 2 (CB2)—that are Gi/o protein-coupled with multiple brain signaling pathways are occupied by AEA and 2-AG. Compared to AEA and 2-AG, DHEA has a lower affinity for binding CB1 and CB2 receptors; instead, it binds GPR receptors, specifically GPR110 in the brain. The percentage of ethanolamides derived from AA and DHA is directly influenced by the dietary omega-3/omega-6 PUFA ratio. In various brain regions, the modulation of eCB is correlated with a decrease in synaptic activity and neuronal CB1R activity. 2-AG and AEA regulate synaptic function by suppressing excitatory and inhibitory synapse neurotransmitter release by acting as retrograde messengers at presynaptic CB1. The importance of brain eCB signaling in the understanding of how altered dietary intake of PUFA correlates with a range of neurological disorders is of high interest. However, other dietary factors may also contribute to improved cognition and prevention of cognitive disorders. Polyphenolic-rich foods are a further example that have been shown to have benefit, particularly in the context of aging.

5.3.5. Dietary Interventions with Polyphenolic-rich Foods Can Improve Neuronal and Behavior Deficits Associated with Aging

By 2050, almost twice as many people in the US as there are now—roughly 20% of the population—will be older than 65. Furthermore, the US is dealing with a growing number of overweight and obese people who are more susceptible to metabolic disorders, which can lead to diabetes, cardiovascular disease, and behavioral impairment. Numerous cognitive and motor deficiencies on tasks requiring balance, fine motor control, short-, and long-term memory or executive function are linked to both aging and metabolic dysregulation. Oxidative stress, inflammation, and poor insulin resistance are common features of obesity, cardio-metabolic, and vascular disease, and age-related declines in cognitive and motor function, according to studies conducted in both human and animal models. Although neuroinflammation only happens locally in the brain, circulating inflammatory mediators and peripheral inflammatory cells (e.g., cytokines) can also enter the brain, and as we get older, this happens more frequently. Therefore, methods for reversing impairments in motor and cognitive function as well as oxidative and inflammatory susceptibility to age-related changes must be developed. Reducing insulin resistance and the infiltration of inflammatory mediators into the brain through targeting peripheral inflammation and insulin signaling may also lower the incidence of various age-related deficits. Increased consumption of fruits and vegetables has been linked to lower fasting insulin levels. Research has revealed that plants, especially vibrant fruit- or vegetable-bearing plants, contain polyphenolic compounds with strong antioxidant and anti-inflammatory properties. There is growing evidence that eating these foods high in polyphenols, especially berries, may help prevent or even reverse age-related neuronal deficits brought on by neuroinflammation. More recently, double-blind, placebo-controlled, randomized human intervention studies have added to the body of evidence showing that flavonoid/polyphenol consumption improves cognitive function.

Preclinical research has led to the theory that modifying the neuronal environment with polyphenol-rich foods, such as berry fruit, will lessen neuroinflammation and oxidative stress and the susceptibility to these conditions, hence lowering the incidence of age-related deficits in behavior. Crude blueberry (BB) or strawberry extracts significantly reduced and reversed age-related motor and cognitive deficits in senescent rodents in early studies using animal models. Additionally, 9-month-old C57Bl/6 mice supplemented with BB were shielded from the negative effects of eating a high-fat diet. The high-fat diet hampered the ability to recognize new objects with memory, but time-dependent blueberry supplementation reversed the effects of recognition memory deficits. After five months on the diets, there was also an improvement in spatial memory as determined by the Morris water maze. Beyond their anti-inflammatory and antioxidant properties, berry fruit polyphenols may have a variety of other effects, according to later studies. Furthermore, it has been demonstrated that the anthocyanins found in blueberries enter the brain, and there is a correlation between their concentrations and cognitive function.

Largely, adequate consumption of fruits and vegetables can prevent cognitive decline, while low intake is linked to increased cognitive decline, according to epidemiological studies that have examined the relationship between fruit and vegetable intake and cognitive function. In particular, consuming more blueberries and strawberries along with higher intakes of total flavanoids and anthocyanidins was linked to a 2.5-year reduction in the rate of cognitive decline. Increasingly, randomized, double-blind, placebo-controlled studies have investigated berry fruit's potential to prevent age-related cognitive decline in humans. Therefore, in older men and women with age-related memory decline who drank blueberry juice, word list recall and paired associate learning significantly improved compared to baseline, and paired associate learning also significantly improved compared to placebo controls. According to a recent study, freeze-dried blueberries (24 g/day, or one cup of fresh blueberries) for 90 days improved two measures of executive function in older adults (ages 60–75). The cognitive tasks measured in the study were similar to those in the rodent studies. Compared to the placebo powder-consuming controls, the blueberry group's participants demonstrated significantly lower switch costs and fewer repetition errors on a task-switching test across study visits. Nevertheless, after consuming blueberries, there was no discernible improvement in balance or gait. Lastly, supplementing with blueberry concentrate for 12 weeks enhanced cognitive function, task-related activation, and brain perfusion (i.e., working memory) in older, healthy adults who drank 30 milliliters of blueberry concentrate, which contained 387 milligrams of anthocyanidins. According to these studies, berry fruit may be a useful tactic to stop, slow down, or even reverse cognitive decline as people age.

Different cognitive domains peak in early adulthood before either declining or reaching a plateau, indicating that cognitive aging does not occur across domains at the same time. Early interventions may therefore produce health benefits that are only quantifiable later in life. Two acute, cross-over studies involving school-aged children (ages 7–10) have demonstrated the positive cognitive benefits of blueberries. In the first study, subjects who drank a 200 g flavonoid-rich blueberry drink performed significantly better on the Rey auditory-verbal learning test in terms of delayed word-list recall when compared to a matched vehicle group.

This suggests that memory items are coded more effectively. The blueberry intervention had a negative effect on proactive interference and no positive effects on measures of attention, response inhibition, or visuospatial memory. In a second study, the same group looked at cognition at baseline, at 1.5, 3, and 6 hours after ingesting blueberry drinks containing 15 or 30 grams of freeze-dried wild blueberry (WBB) powder, or a placebo (vehicle). WBB powder consumption resulted in enhanced recall at 1.15 hours, enhanced delayed word recognition that persisted at all time points assessed, and enhanced accuracy at 3 hours on a difficult interference task. After the 30 g dose, there was an improvement in cognitive performance, especially on tasks requiring more cognitive processing power.

Older adults are more susceptible to neuronal disease and degradation because their bodies' defenses against oxidative stress and inflammation deteriorate with age. By lowering inflammation and oxidative stress in the brain, neuroprotective foods—like berries and other dark-colored fruits—offer a potential defense against this damage to aging brains and prevent cognitive decline in older adults.

5.4. INTERNAL SIGNALS AND COGNITION

Since ancient times, it has been understood that visceral signals affect mental function. Today, treatments for depression, schizophrenia, and bipolar disorder include lifestyle modifications like diet and exercise.

5.4.1. Effects of Vagal Nerve Stimulation on Cognition

The gastrointestinal tract's vagal afferents are essential for tracking several aspects of digestion, including the release of enzymes and the absorption of food. VNS therapy is now a commonly accepted practice for treating partial-onset seizures with refractory symptoms. VNS was thought to be a possible treatment for depression since it was observed that giving the treatment to epileptic patients improved their mood. In humans, a short-term open trial lasting 10 weeks did not result in improvements for depression patients using VNS; however, a longer-term study lasting 12 months showed beneficial effects of VNS that persisted for two years. In the Inventory of Depressive Symptoms self-report, patients treated with VNS specifically doubled their improvement each month when compared to patients receiving treatment as usual (TAU) alone. TAU involved using medication or another therapy that the treating physician thought was suitable to manage depression that was not responding to treatment. The US Food and Drug Administration recently approved the use of VNS for the treatment of chronic (as opposed to acute) resistant depression based on the findings of long-term studies. While the exact mechanisms behind VNS's effects on depression remain unclear, a recent study found that VNS raises noradrenaline levels in the prefrontal cortex and the levels of mRNAs for fibroblast growth factor 2 (FGF2) and brain-derived neurotrophic factor (BDNF) in the rat hippocampus and cerebral cortex. These results shed light on the ways that signals coming from the gut can impact mood, since increases in BDNF and noradrenaline have been linked to the effects of antidepressants. Furthermore, neurotrophins may be involved in sensory and motor signaling from the viscera because neurons in the dorsal motor nucleus of the vagus nerve retrogradely transport BDNF and other neurotrophins. Interestingly, a different line of research found that giving VNS to humans or rodents improved their memory function. This suggests that the information the vagus nerve sends to the brain may have an impact on higher-order cognitive processes.

5.4.2. Gut Hormones Associated with Cognition

Several gut hormones or peptides, including leptin, ghrelin, glucagon-like peptide 1 (GLP1), and insulin, have been found to affect emotions and cognitive processes in addition to the gut's ability to directly stimulate molecular systems linked to synaptic plasticity and learning (Figure 5.1). Adipose tissue synthesizes leptin, which tells the brain to suppress appetite. The hippocampus, cerebral cortex, and hypothalamus are among the brain regions where leptin receptors have been found. BDNF may mediate the effects of leptin on food intake and energy homeostasis because leptin increases BDNF expression in the hypothalamus. Leptin promotes synaptic plasticity in the hippocampal region, just like BDNF does. Long-term potentiation (LTP), long-term depression, and difficulties with spatial learning are all observed in genetically obese rodents with malfunctioning leptin receptors. By delivering leptin into the hippocampal region, these effects were mitigated. It is possible that leptin directly influences hippocampus plasticity because recent research indicates that it stimulates quick alterations in hippocampus dendritic morphology.

An empty stomach releases the adipogenic hormone ghrelin, which stimulates appetite in both humans and mice. The growth hormone secretagogue receptor, which is expressed in the hippocampus and hypothalamus' arcuate nucleus, has an endogenous ligand called ghrelin. Ghrelin increases food intake when administered peripherally to healthy rodents and humans, but long-term administration can cause obesity. Alongside enhanced spatial learning and memory formation, ghrelin also accelerates the reorganization of synaptic terminals in the hypothalamus and the hippocampus. GLP1, which is synthesized by intestinal cells, regulates energy metabolism by stimulating pancreatic insulin secretion and subsequent glucose uptake by muscle cells, as well as by suppressing food intake through actions on the hypothalamus. It has been demonstrated that injecting GLP1 into the brain enhances rats' associative and spatial memory. GLP1 receptors are expressed in neurons. Because ghrelin, leptin, and GLP1 act on both somatic and neural targets, they can integrate processes that affect emotion and thought.

Lastly, it has been discovered that insulin, which was previously thought to be a gut hormone made by the pancreas, also affects cognitive function and synaptic activity. Mentally anticipating meals typically triggers the release of insulin, which is then sustained during digestion and bloodstream absorption. Certain brain regions, like the hippocampus, have distinct signal-transduction receptors that insulin can interact with. Overall, the data seem to point to the possibility that eating itself can affect cognitive processes on two different levels: by releasing gut peptides into the bloodstream and by altering neural circuits that link the brain and the gut (Figure 5.1).

Thus, gut microbiota does affect the molecular mechanisms governing emotion regulation, memory formation ability, and general mental function—just as evolutionary theory predicted. The recognition of visceral signals as critical components in the treatment of psychiatric disorders is hardly surprising.

Gaining more insight into the molecular mechanisms through which peripheral signals can influence mental processes is the current challenge.

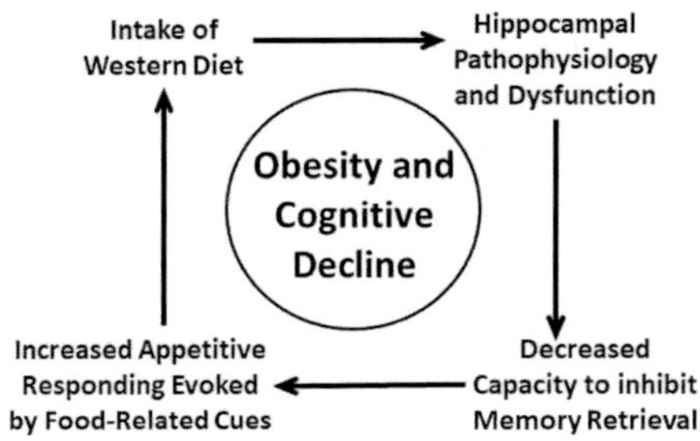

Figure 5.1. *Effects of feeding on cognition.*

Source: *https://encrypted-tbn0.gstatic.com/images?q=tbn:ANd9GcTshpoiM5cS_cYDRoOJRgAvrrHc5sXJwvGTAg&s*

5.4.3. From Energy Metabolism to Cognition

In comparison to other bodily parts, the brain uses a tremendous amount of energy. Therefore, the mechanisms governing the transfer of energy from food to neurons are probably central to the regulation of brain activity. Synaptic plasticity can be influenced by processes linked to energy management in neurons (Figure 5.2), which may help to explain how metabolic disorders can impact cognitive functions. Interestingly, brain processes can affect somatic function at the molecular level through changes in metabolic energy resulting from changes in synaptic function. As a signaling molecule that is closely linked to both energy metabolism and synaptic plasticity, BDNF is a great example. It can activate metabolic signals that impact cognitive function. The brain regions linked to cognitive and metabolic regulation—the hippocampus and the hypothalamus, respectively—have the highest concentrations of BDNF. Task performance enhances BDNF-mediated synaptic plasticity in the hippocampus, and memory formation is compromised by genetic deletion of the BDNF gene. A common genotype in humans linked to aberrant BDNF trafficking and secretion in neuronal cells, the Met variant of the Val66Met BDNF polymorphism is linked to aberrant hippocampal functioning and memory processing. Also, it has been demonstrated that BDNF affects a number of energy metabolism-related factors, including insulin sensitivity, appetite suppression, and glucose and lipid metabolism. Furthermore, the regulation of BDNF expression in the ventral medial hypothalamus by the hypothalamic melanocortin 4 receptor, which is essential for energy balance control, bolsters the connection between synaptic plasticity and energy metabolism. Exercise's effects on downstream effectors of BDNF-mediated synaptic plasticity, including synapsin I, calcium/calmodulin-dependent protein kinase II (CaMKII), and cyclic AMP-responsive element-binding protein (CREB), have been demonstrated to be eliminated in rodents due to a reduction in energy metabolism brought on by a high dose of vitamin D3 infused into the brain. Learning and memory deficits, as well as hyperphagic obesity in humans, have been associated with a de novo mutation in the BDNF receptor TrkB. Even though BDNF-mediated synaptic plasticity and energy metabolism appear to be related, more research is necessary to define the boundaries of this relationship and how it affects cognitive function.

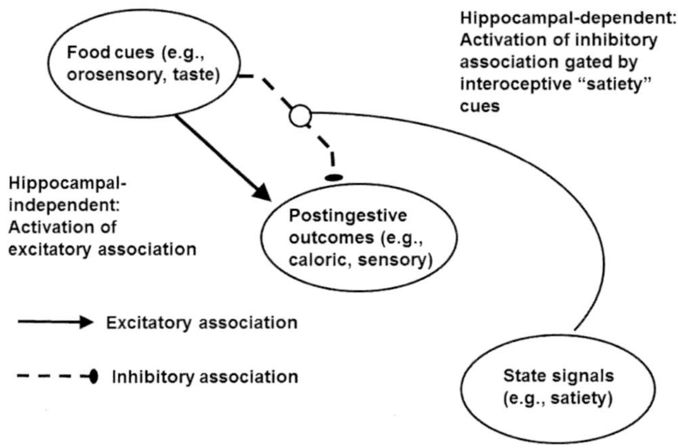

Figure 5.2. *Mechanism underlying the learned control of intake by external food cues.*

Source: Davidson TL, Jones S, Roy M and Stevenson RJ (2019) The Cognitive Control of Eating and Body Weight: It's More Than What You "Think." Front. Psychol. 10:62. doi: 10.3389/fpsyg.2019.00062.

Insulin-like growth factor 1 appears to be a key player in the process by which BDNF influences metabolism and synaptic plasticity (IGF1). The brain, skeletal muscle, and liver all produce IGF1, while the hippocampus is the primary site of brain IGF1 receptor expression. In rodents, a decrease in IGF1 signaling leads to hyperglycemia and insulin resistance, whereas an IGF1 infusion into the brain lowers plasma insulin levels and raises insulin sensitivity. In addition, IGF1 promotes synaptic plasticity, neurotransmitter synthesis and release, nerve growth and differentiation, and may help maintain cognitive function in the face of aging, diabetes, and brain damage. It has been demonstrated that IGF1 stimulates downstream pathways like the Akt signaling system in rodents, which are similar to BDNF. It's interesting to note that the omega-3 fatty acid docosahexaenoic acid (DHA) activates the Akt pathway, which suggests that integrating the effects of signals derived from food on brain plasticity may require activating Akt. The effects of BDNF and IGF1 on energy metabolism, synaptic plasticity, and learning and memory appear to be integrated by the phosphatidyl-inositol 3-kinase (PI3K)/Akt/mammalian target of rapamycin (mTOR) signaling pathway (Figure 5.3) (Wu et al., 2007).

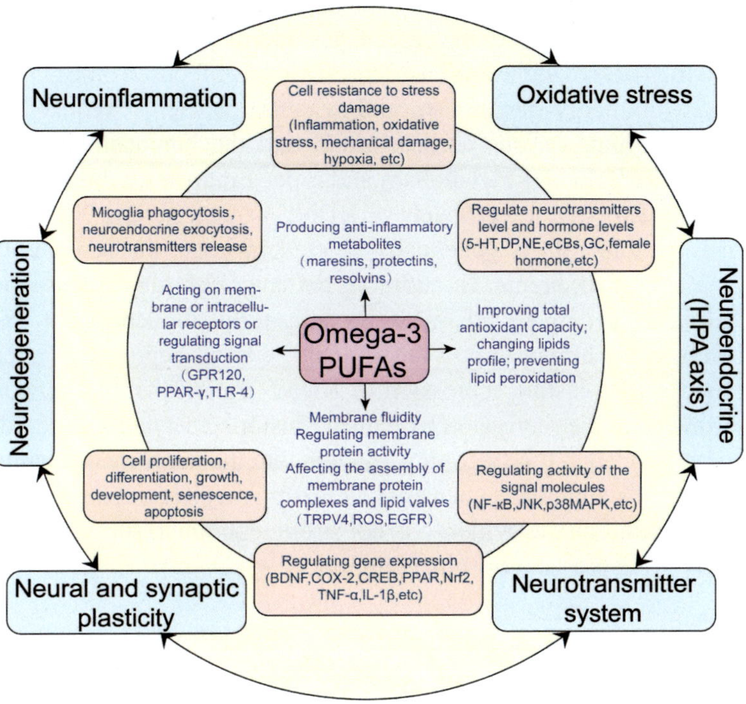

Figure 5.3. *Dietary omega-3 fatty acids can affect synaptic plasticity and cognition.*

Source: Zhou L, Xiong J-Y, Chai Y-Q, Huang L, Tang Z-Y, Zhang X-F, Liu B and Zhang J-T (2022) Possible antidepressant mechanisms of omega-3 polyunsaturated fatty acids acting on the central nervous system. Front. Psychiatry 13:933704. doi: 10.3389/fpsyt.2022.933704.

Dietary management is increasingly being considered a viable treatment option for psychiatric disorders, as disruptions in energy homeostasis have been connected to the pathobiology of several mental diseases. Several studies have suggested a possible link between psychiatric disorders and abnormal metabolism, including obesity, metabolic syndrome, and diabetes type II. Diabetes was found to be more common in patients with manic depression or schizophrenia in a large study than in the general population (1.2% of those aged 18–44 and 6.3% of those aged 45–64). In a group of 95 patients with schizophrenia, the overall prevalence of diabetes was 15.8%, rising to 18.9% by the age of 61. In contrast, in 203 patients with manic depression, the prevalence of diabetes varied, ranging from 2.9% in patients who were roughly 30 years old to 25% in patients who were 75–79 years old. However, considering that schizophrenia, manic depression, and other psychiatric disorders are linked to poor quality of life and the adverse effects of anti-psychotic medication, it is challenging to determine a cause-and-effect relationship between diabetes and psychiatric disorders in these studies. Current theories of schizophrenia and depression have focused on BDNF because of its effects on energy metabolism and synaptic plasticity. Reduced levels of BDNF in the serum, various cortical areas, and the hippocampal regions are linked to impaired glucose metabolism and type II diabetes, while patients with schizophrenia also have lower levels of BDNF in their serum. When the TrkB receptor in the forebrain is genetically deleted in mice, the result is behavior that resembles schizophrenia. In addition, patients with major depression have lower plasma levels of BDNF, whereas long-term antidepressant use raises BDNF levels in the hippocampus. An animal model of depression called defeat stress caused a long-lasting repression of BDNF transcripts, but antidepressant treatment reversed this repression by causing histone acetylation, according to a recent study in rodents. It is becoming evident that the majority of treatments for depression and schizophrenia, such as exercise and medication, involve the action of BDNF, even though the evidence is not strong enough to support the idea that BDNF mediates depression or schizophrenia.

5.4.4. Ultra-Processed Food Can Raise Risk of Cognitive Decline and Stroke

There is a link between an increased risk of stroke and cognitive decline and a diet high in ultra-processed foods, such as deli meats, potato chips, sodas, and sweetened breakfast cereals. Researchers looked at data on eating patterns, cognitive performance, and stroke for about 30,000 adults beginning in their mid-sixties over the course of about 10 years for this study. Before the study began, none of the participants had a history of cognitive impairment. At the conclusion of the follow-up, 768 persons had developed cognitive impairment and 1108 people had suffered strokes. Increased risk of stroke (8%) and accelerated cognitive decline (12%) were associated with higher consumption of ultra-processed foods. The highest intake of unprocessed or minimally processed foods was associated with a 9% lower risk of stroke.

5.4.5. Small Changes in Ultra-Processed Food Consumption Matter

The study also discovered that even modest dietary adjustments can have a significant impact on brain function. There was a 12% decrease in the chance of accelerated cognitive decline for every 10% increase in the amount of unprocessed and minimally processed foods that participants consumed. Furthermore, a 16% increased risk of accelerated cognitive decline was linked to every 10% increase in ultra-processed food consumption. According to the study, people who followed diets known to support brain health, such as the Mediterranean diet, the DASH diet, or the MIND diet, continued to benefit from unprocessed or minimally processed foods and experienced negative effects from ultra-processed foods. All of these diets advise limiting intake of highly processed foods, red meat, and sugary foods and beverages and promoting consumption of more plant-based meals, fruits, vegetables, whole grains, beans, and nuts. In general, the most highly processed packaged goods found in the grocery store are typically ultra-processed foods. These are usually industrial foods, synthesized in factories and labs with little to no ingredients derived directly from natural plant or animal sources, or almost entirely composed of substances extracted from things like oils, fats, sugars, starches, and proteins. While potatoes in cans have undergone minimal processing, whole potatoes from the produce section are not processed at all. However, frozen hash browns and potato chips are highly processed foods (Miller and Spencer, 2014).

REFERENCES

1. Altman, S. E., Shankman, S. A., & Spring, B. (2010). Effect of acute tryptophan depletion on emotions in individuals with personal and family history of depression following a mood induction. *Neuropsychobiology, 62*, 171–176.
2. Anderberg, R. H., Hansson, C., Fenander, M., Richard, J. E., Dickson, S. L., Nissbrandt, H., Bergquist, F., & Skibicka, K. P. (2016). The stomach-derived hormone ghrelin increases impulsive behavior. *Neuropsychopharmacology, 41*(5), 1199–1209.
3. Appleton, J. (2018). The gut-brain axis: Influence of microbiota on mood and mental health. *Integrative Medicine (Encinitas, Calif.), 17*(4), 28–32.
4. Ashton, K., Bellis, M. A., Davies, A. R., Hughes, K., & Winstock, A. (2017). Do emotions related to alcohol consumption differ by alcohol type? An international cross-sectional survey of emotions associated with alcohol consumption and influence on drink choice in different settings. *BMJ Open, 7*(10).
5. Banerjee, N. (2014). Neurotransmitters in alcoholism: A review of neurobiological and genetic studies. *Indian Journal of Human Genetics, 20*(1), 20–31.
6. Benton, D., Haller, J., & Fordy, J. (1995). Vitamin supplementation for one year improves mood. *Neuropsychobiology, 32*(2), 98–105.
7. Bolton, J. L., & Bilbo, S. D. (2014). Developmental programming of brain and behavior by perinatal diet: Focus on inflammatory mechanisms. *Dialogues in Clinical Neuroscience, 16*, 307–320.
8. Boney, C. M., Verma, A., Tucker, R., & Vohr, B. R. (2005). Metabolic syndrome in childhood: Association with birth weight, maternal obesity, and gestational diabetes mellitus. *Pediatrics, 115*(e290–296).
9. Bushman, B. J., DeWall, C. N., Pond, R. S., & Hanus, M. D. (2014). Low glucose relates to greater aggression in married couples. *Proceedings of the National Academy of Sciences of the United States of America, 111*(17), 6254–6257.
10. Dabelea, D., et al. (2000). Intrauterine exposure to diabetes conveys risks for type 2 diabetes and obesity: A study of discordant sibships. *Diabetes, 49*, 2208–2211.
11. de Groot, R. H., et al. (2011). Prenatal famine exposure and cognition at age 59 years. *International Journal of Epidemiology, 40*, 327–337.
12. De Luca, S. N., et al. (2016). Early life overfeeding impairs spatial memory performance by reducing microglial sensitivity to learning. *Journal of Neuroinflammation, 13*, 112.

13. de Rooij, S. R., Wouters, H., Yonker, J. E., Painter, R. C., & Roseboom, T. J. (2010). Prenatal undernutrition and cognitive function in late adulthood. *Proceedings of the National Academy of Sciences USA, 107*, 16881–16886.
14. Dupret, D., et al. (2007). Spatial learning depends on both the addition and removal of new hippocampal neurons. *PLoS Biology, 5*, e214.
15. Elmore, M. R., et al. (2014). Colony-stimulating factor 1 receptor signaling is necessary for microglia viability, unmasking a microglia progenitor cell in the adult brain. *Neuron, 82*, 380–397.
16. Gardner, M. P., Wansink, B., Kim, J., & Park, S. B. (2014). Better moods for better eating? How mood influences food choice. *Journal of Consumer Psychology, 24*, 320–335.
17. Greenwood, C. E., & Winocur, G. (2005). High-fat diets, insulin resistance, and declining cognitive function. *Neurobiology of Aging, 26*(Suppl 1), 42–45.
18. Gugusheff, J. R., Ong, Z. Y., & Muhlhausler, B. S. (2013). A maternal "junk-food" diet reduces sensitivity to the opioid antagonist naloxone in offspring post-weaning. *FASEB Journal, 27*, 1275–1284.
19. Heim, C., Newport, D. J., Mletzko, T., Miller, A. H., & Nemeroff, C. B. (2008). The link between childhood trauma and depression: Insights from HPA axis studies in humans. *Psychoneuroendocrinology, 33*, 693–710.
20. Hidaka, B. H. (2012). Depression as a disease of modernity: Explanations for increasing prevalence. *Journal of Affective Disorders, 140*(3), 205–214.
21. Holford, P. (2007). *Optimum Nutrition for the Mind*. London: Piatkus Books Ltd.
22. Hooper, L. V., Littman, D. R., & Macpherson, A. J. (2012). Interactions between the microbiota and the immune system. *Science, 336*(6086), 1268–1273.
23. Imamura, F., et al. (2015). Dietary quality among men and women in 187 countries in 1990 and 2010: A systematic assessment. *Lancet Global Health, 3*, e132–e142.
24. Innis, S. M. (2008). Dietary omega-3 fatty acids and the developing brain. *Brain Research, 1237*, 35–43.
25. Jacka, F. N. (2017). Nutritional psychiatry: Where to next? *EBioMedicine, 17*, 24–29.
26. Jeon, B. T., et al. (2012). Resveratrol attenuates obesity-associated peripheral and central inflammation and improves memory deficit in mice fed a high-fat diet. *Diabetes, 61*, 1444–1454.

27. Kroes, M. C. W., van Wingen, G. A., Wittwer, J., Mohajeri, M. H., Kloek, J., & Fernández, G. (2014). Food can lift mood by affecting mood-regulating neurocircuits via a serotonergic mechanism. *NeuroImage, 84*, 825–832.
28. Lakhan, S. E., & Vieira, K. F. (2008). Nutritional therapies for mental disorders. *Nutrition Journal, 7*, 2.
29. Levant, B. (2013). N-3 (omega-3) polyunsaturated fatty acids in the pathophysiology and treatment of depression: Pre-clinical evidence. *CNS & Neurological Disorders Drug Targets, 12*(4), 450–459.
30. Locher, J. L., Yoels, W. C., Maurer, D., & van Ells, J. (2005). Comfort foods: An exploratory journey into the social and emotional significance of food. *Food and Foodways, 13*(4), 273–297.
31. MacCormack, J. K., & Lindquist, K. A. (2019). Feeling hangry? When hunger is conceptualized as emotion. *Emotion, 19*(2), 301–319.
32. Mahan, L. K., Escott-Stump, S., & Krause, M. V. (2008). *Krause's Food & Nutrition Therapy* (12th ed.). Philadelphia, PA; Edinburgh: Elsevier Saunders.
33. McNamara, R. K. (2009). Evaluation of docosahexaenoic acid deficiency as a preventable risk factor for recurrent affective disorders: Current status, future directions, and dietary recommendations. *Prostaglandins, Leukotrienes and Essential Fatty Acids, 81*, 223–231.
34. Micha, R., et al. (2014). Global, regional, and national consumption levels of dietary fats and oils in 1990 and 2010: A systematic analysis including 266 country-specific nutrition surveys. *BMJ, 348*, g2272.
35. Miller, A. A., & Spencer, S. J. (2014). Obesity and neuroinflammation: A pathway to cognitive impairment. *Brain, Behavior, and Immunity, 42*, 10–21.
36. Nahas, Z., et al. (2005). Two-year outcome of vagus nerve stimulation (VNS) for treatment of major depressive episodes. *Journal of Clinical Psychiatry, 66*, 1097–1104.
37. Ong, Z. Y., & Muhlhausler, B. S. (2011). Maternal "junk-food" feeding of rat dams alters food choices and development of the mesolimbic reward pathway in the offspring. *FASEB Journal, 25*, 2167–2179.
38. Oomen, C. A., Bekinschtein, P., Kent, B. A., Saksida, L. M., & Bussey, T. J. (2014). Adult hippocampal neurogenesis and its role in cognition. *Wiley Interdisciplinary Reviews: Cognitive Science, 5*, 573–587.
39. Sobesky, J. L., et al. (2014). High-fat diet consumption disrupts memory and primes elevations in hippocampal IL-1beta, an effect that can be

prevented with dietary reversal or IL-1 receptor antagonism. *Brain, Behavior, and Immunity, 42*, 22–32.

40. Spence, C. (2017). Comfort food: A review. *International Journal of Gastronomy and Food Science, 9*, 105–109.
41. Stein, K. (2008). Contemporary comfort foods: Bringing back old favorites. *Journal of the American Dietetic Association, 108*(3), 412–414. https://doi.org/10.1016/j.jada.2008.01.003.
42. Tuulari, J., Tuominen, L., de Boer, F., Hirvonen, J., Nuutila, P., & Nummenmaa, L. (2017). Feeding releases endogenous opioids in humans. *The Journal of Neuroscience, 37*, 8284–8291.
43. Van den Elsen, L. W., Poyntz, H. C., Weyrich, L. S., Young, W., & Forbes-Blom, E. E. (2017). Embracing the gut microbiota: The new frontier for inflammatory and infectious diseases. *Clinical & Translational Immunology, 6*(1), e125.
44. van Strien, T., Cebolla, A., Etchemendy, E., Gutiérrez-Maldonado, J., Ferrer-García, M., Botella, C., & Baños, R. (2013). Emotional eating and food intake after sadness and joy. *Appetite, 66*, 20–25.
45. Wang, Y., & Kasper, L. H. (2014). The role of microbiome in central nervous system disorders. *Brain, Behavior, and Immunity, 38*, 1–12.
46. Wansink, B., Cheney, M. M., & Chan, N. (2003). Exploring comfort food preferences across age and gender. *Physiology & Behavior, 79*(4, 5), 739–747.
47. Wu, A., Ying, Z., & Gomez-Pinilla, F. (2007). Omega-3 fatty acids supplementation restores mechanisms that maintain brain homeostasis in traumatic brain injury. *Journal of Neurotrauma, 24*, 1587–1595.
48. Yam, K. Y., et al. (2017). Exposure to chronic early-life stress lastingly alters the adipose tissue, the leptin system and changes the vulnerability to western-style diet later in life in mice. *Psychoneuroendocrinology, 77*, 186–195.

Chapter **6**

Cancer Risk and Ultra-Processed Foods

Contents

INTRODUCTION

An estimated 14.1 million new cases of cancer were diagnosed globally in 2012. In developed countries, lifestyle, and dietary changes could prevent approximately one-third of the most common neoplasms, according to the World Cancer Research Fund/American Institute for Cancer Research. Therefore, achieving a varied and balanced diet ought to be regarded as one of the most significant modifiable risk factors in the primary prevention of cancer, along with abstaining from tobacco use and reducing alcohol consumption.

Simultaneously, for the last few decades, the diets of numerous nations have changed to dramatically increase the intake of highly processed foods. These food products go through a number of physical, biological, and/or chemical processes before being designed to be inexpensive, convenient, highly palatable, and microbiologically safe. According to several studies (from Europe, the US, Canada, New Zealand, and Brazil) evaluating household food expenditures, supermarket sales, and individual food intake, ultra-processed food items may account for anywhere from 25% to 50% of daily energy intake.

An investigation into this dietary trend is warranted as it may be concerning. Ultra-processed foods have a number of qualities that may contribute to disease, especially cancer. First off, foods that have undergone extensive processing frequently contain lower levels of fiber and vitamin density in addition to higher levels of added sugar, salt, and saturated and total fat. Beyond their nutritional makeup, heat-treated, processed food products contain neoformed contaminants that result from the Maillard reaction. Some of these contaminants, like acrylamide, heterocyclic amines, and polycyclic aromatic hydrocarbons, have the potential to cause cancer. Second, certain materials—like bisphenol A—that have been linked to cancer and endocrine disruption may be present in the packaging of highly processed foods. Lastly, approved but contentious food additives found in ultra-processed foods include titanium dioxide (TiO_2, a white food pigment), which has been linked to cancer in animal or cell models, and sodium nitrite in processed meat.

Research on the possible health effects of ultra-processed foods is relatively new, made possible by the creation of the NOVA product classification system, which rates products based on how much food processing is involved. However, there is still a dearth of epidemiological data, most of which comes from cross-sectional and ecological studies, connecting the consumption of highly processed foods to an increased risk of disease. The few studies that were done found that eating a lot of ultra-processed food increased the risk of overweight, obesity, and hypertension in a prospective cohort of Spanish university students as well as the incidence of dyslipidemia in Brazilian children.

6.1. CONSUMPTION OF ULTRA-PROCESSED FOODS AND THEIR LINK WITH INCREASING RISK OF CANCER

Ultra-processed foods are characterized as food ingredient formulas that are frequently altered chemically before being combined with flavors, colors, emulsifiers, and other cosmetic additions to create ready-to-eat, extremely palatable food and drink items. Savory nibbles, soft drinks, precooked frozen dinners, and reconstituted meat products are some examples of these foods. Simultaneously, during the last few decades, there has been a gradual rise in obesity and other diet-related issues, and these issues now rank among the leading causes of serious diseases. In the US diet, ultra-processed foods account for 89% of added sugars and 58% of daily energy. There is debate regarding the nutritional content of highly processed foods and their place in a balanced diet. In contrast to less processed foods, these foods are more likely to be excessive in saturated fatty acids, added sugar, and sodium, even though they may help Americans consume more of several important nutrients. Foods classified as ultra-processed are thought to have lower nutritional value (Malik, 2013).

Vegetables, beans, nuts, seeds, and fruits are regarded by the WHO as healthful foods, while foods high in sugar, salt, and saturated fat are thought to be disease-causing. Numerous studies utilizing national dietary intake surveys and population-based research have demonstrated that ultra-processed foods are generally high in energy density, high in sugar, unhealthy fats, and salts, and low in dietary fiber, protein, vitamins, and minerals. These foods also contain refined carbohydrates, which can cause cancer, dementia, and mental illness in addition to obesity and diabetes. Additionally, some studies have demonstrated that while moderate caloric restriction slows down the aging process and safeguards the body and brain, excess calories actually shorten lifespans (Sánchez-Villegas, Toledo et al., 2012).

Approximately 2500 different chemicals have been used as food additives to enhance the flavor, color, consistency, quality, and cost of food. Many people consume a variety of food additives on a daily basis, each with pros and cons. Food additives are essential to today's nutrient-dense and plentiful food sources because they enable people to eat a year-round range of palatable, safe, and nutritious foods. Food additives benefit foods in a number of ways. Nevertheless, a number of metabolites, including nitrous compounds and monosodium glutamate, which have been linked to cancer, may be present in food additives.

Understanding the health effects of highly processed foods has become a serious topic as their consumption rises globally. High levels of processing in food may be contributing to rising trends in the prevalence of non-communicable diseases like cancer (Figure 6.1).

Cancer is a serious public health problem that impacts all societal groups. It is regrettably a tissue-level variety illness, and this variation poses significant challenges to both the specific diagnosis and the effectiveness of therapy. Male cancer rates are highest in the prostate, lung, and bronchus, colon, and rectum, and urinary bladder, in that order. The most common cancers in women are those of the breast, lungs, and bronchus, colon, and rectum, uterine corpus, and thyroid. The high rates of cancer-related morbidity and death are associated with rising rates of risk factors like smoking, being overweight, aging more quickly, and population growth.

In the past few decades, research on the connection between nutrition, diet, obesity, and cancer has progressed. In particular, food, and dietary components have a significant impact on cancers of the digestive tract, breast, prostate, and other organs. In the 1980s, the Committee on Diet, Nutrition, and Cancer of the National Research Council (USA) started looking into the possibility that a high-fat diet was associated with a higher risk of cancer in different parts of the body, especially the breast, colon, and prostate. Obesity, nutrition, and diet have all been linked to a higher risk of cancer. Studies have looked at the relationship between the development of neoplasms and calorie intake, lipid type and amount (vegetable or animal), protein, red or processed meat, vitamins, minerals, fiber, and other dietary components.

There is substantial evidence linking the use of food items and products to an increased risk of certain cancers in human organs and body parts. In addition to other unwanted/unavoidable chemicals/ingredients, some common meals and beverages are categorized as carcinogenic because they contain natural toxins, adulterants, artificial colorants, additives, preservatives, and pesticide residues.

Packaged breads and buns, packaged snacks (sweet or savory), industrialized confections and desserts, sodas, sugar drinks, meat balls, chicken, and fish nuggets, and other reconstituted meat products (which may contain preservatives other than salt, like nitrites); instant noodles and soups; frozen or shelf-stable ready meals; and other food products made primarily or exclusively of sugar, oils, and fats, as well as other ingredients not frequently found in culinary preparations, such as hydrogenated oils, modified starches, and protein isolates, are examples of ultra-processed food and beverages.

Various methods of preservation or cooking are employed, along with non-alcoholic fermentation for cheese and breads. Additional Processed Foods: These comprise industrial procedures such as hydrolysis, hydrogenation, molding, reshaping, extrusion, and frying as a pre-processing step. These products frequently contain flavoring agents, colors, humectants, emulsifiers, non-sugar sweeteners, and other cosmetic additives (Augusto MC, et al., 2017).

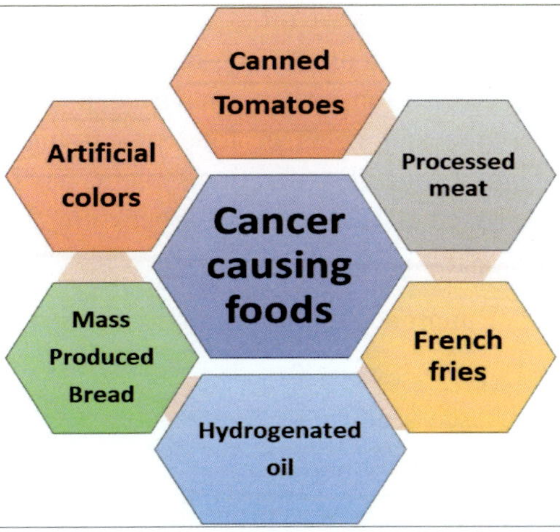

Figure 6.1. *Cancer-causing ultra-processed foods.*

Source: Urvashi Radheshyam Supe, Yash Rajkumar Menghani, Rashmi Vipinchandra Trivedi and Milind Janraoji Umekar, Consumption of ultra-processed foods and their link with increasing risk of cancer, e DOI: https://doi.org/10.30574/gscbps.2022.20.3.0339.

Titanium dioxide (TiO_2), a common food additive with nanoscale particles, is among the many additives found in ultra-processed foods. In addition to being used as a whitening agent, titanium dioxide also has antibacterial and improved texture qualities when it comes into contact with food or beverages in packaging. Titanium dioxide consumption daily may raise the risk of both carcinogenesis and chronic intestinal inflammation. An additional matter of concern pertains to the generation of carcinogenic nitrosamines in meats that contain sodium nitrite upon burning or overcooking. These N-nitroso compounds have been linked to an increased risk of colorectal cancer. Processing meat in ways like overheating, curing, and smoking can produce carcinogenic chemicals, like polycyclic aromatic hydrocarbons (PAH) and N-nitrosocompounds (NOC). These substances damage DNA, which demonstrates how highly carcinogenic processed meat is for humans.

Eating ultra-processed food also exposes one to bisphenols and phthalates found in industrial products. These substances are utilized as fillers in packaging. Because of their capacity to interfere with the endocrine system and because they are endocrine disruptors, phthalates, and bisphenols are linked to a number of harmful health effects, including endocrine cancer.

6.1.1. Food Processing Classification System

Technological developments and increased accessibility to highly processed foods have given rise to the idea of ultra-processed food (UPF). Foods are increasingly being classified using processing-based systems in epidemiological research. New food classification systems have been created as a result of modifications to the scope and goals of food processing. According to the study report, seven distinct food classification systems have been proposed by researchers worldwide based on the degree of processing.

All items were grouped into one of four groups:

1. Non-processed or unprocessed or minimally processed food,
2. Basic or primary processed food,
3. Moderately processed food,
4. Highly or ultra-processed food.

The first classification system was proposed by the National Institute of Public Health (NIPH) in Mexico in 2007. 2009 saw the creation of a second classification in Europe by researchers from the International Agency for Research on Cancer (IARC), utilizing techniques created for the European Prospective Investigation into Cancer and Nutrition (EPIC) study. In order to examine the consumption of highly industrially processed foods, the EPIC study recoded each food report based on the level of processing, dividing it into three primary categories:

1. **Highly processed foods:** Foods that have been prepared industrially, like those from bakeries and catering companies, and only need to be heated and cooked in the home. Examples include commercial cakes, biscuits, sauces, and canned goods like jams, bread, cereals, and cheese.
2. **Moderately processed foods:** This category contains two food groups. First, there are commercial and industrial foods that are consumed raw or cooked after undergoing minimal processing, like dried fruits, raw foods vacuum-packed or prepared in a controlled environment (like salads), frozen foods, extra virgin olive oil, and fruits and vegetables canned in juice or water. Second, foods prepared or cooked from raw or minimally processed foods that have been processed at home.
3. **Non-processed foods:** Foods that are eaten raw and don't require any additional processing or preparation, such as fruits, vegetables, non-processed nuts, fruits, crustaceans, mollusks, and fresh juices.

The distinction between processed and unprocessed food was the basis for a novel classification system created in Brazil by Monteiro et al. They created the NOVA classification system, which underwent revisions in 2015. The NOVA system divides foods into categories based on processing purpose and quantity rather than nutritional content. Foods are categorized into four groups according to this system, which takes processing level into account.

Food processing, according to NOVA, is the use of physical, biological, and chemical techniques following the isolation of food from its natural state but before its consumption or preparation into dishes and meals.

6.1.1.1. Group 1

Unprocessed or minimally processed foods: Natural foods that have been altered by techniques like drying, grinding, crushing, fractioning, filtering, roasting, boiling, pasteurization, refrigeration, chilling, freezing, storing in containers, and vacuum-packaging are considered minimally processed foods. These methods aim to preserve natural foods, prepare them for storage, or enhance their safety, palatability, or enjoyable factor. Examples of foods that fall into this category include fresh milk and meat, grains, legumes, nuts, fruits, vegetables, roots, and tubers. Teas, coffee, herb infusions, tap water, and bottled spring water are also included.

6.1.1.2. Group 2

Processed culinary ingredients: Processes like pressing, refining, grinding, milling, and drying are used to create processed culinary ingredients like oils, butter, sugar, and salt, which are materials obtained from Group 1 foods or from nature. Producing long-lasting goods is the aim of these procedures. They are meant to be mixed with Group 1 items to create freshly made dishes, drinks, and meals rather than being consumed on their own.

6.1.1.3. Group 3

Processed foods: They can be consumed on their own or, more frequently, in conjunction with other foods. In this class, processing aims to modify or enhance the sensory qualities of Group 1 foods, or to make them more durable. Examples include cheeses, freshly baked breads, fruits in syrup, canned fish, and bottled vegetables.

6.1.1.4. Group 4

Ultra-processed foods: Group 4 products are manufactured using a variety of techniques, such as sugaring, baking, frying, deep-frying, curing, smoking, pickling, and canning. Preservatives and cosmetic chemicals are also used, along with intricate packaging designs and the addition of artificial vitamins and minerals.

Based on earlier studies evaluating the contribution of processed foods to the food supply in low-income countries, Asfaw from the International Food Policy Research Institute (IFPRI) developed a food classification system in Guatemala in 2011. Depending on the level of processing, these foods were divided into three categories: highly processed, partially (primary) processed, and unprocessed. Foods categorized as highly processed are those that have undergone additional processing to make them more palatable. These foods probably contain a lot of added salt, fats, and sugars. Examples of such items include baked goods, crackers, sausage, and cooked meats, ice cream, frozen desserts, breakfast cereals, confections (chocolate, sweets), soft drinks with fat spreads and shortening, pasta products, prepared meals (dried soup, for example), formula, and companion foods. Bread, maize products, vegetable oils, and dairy products are examples of primary or partially processed foods.

In order to ascertain the extent to which processed foods contribute to nutrient intake in the US diet, the International Food Information Council Foundation (IFIC) proposed an additional food categorization system in 2012 based on the processing level in the United States of America (USA). Based on the level of processing and the resulting physical, chemical, and sensory changes, every food in the diet can be categorized into one of the IFIC Foundation categories. Items that retain most of their original qualities are classified as minimally processed. Examples of this type of processing include roasted nuts and washed and packed fruits and vegetables. "Foods processed for preservation," The next phase of processing includes canned tuna and beans, as well as frozen fruits and vegetables, for nutritional improvement and freshness. Foods in the "mixtures of combined ingredients" category comprise sugars, spices, oils, colors, flavors, and preservatives intended to promote safety, taste, and attractive appearance.

Cake mix, canned tomato sauce, salad dressing, and rice are among examples. The next stage is "ready-to-eat processed foods," which include morning cereal, crackers, ice cream, yogurt, luncheon meats, fruit drinks, and carbonated beverages. The final category, """prepared foods/meals," includes foods packaged for freshness and ease of preparation such as frozen dinners and entrees as well as prepared deli foods.

In 2015, the University of North Carolina (UNC) created a new classification scheme in the US that featured representative foods from the American diet along with renamed categories. Foods are divided into groups according to how processed they are, from least processed to most processed. A system of classification was devised to divide food into four groups according to the extent of industrial food processing. "Unprocessed and minimally processed: foods and beverages are the most basic type; they are made with just one ingredient and have undergone very little to no processing that has affected the food's inherent qualities when consumed raw or in its natural, unprocessed state. A few examples include unseasoned meat, milk, eggs, fresh fruits, and vegetables.

Basic processed foods and beverages have undergone processing, but they are still separate products. Processed basic ingredients, like sugar, oil, or whole-grain flour, are distinct food ingredients that are taken out of unprocessed or minimally processed foods and refined using physical or chemical methods that change the food's natural properties. Processes include extraction, pressing, milling, refining, purification, and clarifying. Examples include canned fruit or vegetables, white or instant rice, and refined-grain flour or pasta. Foods that have been lightly or simply processed and have flavor additives (sweeteners, salt, flavors, or fats) added to them to improve their flavor are considered to be products with moderate flavor processing. From their original plant or animal origins, they can be easily identified. Almonds salted, fruit in syrup, and vegetables in cans with extra salt are a few examples.

Meals and beverages that are highly processed are industrially produced, multi-ingredient goods that have undergone so much processing that it is impossible to identify the original plant or animal source. Examples include sugar-sweetened beverages (SSBs), cookies, salty snacks, candies, prepared mixed dishes, refined-grain breads, condiments, dips, sauces, toppings, ketchup, margarine, and mayonnaise. Separating foods and beverages into categories of not processed and processed was another proposal made by the Food Standard Australia New Zealand in 2018. However, this classification does not distinguish processed goods according to how much processing they have undergone, which makes evaluating highly or ultra-processed food difficult. It was decided not to include systems that do not distinguish between highly/ultra-processed and processed food. Food classification systems that were mainly applicable locally or regionally were also removed. Consequently, two classifications were disregarded.

6.1.2. Food Processing Techniques and their Health Outcomes

All procedures and methods used by the food, beverage, and related industries to transform whole, fresh foods into food products are referred to as industrial food processing. Ultra-processed food products are made by a variety of methods, such as sugaring, salting, baking, frying, deep-frying, curing, smoking, pickling, and canning. Preservatives and cosmetic chemicals are also frequently used, along with artificial vitamins and minerals and elaborate packaging. All of these industrial procedures are intended to produce products that are attractive, ready-to-eat or ready-to-warm, long-lasting, and easily accessible. Food processing also involves the Maillard reaction. A chemical reaction between amino acids and reducing sugars happens when food products are processed or cooked at high temperatures, giving them unique flavors and a brown color. Thus, it is frequently used in the food industry to impart distinct flavors, colors, and aromas to food. According to reports on the Maillard reaction, some MRPs, especially melanoidins, have positive health effects like antioxidants.

Acrylamide is a carcinogen, and Maillard reaction products like elevated carboxymethyl lysine (CML) encourage diabetes and cardiovascular disorders.

To improve the quality and safety of milk, for instance, ultra-high temperature (UHT) treatment or traditional sterilization are frequently employed. Milk has a lot of sugar and protein. Therefore, it is clear that processing milk at high temperatures has the potential to cause MRP formation. Protein and mineral bioavailability may be impacted by MRPs acting as chelating agents to chelate metal cations by forming various soluble and insoluble complexes.

The high temperatures used in the deep-frying process have the potential to alter the nutritional composition. Deep-frying typically releases harmful fumes that contain two substances that are particularly harmful to human health: aldehydes and polycyclic aromatic hydrocarbons (PAHs). Humans are primarily exposed to these chemicals through ingestion and breathing, which can lead to serious cancers of the skin, stomach, esophagus, and lungs.

High-heat cooking techniques may increase the risk of prostate cancer. The addition of fats or oils, such as those used in deep-frying, raises the possibility that prostate cancer risk may increase. Potentially cancer-causing substances like acrylamide, aldehydes, acrolein, and polycyclic aromatic hydrocarbons can form when food is deep-fried. Deep-fried foods include doughnuts, French fries, fried chicken, fried fish, and snack chips.

Cancer in children may develop if a mother consumes cured and broiled meat while she is pregnant. Five types of meat—ham, bacon, or sausage; hot dogs; hamburgers; bologna, pastrami, corned beef, salami, or lunch meat; and charcoal-grilled dishes—are linked to an increased risk of cancer. After eating bacon or any other food that has been cured with sodium nitrite, there is a higher concentration of nitrate in the cured meats and in the stomach. Because such high nitrite concentrations react most quickly, there is a greater chance that nitrosamides and other N-nitroso compounds will form endogenously. Owing to these compounds' development in cured meats, children are more likely to develop brain tumors. The heat treatments used in food processing include grilling, roasting, smoking, and barbecuing. The main byproduct of these processes is the formation of polycyclic aromatic hydrocarbons, which raises the risk of cancer in humans. PHA contamination results from the direct pyrolysis of food nutrients and the deposition of smoke from incomplete thermal agent combustion.

6.1.3. Hidden Dangerous Additives and Contaminants in Ultra-Processed Foods

Chemicals added to food to enhance its flavor, texture, color, chemical preservatives, taste, appearance, or to help with processing are known as food additives. Certain additives are categorized as carcinogens or agents that cause cancer because it is well-known that they are bad for human health. Our health will be impacted by consuming excessive amounts of food additives over an extended period of time. Food additives and preservatives are sometimes referred to as chemical food contaminants. Antioxidants like butylated hydroxy anisole (BHA), butylated hydroxytoluene (BHT), ascorbic acid, and tocopherols, as well as antimicrobials like benzoic acid, nitrites, and sulfites, have probably influenced food production patterns and eating habits. In addition to helping to preserve and improve the flavor of food, these food preservative chemicals also protect against the harmful effects of reactive oxygen species (ROS), which are linked to aging, cancer, and cardiovascular disease. Sodium nitrite is a preservative used in lunch meats, hams, sausages, hot dogs, and bacon to stop botulism. Sodium nitrite is useful in inhibiting the bacteria that cause botulism, but it can also react with proteins or produce carcinogenic Nnitrosamines when cooked at high temperatures. It has been connected to cancerous growths. Tomato paste in bottles can now have a 40-week shelf life without sacrificing quality thanks to the widely used sodium benzoate. However, when mixed with vitamin C, it can produce the carcinogen benzene.

Two common multipurpose synthetic chemicals found in industrial products are phthalates and bisphenols. They can contaminate food when used in food contact materials or food processing plastics. The ingestion of specific foods, including dairy, meat, spices, flour, wheat, or grains, has been associated with increased levels of certain phthalate metabolites in urine. Greater urine concentrations of bisphenol A have been associated with the consumption of canned foods and beverages. Because certain phthalates and bisphenol A can disrupt the endocrine system, exposure to these substances is linked to a variety of harmful health outcomes.

It has been demonstrated that bisphenol A contributes to the pathogenesis of multiple endocrine disorders, such as polycystic ovary syndrome (PCOS), hormone-dependent tumors like breast and prostate cancer, premature puberty, and infertility in both men and women.

One of the most widely used mineral particle-based food additives in consumer goods is E171, a mixture of titanium dioxide particles that has been shown to contain up to 55% nanosized particles (TiO_2-NPs). TiO_2 has the ability to damage the intestinal barrier, cause inflammation within the intestines, and influence the development of intestinal cancer. The development of the tumor is associated with dysplastic alterations in the colonic epithelium as well as a sharp decline in goblet cells, which are cells that secrete mucus and cause disruption of the intestinal barrier. Citric acid (CA; E-330) is used as an acidulant, pH regulator, taste enhancer, preservative, and antioxidant synergist in a wide range of foods, including soft drinks, jelly-filled candies, baked goods, jam, marmalade, confectionery, and canned vegetables and fruits. In addition to being genotoxic and/or cytotoxic to human dental cells, CA also damages DNA, particularly in human lymphocytes at higher concentrations (Table 6.1).

Table 6.1. *Health Outcomes of Additives Containing Ultra-Processed Food Items*

Additive's	**Food Items**	**Health Outcomes**
Acrylamide	French fries, potato chips, biscuits, bread.	Ovarian and endometrial cancer
Perfluorooctanoic acid	Microwave popcorn	Lung cancer
Sodium nitrates	Process meat Hot dogs	Pancreatic cancer, high risk of leukemia
Parabens	Preservatives	Breast cancer
Omega-6 fats	Refined vegetable oil	Heart disease and cancer
Bisphenols and phthalates	Food packaging material	Endocrine disruptor
Titanium dioxide	Food and drinks packaging material	High risk of chronic intestinal inflammation and causes cancer
N-nitroso and polycyclic aromatic hydrocarbons	Processed meat	Stomach, lungs, Colorectal cancer DNA damage
Tartrazine (coloring agent)	Soft drinks, cereals, sauces	Thyroid cancer, migraine
Citric acid	soft drinks, jelly sweet, jam, marmalade, candy	genotoxic and/or cytotoxic, DNA-damaging

Source: Urvashi Radheshyam Supe, Yash Rajkumar Menghani, Rashmi Vipinchandra Trivedi and Milind Janraoji Umekar, Consumption of ultra-processed foods and their link with increasing risk of cancer, e DOI: https://doi.org/10.30574/gscbps.2022.20.3.0339.

Tartrazine can cause allergies, asthma, and hyperactivity in children, which can be harmful to people's health. Tartrazine is found in a variety of foods and beverages, including sweetmeat, cotton candy, soft drinks, cereals (corn flakes and muesli), flavored chips (Doritos and Nachos), cake mixtures, soups, jam, sauces, ice cream, some rice, candy, chewing gum, marzipan, jelly, gelatins, mustard, marmalade, milk, and noodles. Tartrazine can be found in non-food products like soaps, cosmetics, shampoos, conditioners, hair treatments, and even pastels, crayons, and stamp dyes. Tartrazine consumption has been linked to a variety of immunologic reactions, including general fatigue, anxiety, migraines, clinical depression, purple skin spots, and sleep disturbances.

Many common foods produce acrylamide, a potential human carcinogen when they are cooked at high temperatures. It is common; in the US, foods containing acrylamide account for about 30% of the total calories consumed. Increased dietary acrylamide intake is associated with an increased risk of postmenopausal endometrial and ovarian cancer.

6.2. POTENTIAL BIOLOGICAL MECHANISMS LINKING ULTRA-PROCESSING OF FOODS WITH CANCER RISK

Consuming ultra-processed foods may potentially raise the risk of cancer due to their obesogenic qualities and exposure to contaminants and food additives. When compared to minimally processed foods, unprocessed plant foods (UPFs) may have a lower dietary quality, according to multiple studies. Diets high in unsaturated polyunsaturated fats (UPFs) typically have lower intakes of fiber, protein, sodium, and potassium and higher levels of energy density, free sugar, total fat, saturated fat, and trans fat. When compared to an unprocessed diet, people who eat a diet high in UPFs typically eat more calories and gain more weight. This is according to Hall and collaborators. This study's excess calorie content was primarily found in fats and carbohydrates, demonstrating the hyper palatability and appeal of UPF products. In actuality, diets high in UPFs have been linked to obesity, which has been shown to be a risk factor for at least 13 cancer sites, including the previously mentioned colorectal and breast cancers (Cediel et al., 2020).

Potential mechanisms connecting UPFs to an increased risk of cancer have also been proposed, including the presence of food contaminants such as neoformed processing contaminants. In their unprocessed state, foods undergo a variety of physical, chemical, and biological changes that result in the finished processed goods. As a result of the heat-inducing reactions between the amino group of the amino acid asparagine and the carbonyl groups of glucose and fructose, reactions take place during these operations, changing the food's chemical structure and leading to the formation of neoformed processing contaminants like trans-fat and acrylamide. More industrial trans-fat consumption has been linked to an increased risk of cancer, according to recent research. Other neoformed processing contaminants, which have also been connected to an increased risk of cancer, include oxyhalides, halo acetic acids, heterocyclic amines, and polycyclic aromatic hydrocarbons (Pouzou et al., 2018).

UPFs may also contain certain "indirect" contaminants, such as materials originating from food packaging. In fact, it has been noted that materials that come into contact with food are a hidden source of toxins that may have endocrine-disrupting properties. For instance, consumption of ultra-processed foods has been positively correlated with urinary concentrations of Di(ethylhexyl) phthalate (DEHP), an endocrine disruptor chemical that is widely used as a plasticizer. Research on humans and animals has shown that DEHP exposure may cause cancer by a variety of molecular pathways, including damage to DNA. According to a recent study, DEHP exposure may increase the cancer stemness of colon cancer, which in turn may facilitate its metastasis. Similarly, it has been proposed that exposure to bisphenol-A (BPA), another endocrine-disrupting chemical found in many types of food packaging, raises the risk of cancer. It has been demonstrated that exposure to BPA deregulates signaling pathways linked to head and neck cancers.

A long list of additives, including preservatives and cosmetic additives (like dyes and emulsifiers), may also be present in ultra-processed foods. The majority of the literature is currently derived from animal/cellular experimental models, so even though all food additives used in food production are evaluated and approved for general safety, their cumulative effect on long-term human health is still largely unknown. Numerous food additives have been implicated in cancer through experimental research. Manufacturers currently use sodium nitrate, for example, to preserve processed meat and poultry meats. According to some research, this substance may raise the risk of cancer by forming specific nitroso-compounds that may result in carcinogenic nitrosamines. A meta-analysis comprising 49 studies also revealed that individuals with higher intakes of nitrosamines (OR: 1.34; 95% CI, 1.02–1.76) and nitrites (OR: 1.31; 95% CI, 1.13–1.52) had a higher risk of gastric cancer. There is currently evidence to support the idea that eating processed meat raises the risk of colorectal cancer, with stronger associations found for colon cancer. Similarly, more recent research indicates that titanium dioxide (TiO_2), which was once widely used as a coloring agent in UPFs and as a preservative to improve texture, is no longer safe to use as a food additive. It has been proposed as a potential human carcinogen and to induce altered colonic gene expression in mice models. The impact of artificial sweeteners on the risk of cancer is another contentious issue. While aspartame, the most widely used artificial sweetener, has been deemed safe for human health by the European Food Safety Authority, additional research indicates that aspartame may cause cancer. Furthermore, it has been suggested that some emulsifiers may increase gastrointestinal inflammation, a metabolic change linked to the genesis of cancer.

Alcoholic beverages, particularly those that have been distilled, are regarded as ultra-processed beverages and have been repeatedly linked to the development of cancer. A substantial amount of experimental data indicates that the hazardous alcohol metabolite acetaldehyde may cause cancer by interfering with DNA synthesis and repair. Additional possible underlying mechanisms include the oxidative stress brought on by ethanol consumption and its likely role as a solvent for carcinogens to enter cells (Albano, 2006).

Numerous epidemiological studies examining the relationships between consumption of ultra-processed foods and health outcomes have corrected for nutritional quality, indicating that additional features of UPFs might be responsible for the associations observed. However, additional research is required to confirm that residual confounding cannot be completely ruled out. While the exact processes behind the association between UPF consumption and some cancers remain unclear, available data indicates UPFs may raise the risk of cancer through obesogenic, alcohol-consuming, and dietary quality factors, as well as non-nutritional food processing compounds.

6.3. THE GLOBAL FOOD SYSTEM, ULTRA-PROCESSED FOOD CONSUMPTION AND CANCER RISK

A major factor contributing to the rise in UPF production and consumption as well as the corresponding cancer risk is the global food system. Although there are many theories explaining the complex relationship between the food system and cancer risk, we will focus on two here. The environmental pathway is the first possible pathway. Thanks to modern technology, the food industry can now produce more UPF goods while making use of the limited environmental resources available to it. The land is heavily utilized for non-environmentally friendly agricultural methods, like excessive use of pesticides and deforestation, to meet demand. Local culinary customs and food cultures are in danger due to the globalization of UPFs. It threatens the biodiversity of plants and animals. Massive ingredient production for low-cost UPF puts small-scale farmers in danger and turns agriculture into a monoculture. These goods need to be transported over longer distances, which exacerbates global warming. Furthermore, UPFs are recognized for their elegant and sophisticated packaging, which adds significantly to the waste generated from plastic. Increased pollution of the air, water, and land results from this overproduction, which also strains the available material resources. It is well recognized that these exposures and environmental changes have a detrimental effect on people's health, particularly their risk of cancer.

The consumption pathway is an additional possible pathway. Due to several factors, including their hyper-palatable nutritional profile (high energy, salt, sugar, and low in fiber, vitamins, and minerals), the use of cosmetic additives (colors, and taste enhancers), and their appealing packaging, UPFs are known for their hyperpalatability and attractiveness. Due to these UPFs' qualities, as well as their widespread advertising and low cost, UPFs are consuming more energy globally, which raises the risk of obesity and related cancers. The increased consumption of UPF products may also raise the chance of being exposed to pollutants or other substances that have the potential to cause cancer, such as certain additives that may no longer be deemed safe for use in human health, such as titanium dioxide (TiO_2). Accordingly, it has been proposed that adopting healthier lifestyles and limiting exposure to environmental pollution can prevent between 30 and 50% of cancer cases. This emphasizes how crucial it is for public health policies to support environments that are health-promoting to lower the incidence of cancer.

6.4. FOOD ADDITIVES AND CANCER

Since ancient times, people have been using food additives. Salt, honey, wine, and spices were among the first additives, and they were mostly used as flavorings or preservatives. Subsequent techniques for food preservation involved brining and smoking meats to ensure a year-round supply. The variety and quantity of additives that are available has increased significantly in recent years. These additives serve a variety of functions and, over time, improve global food production and distribution, minimize food waste, and allow the use of convenience foods in urban areas. A few guiding principles apply to the use of food additives: they should be safe to use, technically efficient, and not used in excess of what is necessary. Additionally, the use of non-nutrient additives should be minimized. Finally, the use of food additives should not be motivated by deceiving consumers about the nature or quality of the food. The evaluation of food additive safety encompasses a wide range of factors, one of which is shielding consumers from carcinogen exposure. Cancer is not a rare illness; rather, it is common and has been known about since antiquity. However, cancer is now found to be developing in organs in the industrialized world that were not previously linked to cancer nor are they in the organs where cancer is currently found to be developing in less developed countries. As a result, there is evidence linking certain cancers to environmental changes, and it has been proposed that this may include a rise in the use of food additives. The consumer should be aware, though, that many other naturally occurring substances that may be carcinogenic would still be consumed if food additives were no longer used.

When compared to food additives that have been around for more than 30 years, the risk of carcinogenicity associated with those that have recently been approved is significantly lower. The reason for this is that, while substances used for many years have only recently (the last decade or two) been subjected to carcinogenicity tests deemed to be of adequate design, all recently approved additives must have demonstrated adequate evidence of a lack of carcinogenicity in animal experiments. An excellent illustration of this is the coal tar dyes. They were more concentrated and brighter than the natural colorings used up until 1856 when they were first used as food colors. However, their use was progressively phased out as more and more findings from animal experiments suggested potential carcinogenic risks to the consumer.

It is necessary to evaluate the significance of tumor formation in the experimental animal study before extrapolating the results of a carcinogenicity study. Although short-term *in vitro* screens are now thought to be a prudent step before long-term testing in laboratory animals, long-term testing in animals remains the best method for determining carcinogenicity. If a positive result is found during this screening phase, it is imperative to question the appropriateness of conducting additional long-term testing. The compound is of high carcinogenic risk and should not be used in foods if the results of a long-term carcinogenicity test show that tumor incidence is dose-related, the induction period is shortened, and there is an increase in the number of tumors over control animals. This problem is one for the groups of people involved in safety evaluation. However, if the increase in tumor incidence is only very slightly greater and there is no shortening of the induction period, then the question that needs to be asked is, "What is the significance of these results?" Perhaps more research is necessary, but in the interim, what precautions should be taken? Both the government and industry in the UK are accountable for the general safety of food additives. Independent expert committees assess the safety of using these substances and advise the government based on their evaluation of results from published literature as well as industry. Numerous factors are taken into account, including the goal of the additive, its efficacy, and details about its toxicity and carcinogenicity. Therefore, for the final evaluation to be accurate, good cooperation between the government and industry is essential.

While there are many other ways that food additives can cause cancer besides “direct” carcinogenicity, we will briefly discuss some examples of food additives linked to cancel in animal studies, the testing procedures used, and the issues these findings have raised. The finding of carcinogenicity has been the primary, but not the only, toxicological effect that has prompted legislative decisions, additional research, or both. The following are some suggested examples of this: (1) there might be an impurity present (saccharin was thought to be an example of this for a long time); (2) the additive might react with a food component to form a new compound, such as nitrosamines); and (3) the additive might be metabolically changed by the animal or its gut bacteria to another compound, such as cyclamate to cyclohexylamine.

6.4.1. Furylfuramide (AF-z)

Between 1965 and 1974, AF-z was used as an antimicrobial preservative in Japanese products like fishcakes, meat, and fish sausages, soybean curd, and red bean paste. When AF-z was approved for use as a food additive, no carcinogenicity was found, despite tests being conducted as part of the safety evaluation program, whose design and execution were appropriate at the time. Carcinogenicity studies in animals were started based on mutagenicity results, which were published in 1972, from several test systems (strong mutagenicity with *E. coli* and increased chromosomal aberrations in cultured human lymphocytes). Carcinogenicity was proven two years later. When fed diets containing 0.15–0.45% AF-2, male ddY/SLC mice developed a dose-related increase in forestomachine squamous cell carcinomas.

Since Ikeda and colleagues first discovered carcinogenicity in animals in an unpublished study in 1974, several studies have been conducted to look into this phenomenon. In less than a year, mice, and hamsters fed AF-2 in their diets developed squamous cell carcinomas of the forestomach. The latent period seemed to be marginally shorter in mice. In the rodents administered AF-2 doses insufficient to induce squamous cell carcinomas, stomach papillomas were found. In a three-generation feeding study in mice, AF-2 was also found to cause hepatic enlargement and consistent degenerative hepatic changes in treated dams and their offspring. On the other hand, when rats were given AF-2, they developed mammary adenocarcinomas; however, Takayama, and Kuwabara observed that rats that were killed in the last month of the study had forestomach tumors.

The crucial point to remember in this case is that short-term *in vitro* screening tests were the first to identify its carcinogenic potential, highlighting the importance of paying attention to the findings of these tests. This specific material also highlights the significance of careful pathological techniques—an area that has tragically been neglected in the past—and shows that the same type of tumor would not necessarily be produced in a variety of experimental animal strains.

6.4.2. Safrole

Safrole is a flavoring ingredient that is a minor component of nutmeg and cinnamon leaf oil and a major component of sassafras, star anise, and camphor oils. According to Homburger et al., hepatic tumors were inducted in rats given 1% safrole in their diet for 150–360 days. Cirrhosis and the buildup of large ceroid pigment deposits occurred before the induction of the tumors. At a dietary level of zero%, these effects were not shown. Long et al. confirmed the results utilizing a 0.5% safrole threshold. There were four types of tumors observed in the liver: hepatic cell adenoma, hepatic cell carcinoma, hepatocholangioma, and hepatocholangiosarcoma. No tumors were found in any other organ. Additionally, irregular coarse and fine nodularity, cyst formation, and hepatic cell enlargement were observed. Safrole is converted to 1'-hydroxysafrole in rats and mice. Borchert along with others conducted studies using 1'-hydroxysafrole and 1'-acetoxysafrole by oral and subcutaneous administration in rats and mice. The results indicated that 1'-hydroxysafrole was a proximate carcinogen, more so than safrole itself.

An analysis by Wislocki and colleagues of the mutagenicity of a number of safrole-related compounds showed that the 2'- and 3'-oxides of safrole, as well as 1'-acetoxysafrole, significantly mutagenicized *Salmonella typhimurium* strains TA 1535 and TA IOO. The 2'- and 3'-oxides of 1'-hydroxysafrole, 1'-acetoxysafrole, and 1'-oxosafrole were also demonstrated to be mutagenic. It was discovered that the rodent metabolite 1'-hydroxysafrole is not carcinogenic.

Benedetti and associates examined the metabolism of oral doses of ^{14}C-safrole up to 0.6 mg/kg in rats and humans, finding that the drug was rapidly absorbed and then nearly completely excreted in the urine in less than 24 hours. Raising the dosage caused the compound to accumulate in the kidney and liver and to be eliminated at a slower rate. 1,2-dihydroxy-4-allylbenzene was found to be the primary urinary metabolite in both species, and it was eliminated in a conjugated form. The conjugates of the proximate carcinogen, 1'-hydroxysafrole and 3'-hydroxysafrole, were found in the urine of rats exclusively, not in humans.

Because safrole is a naturally occurring substance, human exposure to it cannot be totally eliminated. Nonetheless, the safrole case highlights the significance of concentrating on the differences in metabolism between humans and test animals. When interpreting the results, common sense should take precedence if it is demonstrated that the recognized carcinogen is not eliminated by humans. More *in vitro* screens should be used to make a valid statement about the ability of short-term *in vitro* screening tests to predict carcinogenicity, even though a comparison of the mutagenic and carcinogenic effects of the rodent metabolite 1'-hydroxysafrole did not appear to show any correlation.

6.4.3. Colors

Over the years, color additives have drawn particular attention, mainly because their use as food additives required more justification than did antioxidants and preservatives. Furthermore, several colors have been phased out over the past 10 years due to findings from research on animals. Butter yellow, which was used in margarine until 1940 when it was discovered to cause hepatomas in rats, was among the first colors to be phased out. Further research revealed that this effect was exclusive to rats.

Toxicologists frequently bring up the issue of how adequate the testing and analytical processes are. Several years ago, subcutaneous injection was used to test a wide range of colors. This method of administration was suggested for compounds that were poorly absorbed because it guaranteed that a sufficient concentration of the test substances would reach the body tissues regardless of how they were processed. Due to this process, some substances—particularly the triphenylmethane dyes—that have not been proven to be carcinogenic when taken orally have occasionally been demonstrated to cause subcutaneous sarcomas at the injection site. Thus, it became crucial to differentiate between cases where the sarcomas were genuinely caused by the material's carcinogenic potential and those where they developed as an artifact resulting from the route of administration. It was widely acknowledged after BIBRA's research that this method of administration was unsuitable for evaluating the safety of food additives.

The color amaranth ban is a good example of a decision made based on studies that were poorly designed and executed. Early in 1976, a statistical reevaluation of previously accepted results revealed a significantly higher number of tumors in female rats only, leading to the prohibition of amaranth in the USA. The results of studies conducted in the USSR early in the 1970s raised concerns about the safety of amaranth regarding its effects on decreased lactation and fertility as well as its carcinogenic potential. Two significant studies were then started in the USA in an attempt to address these questions. The US study that ultimately served as the basis for the ban was of extremely low quality in two areas: (1) post-mortems of dead rats were not performed in a timely enough manner to prevent significant autolysis, and (2) at one point in the study, multiple rats were put in the incorrect cages, resulting in the switching of some treatment and control animals.

The study's contentious statistical analysis showed that while the overall number of tumors did not rise, females fed diets containing 3% amaranth (the highest dose) did see an increase in the ratio of malignant to benign tumors. The US acknowledged that there was no proof of a public health risk, but it did not assert that amaranth was carcinogenic and that safety had not been established. Surprisingly, not many other nations emulated the American move.

As a result, the history of food coloring suggests three crucial considerations for evaluating a material's safety during use: (1) the test substance should be administered using the same route as it is in humans. Thus, although the results of these studies may be pertinent to specific medical and cosmetic uses of these substances, the subcutaneous administration studies with colors are inappropriate for an evaluation of their safety-in-use as food additives; (2) good laboratory practice is of utmost importance in all studies and decisions on safety-in-use of food additives should not be based on poorly conducted studies; and (3) statistical evaluations of tumor results can vary greatly. Significantly different conclusions could be reached depending on how the analysis is chosen and used. For this reason, it's critical to standardize all statistical procedures whenever feasible.

6.4.4. Artificial Sweeteners

Significant doubts concerning the reliability of testing procedures and result interpretation were brought up by the cyclamate ban. It was asserted that feeding rats cyclamates in a 10:1 cyclamate-saccharin ratio increased the incidence of bladder tumors in the rats. The decision to outlaw cyclamates was made in 1969, first in the USA and then in many other countries shortly after. At the time, there was evidence linking cyclamates to bladder cancer, and saccharin had been used for many years without showing any health risks. On the other hand, saccharin was still utilized.

Saccharin was first produced commercially in 1900 after being synthesized from toluene in 1879. With the introduction of diet sodas in response to a population that was becoming more conscious of their weight after World War II, its use in soft drinks dramatically increased. Since 1950, numerous animal studies have been conducted to examine the long-term safety of saccharin. At least 20 studies on rats, mice, hamsters, and monkeys have been conducted to date. The majority of these studies involved single-generation feeding trials with no negative effects observed. However, in three rat studies, wherein saccharin was fed to the pups during pregnancy and then throughout their lives, an increase in bladder tumors was observed in the males. The studies by Munro et al. reported these results as statistically significant along with D. L. Not in the 1972 study by the Wisconsin Alumni Research Foundation Institute, however,

Arnold. The young rats showed a tendency toward proteinuria and microcalculi, even though no other long-term negative findings were found. These results could bolster the theory that physical irritation played a role in the development of bladder tumors in the two-generation studies.

Contrary to most other food additives there have been many epidemiological trials conducted attempting to relate saccharin consumption with bladder cancer. The Howes, etc., and Miller and Howe recently reported increases in bladder cancer in males but a concomitant decrease in females, thus giving rise to a questionable association. One large-scale epidemiological survey being conducted in the USA is now complete but still undergoing analysis of results. Along with 6000 control cases, 3000–4000 newly diagnosed cases of bladder cancer in 1978 were analyzed. Further analysis is needed, but preliminary examination of the data suggests that the conclusions align with the relationships observed in the studies conducted on lab animals.

Therefore, the safety of saccharin after years of use is questioned. Its potential to cause cancer has not been demonstrated or refuted. Furthermore, there is ambiguity in the mutagenicity test results. Saccharin's possible withdrawal in 1977 caused such a stir in the United States that the FDA decided to halt its plans to outlaw the sweetener and instead, at this point, only label foods containing saccharin with hazard warnings. Thus, it appears that the consumer has the final say in whether or not to accept the risk of ingesting a potentially dangerous food additive.

6.4.5. Nitrates, Nitrites, and Nitrosamine

Because nitrate and nitrite, respectively, are common components of a typical diet, the issues raised by the link between nitrosamines and nitrites and cancer are far more complicated than those caused by many other food additives. Plant growth requires the nitrate ion, which can be found in soil naturally or added as fertilizer. Man can ingest soil-derived nitrate through plants, particularly those that concentrate nitrate, like celery, spinach, and beet, as well as through drinking water. Except for potential thyroid effects, nitrate is widely regarded as non-toxic. In humans, the small intestine absorbs nitrate, which is then secreted in the saliva and stomach juice and eliminated in the urine. It is easily converted to nitrite by bacterial action in the human body and during food storage that contains nitrate, particularly when the food is not refrigerated. The body can produce up to 10–12 mg of nitrite/d just from saliva. All of these sources of nitrite, along with nitrite that is purposefully added to food in its capacity as a preservative, pose a toxic risk to human health because (1) nitrite is directly toxic to humans (causing hemoglobinemia in children); and (2) nitrite reacts with secondary tertiary amines and amides to form N-nitroso compounds, such as nitrosamines and nitrosamides. Methemoglobinemia is known to be caused by nitrite, however, its carcinogenic potential is uncertain. In a 1978 study, rats were given nitrite concentrations as high as 2000 parts per million in a variety of food and water media. The treated rats had a statistically significant increase in "lymphomas" (10.2% incidence compared to 5.4% in the controls), which was not dose-related. This is the first study that links the carcinogenicity of nitrite to itself. Unfortunately, UAREP (Universities Associated for Research and Education in Pathology) is currently re-evaluating the study due to significant criticism.

However, nitrosamines' carcinogenic effects have been well-documented in recent years. Many studies, nevertheless, have included dosages of administration that are significantly higher than the estimated likelihood of human exposure to these substances. However, recent research using similar dosages of administration has shown that male mice and rats seem to be more vulnerable to the cancer-causing properties of several nitrosamines. One study's findings are currently being statistically evaluated, and a time-related analysis is intended to be conducted to calculate the risk associated with a daily exposure of one kilogram of nitrosamines to humans.

Numerous researchers have attempted to link nitrate and nitrite exposure and subsequently nitrosamine exposure to the incidence of stomach cancer. In Chile and Japan, stomach cancer rates are highest. The world's largest nitrate mines are located in Chile, and Japanese people consume a high concentration of dried and salted foods, many of which may concentrate nitrite. These two findings are correlated. However, it will require many more years of work to establish a positive causal relationship.

With the aforementioned in mind, legislative bodies should specifically ask themselves, "Where do all these studies leave us?" Since nitrates are naturally occurring in the environment, prohibiting nitrite as a food additive won't stop nitrate from being reduced in the body to nitrite and interacting with secondary or tertiary amines derived from naturally occurring proteins in the diet, nor will it stop the potentially cancer-causing effects of nitrite itself. Meanwhile, the important thing to think about is how much the normal body burden of nitrite is increased by nitrate and nitrite used as food preservatives. If this is demonstrated to be negligible, the advantages of reducing the risk of botulism as a food additive must be greatly greater than the "risks."

6.4.6. Carcinogenicity Testing

In recent years, there has been a significant evolution in the methods used to test food additives for potential carcinogenicity. To determine whether there were any effects on the gross and histopathology, as well as the appearance, behavior, survival, growth, and clinical biochemistry (blood, serum, and urine), long-term tests were first designed.

Of all the parameters observed, pathology received the least attention, and frequently only a small percentage of the animals were examined. However, as pathological techniques and statistical analysis methods advanced, more animals were examined and the likelihood of finding a carcinogenic effect increased. The design of specific carcinogenicity studies occurred during the past 10 years. Among these was the use of sizable animal groups, some of which were put down periodically throughout the investigation to see if there were any effects on tumor latency. These days, long-term animal experiments are typically made to examine a compound's possible carcinogenic effects as well as its chronic toxic effects; as a result, they are very costly and large-scale studies. The amount of test materials given to the experimental animals in comparison to the known or probable human intake has recently come under heavy fire (as with saccharin for example). Drugs ought to be given at reasonable dosages; one low dose level should roughly correspond to the average human intake, and subsequent dose levels should be multiples of this. "Maximum tolerated doses" may yield results that are challenging to interpret. It is important to remember that feeding an animal excessively high concentrations of a compound can cause it to follow a different metabolic pathway than it would after low-dose administration. This latter point is crucial when the test species' metabolic pathway is shared by humans.

There are many problems involved in the design and interpretation of results from carcinogenicity studies, particularly those relating to food additives.

1. The possibility that young animals may have an increased susceptibility to carcinogens has led to some study designs including exposure in utero followed by lifetime feeding of the pups, thus giving transplacental exposure for compounds with the ability to cross the placental barrier. The need for this type of approach has not yet been established although it would mimic the human situation if the levels of administration in the animal studies were comparable with those in man.
2. The interpretation of results obtained from inbred strains of animals poses many questions. The main problem arising from the use of such animals is the possibility of selection of a line that may be particularly resistant to the test material. Therefore, more than one line should be used in the animal experiments. There is also much debate on whether strains with a high or low "spontaneous' tumor incidence should be used.
3. The question of the extent to which short-term *in vitro* screening

tests should be used in predicting potential carcinogenicity is very important at the present time. Currently, the use of a range of screening tests serves as a useful guide in this prediction but as yet there is no group of *in vitro* tests available which could reliably replace the conventional long-term testing in animal species. Vast quantities of time and resources are being devoted to this particular area and every encouragement should be given to the development of suitable screens.

6.5. CANCER PREVENTION WITH NUTRITION AND LIFESTYLE

For the most part, ancient healing theories and methods heavily emphasize nutrition. Nutrition-based cancer prevention is a major field of study and concern for public health. A lower risk of developing cancer has been linked to specific dietary patterns and nutrients, although no single food or diet can ensure cancer prevention.

6.5.1. Facts and Figures – Cancer Prevention

Cancer is one of the most common causes of death, as is well known; it accounts for 25% of all deaths in Germany, making it the second most common cause of death there. Both endogenous and exogenous factors have the potential to start the development of cancer. Endogenous factors are unaffected by behavior or way of life, such as age or genetic predisposition. Smoking and diet are the two main exogenous environmental and lifestyle factors that cause cancer. Worldwide, exposure to tobacco smoke and smoking cigarettes continue to be the leading causes of preventable deaths, despite a decline in the overall number of smokers. An estimated 13.5% of deaths in Germany are thought to be related to smoking, and 28.8% of cancer deaths in the US are caused by smoking. Furthermore, the primary risk factor for lung cancer is smoking. Smoking is responsible for about 80% of lung cancer deaths. Therefore, it is undeniable that quitting smoking is the best way to prevent lung cancer. Numerous international investigations have been carried out to look into how diet and nutrition affect cellular processes involved in carcinogenesis, including immunity, inflammation, apoptosis, and the cell cycle. These studies demonstrate that a nutritious, well-balanced diet helps the body receive enough nutrients at the cellular and molecular levels as well as for the entire body. At these levels, improper diet has an effect on metabolism and the microenvironment. This creates an environment in cells that is favorable for harmful alterations like the buildup of DNA damage and, consequently, the development of cancer. A low intake of fruits and vegetables combined with other dietary risk factors resulted in 11 million deaths and roughly 255 million disability-adjusted life years in 2017, according to startling statistics published in a recent Lancet paper. A healthy lifestyle, avoiding occupational carcinogens and environmental toxins, and other preventive measures are thought to be responsible for between 30 and 50% of cancer cases that can be prevented. The type of cancer determines how different this number is. The two cancers with the highest rates of potentially avoidable cases are lung and colorectal cancer.

6.5.2. Cancer Prevention Recommendations of the World Cancer Research Fund

The World Cancer Research Fund (WCRF) is one of the most significant scientific organizations when it comes to cancer prevention and lifestyle. This non-profit organization is made up of independent, well-known experts from across the globe who evaluate and assess the data regarding the impact of lifestyle factors on cancer risk. These researchers evaluate the data and offer dietary, nutritional, and physical activity recommendations that can prevent cancer and improve prognosis following diagnosis. The Third Expert Report, an incredibly thorough summary of the most recent research in this field, published the most recent expert insights in 2018. The 10 WCRF cancer prevention recommendations provide an overview of the main conclusions. Even though observing each recommendation separately may help prevent cancer, it is best to follow them all and take them into consideration as a way of life (Hastert, White., 2016).

6.5.2.1. Be a Healthy Weight

The first and most crucial WCRF recommendation is to keep body weight as low as possible and within a healthy range, as there is compelling evidence that being overweight increases the risk of cancer. BMI measurements indicate that a weight range of 18–24 points is considered "normal" or healthy. A recent survey conducted among German households found that 43.1% of women and 62.1% of men are overweight or obese. Since childhood is when overweight people first gain weight, it is important to stop weight gain from an early age. Maintaining a healthy weight and energy balance throughout life is also recommended for adults. Strong evidence suggests that being overweight increases the risk of developing many cancers, including those of the mouth, throat, larynx, esophagus, stomach, pancreas, gallbladder, liver, colorectal, breast, ovarian, endometrial, prostate, and kidney.

6.5.2.2. Be Physically Active

There is compelling evidence to suggest that physical activity lowers the risk of various cancer types. Thus, the WCRF's recommendation for exercise is to "walk more and sit less." Though there is more evidence for endometrial, colon, and (postmenopausal) breast cancers, physical activity may also lower the risk of other cancers.

Insufficient physical activity may be the cause of 9–19% of cancer cases in Europe. Exercise affects the body's immune, metabolic, and endocrinological systems, among other systems, which in turn affects the risk of cancer. Additionally, it lowers body fat, which is a separate risk factor for many cancers, and aids in maintaining a healthy weight. The World Health Organization advises minimizing sedentary behavior and engaging in daily physical activity. Adults are advised to engage in at least 150 minutes per week of moderate-intensity exercise (such as walking, dancing, or gardening) or at least 75 minutes of vigorous exercise (such as team sports, aerobics, or jogging). What kind of activity you select is irrelevant. "Any movement that uses skeletal muscles and requires more energy than resting" is the simple definition of physical activity. Recreational activities, certain jobs, gardening, and household chores all support cardiometabolic health. People around the world spend a lot of time in front of screens, even at work, but they engage in less physical activity than is healthy. These behaviors can readily lead to weight gain or overweight, which raises the risk of developing cancer.

Five recommendations for a healthy diet complete these two WCRF guidelines for weight and physical activity:

(1) Eat a Diet Rich in Wholegrains, Vegetables, Fruit, and Beans: Plant-based foods should make up a large portion of a diet to prevent cancer. Vegetables, fruits, whole grains, and pulses (legumes) like beans or lentils offer a variety of dietary fiber and important nutrients. 30 g of fiber and five servings (or at least 400 g) of fruit and vegetables should be consumed daily, according to the WCRF. This diet works well at controlling energy intake, preventing weight gain and becoming overweight. Furthermore, dietary fiber contributes to the prevention of colorectal cancer.

(2) Limit Consumption of "Fast Foods" and Other Processed Foods High in Fat, Starches, or Sugars: A "Western type" diet includes "fast foods" and processed foods that are high in fat, starches, or sugars; these foods should be avoided. These foods are typically tasty, accessible, and reasonably priced. Generally speaking, they have less micronutrients but more energy than unprocessed foods.

As a result, they make it easier for excessive energy intake to occur relative to energy expenditure, which can quickly result in weight gain and overweight. Highly processed foods account for more than half of the calories consumed in the diet of Americans. According to a recent pilot intervention study, eating more ultra-processed food than unprocessed food increases energy intake, which in turn contributes to weight gain.

(3) Limit Consumption of Red and Processed Meat: The WCRF recommends consuming no more than 350–500 g (cooked weight) of red meat per week. Beef, pork, veal, lamb, goat, mutton, and horse are examples of red meat. Eat as little processed meat as possible, such as sausages, ham, bacon, and salami. The advice regarding red meat finds a balance between the benefits and drawbacks of meat consumption. This means that while eating meat is not necessary for a healthy diet, it is preferable to not completely avoid it. Meat is a good source of micronutrients like zinc, iron, and vitamin B12. It is also a highly bioavailable and valuable source of protein. However, excessive red meat consumption appears to raise the risk of noncommunicable diseases (NCDs), including colorectal cancer. Therefore, it is recommended to limit the consumption of red meat to moderate amounts. It is better to have lean cuts than fatty ones. Additional excellent sources of micronutrients and protein include fish, poultry, eggs, and dairy products. Their plant-based alternatives to meat can round out a daily diet that is high in nutrients.

(4) Limit Consumption of Sugar-Sweetened Drinks: Drinks with a lot of sugar added can make people gain weight and become overweight. Drinks with added sugar that don't trigger a satiety signal but are high in calories allow for an excess of energy intake over energy expenditure. People may then become overweight or obese as a result, raising their chance of developing many types of cancer. Therefore, it is best to stick to drinking water and tea or coffee without added sugar to stay properly hydrated. Coffee, incidentally, appears to offer some protection against endometrial and liver cancer. Consuming sugar and foods sweetened with sugar, such as fruit soups and soft drinks, may increase the risk of pancreatic cancer.

(5) Limit Alcohol Consumption: Avoiding alcohol is the best way to prevent cancer. This stringent WCRF recommendation is supported by the fact that alcohol consumption, even in tiny doses, may raise the risk of at least some malignancies. There is compelling or strong evidence that the risk of developing specific cancers, i.e., alcohol

consumption is strongly linked to cancers of the mouth, throat, larynx, esophagus, liver, colon, breast, and stomach. Although a significant portion of the population still associates alcohol use with their quality of life, the WCRF advises against going over the national recommendations. In Germany, this translates to a limit of two alcoholic drinks (20 g of ethanol) for men and one drink (10 g of ethanol) for women per day.

6.5.3. Do Not Use Supplements for Cancer Prevention

Supplemental nutrition has not been shown to be able to prevent cancer. Generally, one must assume that all nutritional needs are met by a healthy, balanced diet that includes foods from every food group. Supplements should only be prescribed by licensed healthcare providers to individuals who have deficiencies that have been demonstrated.

6.5.4. Eating Behaviors and Cooking

A few examples of lifestyle suggestions for cancer prevention are the WCRF guidelines. The International Agency for Research on Cancer's European Code against Cancer, which offers 12 strategies to lower cancer risk, is another example. The most renowned and expert organization in Germany for dietary advice is called the "Deutsche Gesellschaft für Ernährung." Two more items that they list in their 10 recommendations for a healthy diet and lifestyle are, in our opinion, crucial for leading a balanced life. Eating a varied selection of food and appreciating its diversity is the first step. This suggests that to meet all nutritional needs, a varied diet is necessary. No single food, like kale or raspberries, has all the necessary nutrients.

"Take a break while you eat and allow plenty of time for eating" is the next suggestion we would like to add to the WCRF guidelines. Eating slowly makes sure that there is adequate time for chewing, which improves the meal's acceptability. It also makes the feeling of satiety easier to perceive. This assists in preventing overweight, which is one of the primary risk factors for cancer as previously mentioned. Moreover, conscious enjoyment is another benefit of mindful eating. This includes cooking your own food and sharing meals with others, both of which are easy ways to unwind and incorporate relaxation into daily life. Reducing stress and finding relaxation are essential to preserving health and wellness. For example, breast cancer survivors' psychological well-being and quality of life are enhanced by mindfulness-based stress reduction (Huang et al., 2016).

6.5.5. Recent Data on Nutrition and Physical Activity Recommendations and Cancer Prevention

Following these lifestyle suggestions can help prevent cancer. The risk of cancer is reduced by approximately 36% by, for instance, never having smoked, having a BMI <30, engaging in physical activity for at least 3 points 5 hours per week, and meeting healthy eating criteria like eating a lot of fruits and vegetables or whole grains. A different analysis showed that the incidence of colorectal cancer is reduced by 34–45% or 58%, respectively, by meeting 1–3 of the WCRF recommendations or meeting 4–6 recommendations as opposed to following none of these guidelines. According to a sizable population-based cohort study, following dietary and lifestyle guidelines like those recommended by the WCRF lowers the risk of developing breast and prostate cancer in general. According to a systematic review, there is a 10–61% reduction in overall cancer incidence and mortality when cancer prevention guidelines are followed. The WCRF guidelines are also relevant and lower the elderly's risk of developing cancer. Following an extra recommendation delayed the onset of cancer in individuals 60 years of age or older by approximately 1.6 years.

6.5.6. Public Health Policies

Currently, there is a lack of awareness and participation in prevention programs. Furthermore, very few people in the population follow the recommendations for cancer prevention. For example, only 8.5% of African American women followed >4 guidelines. Furthermore, the global burden of cancer and NCDs is increasing. The WCRF and other similar recommendations, which were developed based on scientific evidence, serve as the foundation for lowering the risk of cancer and other NCDs because they share common underlying risk factors. These prevention guidelines also benefit the environment by encouraging dietary patterns that emphasize foods with a plant-based origin. It is now necessary to take immediate policy action to enhance public communication of these recommendations.

However, since environmental, economic, and social factors also impact people's behavior, what can policymakers and health professionals do to encourage adherence to the WCRF recommendations? First and foremost, more needs to be done to address the motivational aspects of the recommendations. The first and most important step in altering behavioral patterns is raising people's awareness of how their habits affect their health. This can be done in a number of ways, such as by educating people about disease prevention or stressing the benefits of screenings for diseases early on and follow-up care afterward. A paradigm shift from dietary restriction to the promotion of healthy lifestyle behaviors and from nutrient-based to food-based guidelines is currently taking place in public health organizations. In addition, to prevent cancer and other NCDs, strong public health policies in the form of laws, rules, or guidelines are needed. Policymakers are in charge of establishing social contexts that support health. In comparison to unhealthy foods, beverages, alcohol, and physical inactivity, there should be improvements made to the accessibility of environments for active lifestyles and the availability and affordability of healthy foods. This could involve limiting food advertising, enforcing stricter food labeling regulations, or raising the general nutritional value of food. Another strategy is to provide everyone with anytime access to dietary guidance and advice in medical settings (Hastert & White, 2016).

REFERENCES

1. Albano, E. (2006). Alcohol, oxidative stress and free radical damage. *Proceedings of the Nutrition Society, 65*, 278–290.
2. Augusto, M. C., Geoffrey, C., Jean-Claude, M., Bertazzi, L. R., Louzada, M. L. C., & Constante, J. P. (2017). The UN Decade of Nutrition, the NOVA food classification and the trouble with ultra-processing. *Public Health Nutrition, 21*(1), 5–17.
3. Cediel, G., Reyes, M., Corvalán, C., Levy, R. B., Uauy, R., & Monteiro, C. A. (2020). Ultra-processed foods drive unhealthy diets: Evidence from Chile. *Public Health Nutrition, 27*, 1–10.
4. Eicher-Miller, H. A., Fulgoni III, V. L., & Keast, D. R. (2015). Processed food contributions to energy and nutrient intake differ among US children by race/ethnicity. *Nutrients, 7*(12), 10076–10088.
5. Gangwisch, J. E., Hale, L., Garcia, L., Malaspina, D., Opler, M. G., Payne, M. E., Rossom, R. C., & Lane, D. (2015). High glycemic index diet as a risk factor for depression: Analyses from the Women's Health Initiative. *The American Journal of Clinical Nutrition, 102*(2), 454–463.
6. Hassanpour, S. H., & Dehghani, M. (2017). Review of cancer from perspective of molecular. *Journal of Cancer Research and Practice, 4*(4), 127–129.
7. Hastert, T. A., & White, E. (2016). Association between meeting the WCRF/AICR cancer prevention recommendations and colorectal cancer incidence: Results from the VITAL cohort. *Cancer Causes & Control, 27*(11), 1347–1359.
8. Huang, H. P., He, M., Wang, H. Y., & Zhou, M. (2016). A meta-analysis of the benefits of mindfulness-based stress reduction (MBSR) on psychological function among breast cancer survivors. *Breast Cancer, 23*(4), 568–576.
9. Kamal, A. A., & Fawzia, S. A. (2018). Toxicological and safety assessment of tartrazine as a synthetic food additive on health biomarkers: A review. *African Journal of Biotechnology, 17*(6), 139–149.
10. Ke, X., & Shen, L. (2017). Molecular targeted therapy of cancer: The progress and future prospect. *Frontiers in Laboratory Medicine, 1*(2), 69–75.
11. Lauby-Secretan, B., Scoccianti, C., Loomis, D., Grosse, Y., Bianchini, F., Straif, K., et al. (2016). Body fatness and cancer-viewpoint of the IARC working group. *New England Journal of Medicine, 375*, 794–798.
12. Malik, V. S., Willett, W. C., & Hu, F. B. (2013). Global obesity: Trends,

risk factors and policy implications. *Nature Reviews Endocrinology, 9*(1), 13–27.

13. Modica, C., & Hoenig, K. (2018). Mindfulness in follow-up care after breast cancer: Can it prevent recurrence? *Breast Care (Basel), 13*(2), 102–108.
14. Moore, S. C., Lee, I. M., Weiderpass, E., Campbell, P. T., Sampson, J. N., Kitahara, C. M., et al. (2016). Association of leisure-time physical activity with risk of 26 types of cancer in 1.44 million adults. *JAMA Internal Medicine, 176*(6), 816–825.
15. Moubarac, J. C., Martins, A. P., Claro, R. M., Levy, R. B., Cannon, G., & Monteiro, C. A. (2013). Consumption of ultra-processed foods and likely impact on human health: Evidence from Canada. *Public Health Nutrition, 16*, 2240–2248.
16. Poti, J. M., Mendez, M. A., Ng, S. W., & Popkin, B. M. (2015). Is the degree of food processing and convenience linked with the nutritional quality of foods purchased by US households? *The American Journal of Clinical Nutrition, 101*(6), 1251–1262.
17. Pouzou, J. G., Costard, S., & Zagmutt, F. J. (2018). Probabilistic assessment of dietary exposure to heterocyclic amines and polycyclic aromatic hydrocarbons from consumption of meats and breads in the United States. *Food Chemistry and Toxicology, 114*, 361–374.
18. Proquin, H., Jetten, M. J., Jonkhout, M. C. M., Garduno-Balderas, L. G., Briede, J. J., de Kok, T. M., et al. (2018). Gene expression profiling in colon of mice exposed to food additive titanium dioxide (E171). *Food Chemistry and Toxicology, 111*, 153–165.
19. Sánchez-Villegas, A., Toledo, E., De Irala, J., Ruiz-Canela, M., Pla-Vidal, J., & Martínez-González, M. A. (2012). Fast-food and commercial baked goods consumption and the risk of depression. *Public Health Nutrition, 15*(3), 424–432.
20. Soffritti, M., Padovani, M., Tibaldi, E., Falcioni, L., Manservisi, F., & Belpoggi, F. (2014). The carcinogenic effects of aspartame: The urgent need for regulatory re-evaluation. *American Journal of Industrial Medicine, 57*, 383–397.
21. Sommerfield, A. J., Deary, I. J., & Frier, B. M. (2004). Acute hyperglycemia alters mood state and impairs cognitive performance in people with type 2 diabetes. *Diabetes Care, 27*(10), 2335–2340.
22. Song, M., Fung, T. T., Hu, F. B., Willett, W. C., Longo, V. D., Chan, A. T., & Giovannucci, E. L. (2016). Association of animal and plant protein intake with all-cause and cause-specific mortality. *JAMA Internal Medicine, 176*(10), 1453–1463.

23. Townsend, N., Wilson, L., Bhatnagar, P., Wickramasinghe, K., Rayner, M., & Nichols, M. (2016). Cardiovascular disease in Europe: Epidemiological update 2016. *European Heart Journal, 37*(42), 3232–3245.
24. Viennois, E., Merlin, D., Gewirtz, A. T., & Chassaing, B. (2017). Dietary emulsifier–induced low-grade inflammation promotes colon carcinogenesis. *Cancer Research, 77*, 27–40.
25. Wilson, L. F., Antonsson, A., Green, A. C., Jordan, S. J., Kendall, B. J., Nagle, C. M., et al. (2018). How many cancer cases and deaths are potentially preventable? Estimates for Australia in 2013. *International Journal of Cancer, 142*(4), 691–701.

Chapter 7

Effects on Children and Adolescents

Contents

INTRODUCTION

Children and adolescents who work have a variety of effects on them, including effects on their social development, academic performance, emotional stability, and physical health. Teenagers are juggling more and more part-time jobs with their academic obligations, and how they manage work and other commitments can have a big impact on their whole development. The rigors of part-time employment can make it difficult for many teenagers to maintain their academic standing. Extended workdays or erratic shift patterns may impede the ability to study, finish homework, and attend classes, which could lead to poorer grades and stress related to the classroom. Having to balance work and school obligations can also result in exhaustion, which impairs learning and cognitive function. There are several effects of work on physical health.

Physically demanding jobs may increase your risk of injury and physical strain. Extended work hours or hard assignments can cause back pain, repetitive strain injuries, and overall exhaustion. Conversely, physical benefits from active employment may exist, though they are less common in more sedentary roles. Long work hours can also interfere with sleep cycles, which is bad for general health and wellbeing. Teenagers who work frequently feel more stressed out because of the demands of their jobs and their studies. Anxiety, sadness, and burnout are some of the ways that stress can show up. Mental health may be impacted by the emotional toll that juggling many obligations and possibly difficult work environments can take. However, as teenagers gain independence, financial autonomy, and important life skills, working can also help them feel more confident and good about themselves.

By giving teenagers the chance to engage with a wide variety of people, practice teamwork, and navigate professional settings, work can improve social skills. These encounters may have a favorable effect on their social growth. Extensive work schedules, however, may make it difficult to socialize with peers, which could result in social isolation and a decline in participation in extracurricular activities. Teenagers who work might be able to support their families financially, which could change the roles and responsibilities within the family. Although this contribution can ease financial strains, it might also put extra weight on young people's shoulders, which could have an impact on their general development and family dynamics. Work has a variety of complex effects on kids and teenagers, both good and bad. To optimize advantages and minimize disadvantages, one must strike a harmonious equilibrium between professional, academic, and personal obligations, with the backing of family and educational establishments.

7.1. EFFECTS ON CHILDREN OF ULTRA-PROCESSED FOODS

Ultra-processed foods (UPFs) have been linked to several serious and varied health, cognitive, and emotional problems in children. Fast food, sugary snacks, and pre-packaged meals are examples of ultra-processed foods. These foods are usually low in vital nutrients and high in unhealthy fats, artificial additives, and added sugars. When these foods are widely consumed during crucial stages of growth and development, there can be many negative effects.

7.1.1. Physical Health Impact

An elevated risk of obesity in children is among the most notable and immediate effects of UPFs. UPFs frequently have low nutritional value and a high-calorie content, which contributes to overeating and insufficient satiety. This imbalance has the potential to cause weight gain and the emergence of obesity, a significant risk factor for a number of illnesses. Obesity in children is linked to an increased risk of type 2 diabetes, hypertension, and dyslipidemia. Due to bad eating habits, these disorders—which were formerly thought to be adult-onset diseases—are becoming more common in younger populations.

Academic performance and cognitive function are also impacted by UPFs. Diets heavy in processed fats and sugars can have an impact on cognitive development and brain health. Excessive consumption of sugar is associated with blood glucose fluctuations, which can cause cognitive fatigue, memory issues, and concentration issues. Additionally, UPFs frequently replace healthier food options, which can result in dietary deficiencies in important nutrients like vitamins, minerals, and omega-3 fatty acids that support brain health. These deficits may affect cognitive abilities, such as learning and problem-solving techniques, which may result in subpar academic performance and lower academic achievement.

7.1.1.1. Behavioral and Emotional Well-being

UPF consumption may also have an impact on behavior and mental well-being. Due to the quick fluctuations in blood sugar levels caused by the high sugar content of these foods, mood swings and irritability may be exacerbated. Studies have indicated a connection between diets rich in unsaturated polyphenols (UPFs) and a heightened likelihood of behavioral issues like hyperactivity and attention deficits, which are frequently linked to attention deficit hyperactivity disorder (ADHD). Additionally, research has linked high UPF consumption to an increased risk of anxiety and depression, suggesting that the poor nutritional quality of UPFs may have an impact on mental health. Mood swings and nutritional deficiencies together have the potential to negatively affect social interactions and general emotional health.

7.1.1.2. Long-Term Health Consequences

The long-term effects of early UPF exposure are possible. Regular UPF consumption in childhood increases the risk of chronic health conditions in later life. Childhood-established patterns of inadequate nutrition frequently carry over into adulthood, resulting in lifelong health issues. Chronic illnesses that impact health throughout life, including metabolic disorders, psychological disorders, and cardiovascular diseases, can be caused by early dietary habits.

7.1.1.3. Addressing the Issue

A comprehensive approach is necessary to address the impact of UPFs on children. It is crucial to teach parents and other caregivers the value of a diet rich in nutrients and well-balanced. Communities and schools can contribute by encouraging better eating options and limiting the availability of UPFs. Better food labeling and stronger restrictions on food marketing to children are two examples of policy interventions that can help inform healthier eating choices. Children's general health and development can be supported by minimizing the harmful effects of UPFs and encouraging healthier eating habits from an early age.

7.1.2. Work's Effects on Children and Adolescents

Children and adolescents who work may experience significant effects on their cognitive, emotional, and physical development. Adolescents who manage part-time employment and school obligations face significant consequences to their academic achievement and overall well-being. Due to work's tendency to interfere with study and rest time, long work hours and demanding jobs can cause exhaustion, stress, and a decline in academic achievement. Their capacity to finish their homework and do well in class may be impacted by this. But employment can also impart important life skills.

Adolescents who work can improve their personal and professional development by learning time management, communication, and problem-solving skills. Furthermore, as teenagers become more independent and feel like they've accomplished something, work experiences can increase confidence and self-worth. The effects of work on physical health can vary. While more active jobs may improve physical fitness, demanding or long hours can cause weariness and possibly injuries. Work can occasionally cause emotional stress and burnout, especially if the demands of the job are great and there are no support systems in place. On the other hand, social benefits from work can also include enhanced collaboration and interpersonal skills. Working teenagers may make financial contributions to their homes, which changes the roles and responsibilities within the family.

This has an impact on family dynamics. The overall impact of employment on young people is heavily influenced by how well they manage their personal, professional, and educational obligations as well as how supportive their families and employers are. Maintaining a healthy balance can maximize the positive effects of early work experiences while reducing any potential negative effects.

7.1.3. Role of Work Effects on Children and Adolescents

Teenagers are typically the age at which most people enter the workforce for the first time. Therefore, it's critical to analyze how employment affects adolescent development and how working experiences can help or hinder young people's socioeconomic achievements as well as their emotional, intellectual, and physical growth. To respond to these queries, one must comprehend the characteristics of teenage development in contemporary culture.

The majority of experts concur that adolescence plays a significant role in modern human development for many interconnected reasons. First of all, adolescence is a time when a person's long-term development is influenced by experiences in the family, school, and other environments. To put it simply, adolescent experiences have an impact on future performance. Adolescence is a time of great variability, to start with. During those years, people's life paths start to diverge significantly, partly due to the abundance of diversity and flexibility afforded by contemporary society. Lastly, adolescence is a particularly crucial formative stage when a lot of developmental trajectories solidify and become more challenging to change. Due to the malleable, variable, and formative aspects of adolescence, schools, and other organizations that have an impact on children during this time must be set up to maximize the children's chances of having happy, healthy adolescent years and successful adulthood.

Generally speaking, adolescent development is heavily centered on getting ready for adult responsibilities such as family, employment, and citizenship. The development of specific personal competencies—such as the ability to be self-reliant and behave responsibly—interpersonal competencies—such as the ability to establish and maintain fulfilling relationships with others—and social competencies—such as the ability to participate in a larger community—is fundamentally necessary for success in each of these domains. It is during adolescence that these abilities both grow and stabilize. Like any other experience, work experience can be assessed based on how much and in what ways it aids in young people's development as individuals, as well as their interpersonal and social skills.

It is impossible to consider the growth of these psychosocial competencies in isolation from the larger context of a young person's or a group of young people's coming of age. Societal definitions of social, interpersonal, and personal competence are shaped by contextual factors. Therefore, the environment in which teenagers grow determines not only the paths toward maturity but also the very skills that characterize it. The preparation that today's teenagers need for a successful adulthood is very different from that of their great-great-grandparents because the elements of competent adulthood in the twenty-first century will differ from what they were a century ago.

The function of work in a young person's transition to adulthood is contingent upon the characteristics of adolescence, or the developmental needs of the age group, as well as the characteristics of adulthood and the abilities that the adult population is expected to possess. Because of this, laws, and policies of youth employment must be periodically reviewed in light of how adolescence and the transition to adulthood are evolving.

More specifically, given the circumstances that characterize adolescence and adulthood in contemporary society, what role does work play in the development of personal, interpersonal, and social maturity? Answers to these and other questions regarding the nature of the work adolescents perform, the features and conditions of the settings in which they work, the amount of time they dedicate to their jobs, and how experiences at work complement or conflict with other demands of the adolescent years are necessary to comprehend the role of work in the lives of adolescents.

7.1.4. The Nature of Work by Young People

When analyzing the effects of working during adolescence on development during this stage of life, specific characteristics of this work must be considered. When compared to adult labor, youth labor is typically more voluntary (i.e., not financially necessary), part-time as opposed to full-time, and erratic, since young people enter and exit the labor force in response to changes in employer demands, labor market dynamics, and life events that alter in other domains. The quantity of time spent working and the order in which those hours are worked are likely to vary regularly. High school students are more employable on the weekends than they are on weekdays, and during the summer and break than they are during the academic year. However, during the school year, about 80% of all students work for pay at some point during their high school careers. Apart from these temporal aspects of work, certain characteristics of work experience might be especially significant for young people. These characteristics might include, for instance, how much work allows teenagers to apply what they learn in school or offers them additional learning opportunities; how their money is spent; how stressful the work is; and how well the young people get along with their managers or other adults at work. The majority of teenagers in the United States work in jobs that are not aligned with school curricula, do not consistently impart the skills needed for career advancement, and offer minimal opportunities for meaningful interaction with adult supervisors.

The quality of jobs held by young people has not been the subject of much systematic research. During their high school years, teenagers transition from casual jobs like babysitting, yard work, and snow shoveling to more formal jobs, especially in the fast-food and retail/service industries. They also become more evenly distributed across different job categories. As they get older, they are also more likely to be given supervisory responsibilities and to receive additional training from their employers.

7.1.5. Consequences of Working

Selection effects present a problem for practically all studies on the effects of working with young people. That is to say, individuals who work long hours may differ from those who work fewer hours, and young people who work may differ before they started from those who do not. Adolescents who are disinterested in school, for instance, might decide to work longer hours than those who do. Additional distinctions between the groups could be attributed to prior academic achievement, career aspirations, family income, parents' educational attainment, motivation, and a variety of other variables that are frequently not measured specifically. Due to these variations, it is very challenging to determine whether working itself results in any specific outcome (positive or negative), or whether those outcomes might have happened even in the absence of the young people who were employed.

The committee drew upon a thorough analysis of research that tracks youth over time (longitudinal studies) and includes pre-existing disparities among youth who participate in different work patterns (statistically controlling for differences). In the context of pre-existing differences, these studies quantify the statistical correlation between working and different outcomes, i.e., the extent to which the occurrence of certain outcomes varies with different work patterns. Correlational studies do not directly link work to outcomes, but the best information may come from a pattern of consistent findings from studies with strong statistical controls. Selection effects, however, have to be taken into account as a different way to interpret the data in the absence of random assignment.

Research using information from cross-sectional studies (with data from a single point in time) and surveys of regional samples are also discussed when they supplement national longitudinal studies and provide information not found in national studies. Cross-sectional studies include first-year data from the National Longitudinal Study of Adolescent Health and Monitoring the Future, which annually surveys a national sample of high school seniors. The Youth Development Study, one of the regional studies, tracks a random sample of St. Louis students who were enrolled in the ninth grade in 1988. Paul, Minnesota, and a 1993 study by Steinberg and colleagues that tracked a group of students from nine Wisconsin and northern California high schools for a full year. Regional studies may have a limitation in that findings drawn from these samples may not apply to the entire country. Studies on the impact of working on teenagers have concentrated on a range of outcomes, such as problem behaviors, relationships, education, and careers.

These studies examine a wide range of outcomes, but they typically approach work in a fairly one-dimensional way. The majority of studies look at the effects of work intensity, which is typically determined by the average number of hours worked per week during the academic year. Few research studies have attempted to account for changes in work intensity over time. Work schedules, such as the number of hours worked on school days compared to non-school days or the start and end times on school days, have not been taken into account in any research on the effects of work. Given the unpredictability of young people's work schedules, gathering these kinds of information is more challenging, and the datasets utilized to research work outcomes do not contain them.

7.1.5.1. Relationships

One of the main developmental tasks during adolescence is the development of interpersonal competencies, which include the ability to establish and maintain satisfying relationships. The majority of nationally representative surveys, however, do not gather data regarding the nature of the connections that youth have with their peers and families. Furthermore, the majority of research on the impact of work on relationships uses non-representative or local samples, except Monitoring the Future and the National Youth Survey, making it challenging to extrapolate the findings to the U.S. youngsters in general. However, despite their limitations, research on the significance of interpersonal relationships during adolescence offers helpful insights into the possible consequences of adolescent employment on relationships.

Teenagers are frequently employed in age-segregated environments. Young workers may interact with people their own age a lot, but their relationships with their coworkers may be superficial, and work environments may make it difficult for them to form lasting friendships. On the other hand, employment can elevate young employees' status and strengthen their bonds with their peers. Strong bonds with coworkers or superiors can balance out tense relationships with friends or family.

According to the National Youth Survey, students who worked longer hours—for a full year during their high school years—spent less time with their parents than their peers. Students who were employed spent less time with their parents than students who were not employed, and this difference decreased as the number of work hours rose. There was a correlation between the amount of time high school students in Wisconsin and California worked and their time spent with their families. Research shows that children who work more during high school become less dependent on their parents, which is a positive outcome that most parents consider. While being independent from parents can be harmful if not accompanied by adequate parental supervision, this is a typical developmental stage for teenagers.

Due to the high number of young people who work long hours, their employment may cause inconvenience to their families or, in the worst-case scenario, may lead to arguments or confrontations with their parents. Teenagers who work, for instance, might not be as available for household chores; parents of such adolescents might be asked to do additional duties that they wouldn't normally have to (e.g., escorting the teenagers to work, driving the paper carriers during inclement weather, and so on). Research utilizing information from the National Survey of Families and Households shows that parent-child arguments escalate when the kids have jobs. Parents of workers reported more disagreements with their spouses about money, school, getting along with the family, smoking, drinking, and drug use than did parents of nonworkers. According to data from Monitoring the Future, people who are not employed reported having the fewest arguments with their parents. Arguments about the number of hours worked rose to roughly 20 hours, after which they either decreased (for girls) or assumed an irregular pattern (for boys) for additional hours worked. The boys and girls who worked more hours also reported having more arguments with their parents in the Youth Development Study.

Though it doesn't seem to be the case, the data on the amount of time spent with the family and the frequency of arguments could imply that the overall quality of parent-child relationships is negatively impacted by teenage employment. According to Greenberger and colleagues (1980), there are no appreciable effects of adolescents' employment intensity or status on the caliber of their family relationships. The same result was obtained from an analysis of data from the Youth Development Study. However, there was some evidence in the later study that boys who were able to learn new skills for their jobs eventually grew closer to their fathers; this pattern was not observed for girls. Additionally, according to Schulenberg and Bachman (1993), developing, and applying skills at work was linked to a decrease in the perception of conflict between work and family life. As a result, high-quality work mitigated the challenges brought on by intense work. Conversely, as the intensity of the work increased, adolescents were more likely to perceive their jobs as interfering with other aspects of their lives if they did not believe that the jobs made good use of their talents and skills, and were unrelated to their future careers, or were the kinds of jobs that people only did for the money.

Positive work environments may mitigate the negative effects that stress in parent-adolescent relationships may have on the adjustment and mental health of adolescents, according to some research. Teens' self-esteem, mastery, and general well-being were all negatively impacted by parent-child relationship strain when they reported not receiving any support from their supervisors. However, such strain had no appreciable impact on these aspects of the adolescents' mental health when supervisory support was available. Interestingly, in view of these findings, more than one-third of employed students reported that they were "quite close" or "very close" to their supervisors and that the supervisors were "often" or "always" willing to listen to their problems and help them find solutions.

Examining how adolescent employment affects parent-child relationships requires consideration of the social context and significance of work. the correlations between parent-child relationships in rural and urban environments and the earnings of adolescents. In rural areas, the income of teenagers encouraged parents and teenagers to share advice more, and the teenagers' emotional relationships with their parents either stayed the same or got better. Earnings in urban environments were not associated with these favorable results. Young people in rural areas who worked more than 10 hours a week were more likely to spend time with their families than those who worked fewer than 10 hours. In cities, the amount of time spent with families fell as work hours increased. Although the majority of rural families are not farming families, young people who live and work with their parents on family farms may account for a portion of the rural effect. 34% of the families in this study lived on farms, 12% in rural areas without farms, and 54% in towns with fewer than 6500 residents.

The pattern is interpreted by the researchers in terms of ecological variations in the significance of the work produced by adolescents. Because rural work functions in distinct ways that benefit families, productive activities are more likely to be interpreted as adult-like behaviors in rural settings. Adolescent labor is more likely to be a part of the family economic strategy in rural areas;

in fact, earnings for adolescents were four times more likely to be allocated to family-supporting activities (e.g., donating money to parents, setting aside funds for further education, or covering tuition costs) in rural as opposed to urban areas. These kinds of spending were positively correlated with children's relationships with their parents in both urban and rural contexts, although these uses of earnings were less common in urban areas.

There is little evidence that an adolescent's work status or the intensity of their work has a positive or negative impact on their peer relationships, despite the fact that orientation to peers is a factor that influences whether or not they choose to work and that there is some indication that employed adolescents are more involved in dating. Good work experiences, on the other hand—those that promote job skills and work involvement—are linked to improved peer relationships and the conviction that one's work grants status within the peer group. It has also been discovered that strong links between the workplace and education support this idea. According to this data, adolescent employment that implies success and a move toward adult status also fosters the idea that employment elevates one's standing among peers.

7.1.6. Personal Development

One of the most important indicators of becoming an adult is the ability to work; assuming and keeping a job as an adult leads to financial independence and forms a significant part of one's adult identity. Since almost all teenagers—boys and girls alike—expect to work as adults, the adult worker role is probably going to be a prominent part of an adolescent's imagined future self. If working part-time represents a step closer to achieving that objective, one might anticipate that it would benefit mental and personal growth.

Teens who work have to balance being a worker, student, friend, and family member. If this is done well, the child may feel more capable overall and be better equipped to handle the demands of playing many roles in the future. Given the potential advantages of working for one's own growth, it is not surprising that employment for young people is widely supported. Many people, including parents of working teenagers, think that having a job during this stage of life fosters self-confidence, responsibility, and time management abilities. Indeed, young people who work report that they are more trustworthy, accountable, and on time than those who do not.

The possibility of role conflict arises when working students take on more roles, particularly when their work hours rise. Conflicts between work and other roles may cause adolescents to lose confidence in their abilities if they are unable to manage the demands of their roles. The likelihood of this unfavorable result would seem to rise with increased work hours. In the Monitoring the Future studies, high school seniors who worked six to 10 hours a week reported being the happiest in their lives, surpassing both nonworkers and those who worked more than 10 hours a week, even after adjusting for background variables and academic achievement.

While it is generally accepted that employment fosters better money management abilities, some have questioned whether it is advantageous to have comparatively high levels of discretionary income during this stage of life. Bachman (1983) issues a warning that taking a paid job early on may lead to later dissatisfaction with certain lifestyles and living standards. Young workers mostly buy items they want but may not actually need with their earnings.

Moreover, work may restrict the opportunities available to young people. According to Greenberger and Steinberg (1986), teenagers lose out on a crucial moratorium period during which they can explore alternate identities and interests because work frequently takes up a significant amount of time. This argument holds that employment diverts youth from activities that are better for their development. Schoenhals and colleagues (1997) conducted an analysis of the effects of working and working hours on how adolescents use their time.

They found that there was a strong negative correlation between the number of hours worked per week and the amount of time children spent watching television during the 10th grade—that is, the more hours worked, the less time they spent watching television. If watching television is the activity that is primarily sacrificed by employment—a task that involves little challenge, low demand, and typically little educational benefit—then this could explain the null findings about the impact of working on children's homework completion time.

Additionally, the investigators note that people who work longer shifts are less likely to eat breakfast. Additionally, if they work up to 25 hours a week, they are less likely to exercise vigorously; beyond that, their likelihood of exercising increases. Carskadon and colleagues have also discovered that high school students who work have less sleep and more daytime drowsiness; the effects intensify with the number of hours worked.

Researchers have looked into whether teenage psychological functioning is similarly responsive to the quality of work because adult psychological functioning has been found to be significantly impacted by the quality of adult work. Young workers might encounter stressors related to their jobs, particularly if they are expected to take on adult responsibilities before they are ready or if their coping mechanisms are still developing. Furthermore, juggling work and school may not be simple; teenagers generally view it as stressful, and this perception grows as they move through high school (Mortimer et al., 1994).

This is particularly true for the boys who work very hard and almost nonstop throughout high school. Resnick and colleagues (1997) discovered that working more than 20 hours per week during the school year was linked to emotional distress (defined as physical or emotional symptoms of distress as reported by the young people themselves or by their parents). This analysis was conducted using data from the first year of the National Longitudinal Study on Adolescent Health.

There is proof that children's mental health is, in fact, greatly impacted by the caliber of their work. In the Youth Development Study, girls' and boys' good pay and prospects for advancement and a work-school balance over time increased their sense of self-efficacy (Finch et al., 1991). Girls' sense of efficacy increased when they were given early opportunities to be helpful to others at work, whereas boys' sense of efficacy increased when their supervisors included them in discussions about work tasks and did not subject them to close supervision. Stressors related to the job (e.g., time constraints, overload, and early decision-making autonomy at work increased their distress, but learning practical skills improved their melancholy attitudes. Stress at work and taking accountability for circumstances beyond their control were linked to higher rates of depression and lower levels of efficacy in females. The Youth Development Study provides more proof that the caliber of a student's work during high school continues to affect their personal outcomes (e.g., for coping strategies for issues at work, wellbeing, and depression) four years after high school.

Furthermore, the effects of working hours may change depending on the caliber of the work experience. That is, the possible detrimental effects of working long hours on personal development may be mitigated when the work is of a high caliber. Therefore, when autonomy and role clarity were high, hours worked were positively associated with self-esteem in a cross-sectional study of urban white Canadians in grades 10 through 12. Compared to other participants, senior participants in the Monitoring the Future studies who had jobs that allowed them to use their skills and learn new ones expressed higher levels of life satisfaction and optimism about the future. Individuals whose occupations aligned with their aspirations for the future were less vulnerable to challenges associated with extended work hours. Conversely, individuals who perceived minimal correlation between their current and prospective occupations linked a rise in work intensity to declines in their overall health and well-being. Low-quality work can therefore compound with long hours to have detrimental effects on personal growth. For instance, Shanahan (1992) found that boys who worked more than the median number of hours in the tenth and twelfth grades experienced higher levels of depression due to work stress and a lack of supervision.

7.1.7. Problem Behaviors

Studies find the strongest evidence of the negative effects of teen employment in the group of behaviors commonly referred to as problem behaviors. The overwhelming body of research—from cross-sectional studies using representative sampling designs to national longitudinal studies—has revealed that young people who work—especially those who work at high intensity—have higher rates of problem behaviors, such as minor delinquency and the use of alcohol and other drugs. Even after statistically adjusting for additional problem behavior correlates, these results hold true. As previously mentioned, these studies do not lend themselves to experimental designs because they were carried out in natural settings.

Thus, it is impossible to draw firm conclusions about the relationship between high-intensity work and behavioral issues in young people. But considering the findings' consistency across different youth populations and settings, it appears highly likely that high-intensity work actually plays a role in young people's problem behaviors. Students who have jobs are more likely to participate in deviant behavior and school misconduct than youth without jobs. Wofford (1988) discovered, using data from the National Youth Survey, that minor delinquency is higher among teenagers who work full-time compared to those who work part-time, and higher among those who work part-time compared to those who do not work at all. On the other hand, teenagers who don't work at all typically commit the more serious crimes.

Long work hours have also been connected in males to aggressive behavior, theft, and run-ins with the law, particularly for those who put in more than 30 hours a week. Using data from Monitoring the Future, these researchers statistically adjusted for the impacts of parental education, race, region, high school grade point average, 4-year college plans, and high school curriculum. Higher-level employees (i.e., found that high work intensity increased the likelihood of delinquent involvement among males who were already at risk for delinquent behavior: that is, the more hours such high-risk adolescent males worked, the more likely they were to engage in delinquent behavior. However, the majority of the incidents occurred in or near schools. For low-risk males or females, the number of work hours did not directly affect the likelihood of involvement in delinquency. Using data from the 1988 National Survey of Families and Households, a cross-sectional, nationally representative sample of 13,079 people living in 9643 households, the researchers examined the connection between work intensity and delinquency. The parents of the teenagers who participated in this study were the respondents.

Research has shown that teenagers who work more than 20 hours a week are more likely to use alcohol, tobacco products, and illegal drugs like cocaine and marijuana. Learners in St. Paul, Minnesota, who put in more than 20 hours a week at work consumed more alcohol annually than their peers. One of the most compelling findings in this field is the correlation between intensive work and substance use, which holds true even after applying numerous statistical controls to account for background variables and pre-existing variations in substance use among the groups. For instance, Mortimer, and colleagues (1996) discovered that the number of hours worked in high school was linked to alcohol consumption. This relationship persisted even after statistical adjustments were made for sex, parental socioeconomic status, race, family composition, and nativity, as well as the frequency of prior alcohol consumption. Statistically controlling for sex, age, parental socioeconomic status, place of residence (urban/suburban, rural), ethnicity, and prior drug and alcohol use (at first interview), the authors examined the short-term effects of work hours on the use of alcohol, marijuana, and other drugs among 11–17-year-olds from 1976 through 1980. They discovered that employment had an impact on marijuana and alcohol use above and beyond any preemployment differences.

While long hours are linked to higher rates of substance abuse and other problem behaviors, not much research has looked at how work experience directly impacts these outcomes or modifies the effects of employment. In a sample of California high school students, Steinberg, and Cauffman (1995) found a correlation between the use of drugs and alcohol by adolescents and work-related stressors. Schulenberg and Bachman (1993) discovered that skill utilization at work was linked to a decrease in alcohol use for female adolescents and a decrease in cigarette and marijuana use for adolescents as a whole, based on data from Monitoring the Future. Furthermore, as the intensity of their work increased, adolescents who described their jobs as not requiring the use of their skills, as being unconnected to the future, and as the kind of work that people do just for the money smoked more cigarettes. Young people who said their jobs were relevant to their futures were found to be less affected by work intensity.

7.2. ADOLESCENT DEVELOPMENT

Teenagers nowadays have a difficult time making healthy eating choices because ultra-processed foods (UPFs) are so widely consumed. Due to their industrial manufacturing processes and use of ingredients, these foods frequently have high added sugar, fat, and artificial additive content. Considering their possible detrimental effects on health and development, the frequency of UPFs in adolescent diets is concerning. Research has indicated that diets rich in unsaturated polyphenols (UPFs) are associated with several health problems, such as obesity, type 2 diabetes, cardiovascular disorders, and cognitive decline. Your child's exciting but stressful transition from childhood to adulthood occurs during adolescence. It's not always a smooth ride during this period of rapid changes and developments.

As your child goes through physical, cognitive, and psychological changes, they will start to figure out who they are and where they fit in. There are multiple reasons for the increase in UPF usage among teenagers. First off, these foods are a great option for busy teenagers due to their affordability and ease of use. Ultra-processed snacks and meals are frequently aggressively marketed to younger consumers with eye-catching packaging and branding. These products are also made to be extremely tasty, with ratios of sugars, fats, and salts that cause the brain to respond pleasurably, making it hard to resist them. Teenagers may consequently discover that they prefer these foods over wholesome alternatives like fruits, vegetables, and whole grains.

Adolescents who use UPFs are affected in ways other than their physical health. A high intake of these foods may also have negative effects on academic performance and mental health. High UVF intake has been linked in studies to an increased risk of mood disorders, such as anxiety and depression. Furthermore, UPFs' lack of nutrients can result in vitamin and mineral deficiencies, which may have an impact on academic achievement and cognitive development. Adolescents' dietary decisions are especially important because they are going through a crucial stage of growth and development. To tackle the problem of UPF consumption, a comprehensive strategy is needed.

It is essential to teach teenagers the advantages of eating whole, minimally processed foods. Better food options and the promotion of healthier eating habits are important roles that parents and schools can play. Adolescents can also be guided toward healthier options by policy changes like banning the marketing of UPFs to minors and enhancing food labels. Encouraging teens to cook and raising their knowledge of the health risks associated with UPFs can equip them to make healthier food choices. Adolescents who frequently consume highly processed foods run the risk of developing major health problems that could impair their physical and emotional wellbeing. In order to address this problem and improve young people's general health outcomes, a combination of supportive environments, policy changes, and education are needed to encourage healthier eating habits.

7.2.1. What is adolescence?

The transitional stage between childhood and adulthood is known as adolescence. As they approach adolescence, children's bodies and minds undergo numerous changes. These encompass challenges that are social, psychological, intellectual, and physical in addition to the formation of their own moral compass. The changes happen quickly and frequently at various rates. Adolescence can be an exciting and demanding time in a person's life. Your child starts to explore their identity and grows more self-reliant during adolescence.

7.2.2. The Physical Changes of Adolescence

During adolescence, physical development involves changes brought on by the puberty process. Certain hormones are released by your child's brain during puberty. Your child's body changes physically as a result of the hormones, and their sexual organs mature. Your child is probably going to go through a growth spurt.

They will experience rapid weight and height growth during this time. Additional physical alterations could be an increase in body hair, acne, and body odor. In comparison to boys and adolescents assigned male at birth (AMAB), girls, and adolescents assigned female at birth (AFAB) typically experience growth spurts earlier. The majority of AFAB girls and teenagers experience growth spurts between the ages of 10 and 14. Growth spurts typically occur in boys and adolescents AMAB between the ages of 14 and 17. Breast development will start in girls and adolescents who are AFAB.

This should begin by the age of 14, but it can occur as early as age 10. Additionally, they will go through their first menstrual cycle, which typically occurs two years after breast and pubic hair become noticeable. The penis and testicles of boys and adolescents AMAB will grow. They'll start having ejaculations and erections. As demonstrated by an ultrasound during pregnancy, erections can also occur regularly from early infancy to old age.

Everyone experiences these physical changes, though each person's experience will differ in terms of timing and sequence. While some teenagers mature sooner than others, others mature later. The additional stress of standing out among their peers can arise for those who fall on either end of this spectrum.

An adolescent medicine specialist if puberty is occurring early (before age 8 for girls and adolescents AFAB and before age 9 for boys and adolescents AMAB) or late (after age 14 for girls and adolescents AFAB and after age 15 for boys and adolescents AMAB). They can assist in treating and managing this puberty-related issue. Ignoring these issues may have an effect on the growth and development of bones.

7.2.2.1. What are the Cognitive Changes of Adolescence?

Adolescent brain development is more advanced than that of a child's. Youngsters can only reason about the tangible, the present moment. Adolescents can think beyond these constraints and consider possibilities for truth rather than simply accepting things as they are. They are able to see infinite possibilities, test hypotheses, and work with abstractions. Adolescents nevertheless frequently exhibit egocentric attitudes and behaviors. Adolescents' cognitive development is characterized by the rapid growth of many neurons. There is an increase in the connections between these bundles of nerves in your child's body. This makes it possible to think more deeply and intelligently.

7.2.2.2. Which Part of the Brain Develops Last in Adolescence?

One of the last areas of your child's brain to fully develop is the frontal cortex, located in the front of the brain. It will continue to develop until your child is in their mid- to late-20s. Planning, setting priorities, and impulse control are examples of executive functions that are governed by this region of the brain. Your teenager might have made mistakes in judgment because it manifests itself so late. Mood fluctuations and risk-taking behaviors might increase.

This type of thinking is known as "hot cognition," and it occurs when a teen acts impulsively without using their frontal cortex. Using your logical mind is what is meant by "cold cognition," not being "cold." A young person can be helped by parents to shift from "hot" to "cold" thinking by asking questions instead of giving lectures, showing empathy, and not having high expectations.

7.2.3. Which Mental Characteristic Develops Over Adolescence?

Mental characteristics that develop during adolescence include improved:

- Abstract thinking.
- Reasoning skills.
- Impulse control.
- Creativity.
- Problem-solving abilities.
- Decision-making skills.

7.2.3.1. What Are The Emotional Changes of Adolescence?

Your child will learn to recognize, quantify, and control their emotions during adolescence. This implies that they'll start to become more conscious of both their own and other people's feelings. Your child's emotional development process will enable them to hone their abilities and find their special traits. Some teenagers embrace these new challenges as they grow more self-reliant. Others might require extra assistance in order to boost their confidence.

7.2.3.2. How Does Self-Esteem Develop in Adolescence?

Your child's self-esteem may be impacted by the physical, hormonal, and emotional changes they go through during adolescence. Adolescents who exhibit delayed or early development in comparison to their contemporaries may experience body dysphoria. Their self-esteem grows increasingly dependent on their ability to fit in. Self-worth is not always simple. In the presence of their families, some teenagers may exhibit high self-esteem, but not in the presence of their peers.

Adolescents benefit most from having a parent who is a "lighthouse" rather than a "snowplow parent" who clears the path of obstacles for them or a "helicopter parent" who comes in and saves the day. When it comes to matters of safety or ethics, this type of parent sets boundaries and lets their child practice making their own decisions. For teenagers, being a lighthouse-wielding adult with compassion can change their entire lives.

While a challenging part of adolescence, it's important that your child learns to accept who they are and gains a sense of capability. They can develop their self-esteem by:

- Making mistakes.
- Learning from their mistakes.
- Holding themselves accountable for their actions.

7.2.3.3. What Are The Social Changes of Adolescence?

During this period, adolescents are also developing socially. The quest for identity is the most significant social development task during adolescence. This is frequently a journey that begins in adolescence and lasts a lifetime.

Along with the search for identity comes the struggle for independence. Your child may:

- Develop an interest in their sexuality and romantic relationships.
- Turn to you less in the midst of a challenge.
- Show more independence from you.
- Spend less time with you and more time with their friends.
- Feel anxious, sad or depressed, which can lead to trouble at school or risk-taking behaviors.

7.2.4. Identity Development in Adolescence

When your child finds a strong sense of self and personality, as well as a connection to others, identity development takes place. A healthy sense of self-identity is crucial because it affects how your child will always feel like they fit in.

A positive self-identity is also associated with higher self-esteem. You can help reinforce a positive self-identity in your child by:

- Encouraging their efforts.
- Praising their good choices.
- Inspiring perseverance.

7.2.4.1. How Does Social Media Affect Adolescent Development?

Social media can have a detrimental effect on the growth and well-being of your child. Teenagers report being exposed to offensive content online and experiencing cyberbullying. Furthermore, socializing in person differs from socializing online. Adolescents are blind to important body language cues and facial expressions that are only visible when they interact in person. Adolescents who compare themselves to people they see online may also experience self-esteem issues. Depression, anxiety, and low self-esteem can all result from these factors.

7.2.4.2. What Are the Moral Changes of Adolescence?

Your child may begin to think more deeply and abstractly about the world during adolescence. This influences your child's perspective on the world and their desired interactions with it. Additionally, morals, and values that your child will uphold throughout their life will start to develop. Your youngster might start to realize that not all choices are clear-cut.

Empathy will grow in them as they start to understand why other people make decisions that are different from their own. Additionally, they'll start to comprehend the rationale behind the world's rules on a deeper level. They will begin to develop their own ideas about what is right and wrong. They might also ponder over their spirituality and religious convictions. When you have the chance, encourage your child to have these discussions. As your child practices problem-solving and thinking through situations, practice listening and learning.

7.2.4.3. Can ADHD Develop in Adolescence?

Children with attention-deficit/hyperactivity disorder (ADHD) are usually diagnosed by medical professionals. However, some people don't get diagnosed until they're in their teens or even older years. Adolescents who have undiagnosed ADHD may experience worsening symptoms as they approach puberty. Undiagnosed learning disorders or abnormal ways of processing information are present in other teenagers. Dropping grades could indicate undiagnosed ADHD, learning disabilities, or stress-related symptoms like depression.

7.2.4.4. How Can Parents Support Healthy Adolescent Development?

Adolescence can be a trying period for both you and your child. But your home doesn't have to become a battleground if you make special efforts to understand one another. The following guidelines may help parents:

1. Give your child your undivided attention when they want to talk. Don't read, check your email, watch television or busy yourself with other tasks.
2. Listen calmly and concentrate on hearing and understanding your child's point of view. Reflect back on what you have heard.
3. Speak to your child as courteously and pleasantly as you would to a stranger. Your tone of voice can establish the mood of a conversation.
4. Understand your child's feelings, even if you don't always approve of their behavior. Try not to make judgments. Keep the door open on any subject. Be an "open/approachable" parent.
5. Avoid humiliating your child. Don't laugh at what may seem to you to be naive or foolish questions and statements.
6. Encourage your child to "test" new ideas in conversation. Don't judge their ideas and opinions. Instead, listen, and then offer your own views as plainly and honestly as possible. Love and mutual respect can coexist with differing points of view.
7. Help your child build self-confidence. Encourage their participation in activities of their choice (not yours).
8. Make an effort to commend your child frequently and appropriately. Too often, we take the good things for granted and focus on the bad. Your child needs to know you appreciate them.
9. Catch them doing something right and encourage them to do more of that.
10. Encourage your child to participate in family decision-making and to work out family concerns together with you. Understand your child needs to challenge your opinions and your ways of doing things. This is how they achieve the separation from you that's essential for their own adult identity.

7.2.4.5. What Can Adolescents Do During This Time?

1. Avoid looking at your parents as the enemy. Chances are they love you and have your best interests in mind, even if you disagree with their way of showing that.
2. Try to understand that your parents are human beings with their own insecurities, needs, and feelings.
3. Listen to your parents with an open mind. Try to see situations from their point of view.
4. Share your feelings with your parents so they can understand you better.
5. Live up to your responsibilities at home and in school. That way, your parents will be more inclined to grant you the kind of independence you want and need.
6. Bolster your criticisms of family, school, and government with suggestions for practical improvements.
7. Be as courteous and considerate to your parents as you would be to your friend's parents (Blum and Nelson-Mmari, 2004).

7.3. IMPACT ON GROWTH, DEVELOPMENT, AND LEARNING

Considering that adolescents are in a critical stage of physical and cognitive development, ultra-processed foods (UPFs) have a significant effect on learning, growth, and development. These foods can seriously harm teenage health in several ways because of their high added sugar, unhealthy fat content, artificial additives, and low nutritional value.

1. **Growth and Development**: The body experiences quick physical changes during adolescence, such as growth spurts and the expansion of bone and muscle mass. Consuming a lot of UPF foods can impede these procedures. UPFs are generally deficient in vital nutrients, which are necessary for normal growth and include vitamins, minerals, and proteins. For example, insufficient consumption of calcium, vitamin D, and protein can have an impact on bone density and overall development of the body. Furthermore, UPFs' high added sugar and unhealthy fat content can exacerbate weight gain and obesity. Adolescent obesity can cause long-term health problems, such as metabolic syndrome and cardiovascular diseases, which can make healthy growth even more difficult.

2. **Cognitive Development**: Adolescents' cognitive development is extremely responsive to dietary influences. Nutrients that promote brain health, like antioxidants, different vitamins, and omega-3 fatty acids, are frequently lacking in UPFs. Research indicates that diets high in UPFs may have a deleterious effect on executive functioning, memory, and attention. For instance, a high-sugar diet has been connected to memory loss and learning challenges. Furthermore, consuming UPFs is linked to unhealthful eating habits that might deprive the body of vital nutrients needed for the best possible development of the brain, which could have an impact on cognitive function and academic achievement.
3. **Learning and Academic Performance**: Dietary decisions can have an impact on academic performance, and UPFs are important in this context. Many UPFs have a high glycemic index, which can cause blood sugar fluctuations. These fluctuations can impair focus and cause fatigue, which can negatively impact academic performance. In addition, consuming UPFs frequently takes the place of better food choices that offer long-lasting energy and concentration. A diet high in UPF can cause nutritional deficiencies that can affect mood and brain function. These deficiencies can cause problems like irritability and low motivation, which can further hinder academic success. Ultra-processed foods negatively impact adolescents' learning, development, and growth. Their poor nutritional value and high content of unhealthy ingredients can be detrimental to one's physical and mental well-being as well as academic achievement. In order to address these issues, we must move toward advocating nutrient-rich, balanced diets that support both developmental needs and general well-being.

7.3.1. Risks of Childhood Obesity and Related Diseases

The rising prevalence of childhood obesity is largely attributed to the consumption of UPFs, which is a serious public health concern with substantial long-term health implications. UPFs, or unhealthy processed foods, are products that are low in vital nutrients and high in added sugars, unhealthy fats, and salt. Examples of these products include sugary cereals, snack foods, fast foods, and sugary beverages. They are a major contributor to excessive calorie intake and poor dietary quality due to their widespread availability and kid-friendly appeal. There is ample evidence linking UPF consumption to childhood obesity.

These foods frequently have high-calorie content but low nutritional value, which causes an imbalance between energy intake and use. Kids who eat a lot of UPFs are more likely to eat more than they need each day, which could result in weight gain. These foods' high sugar and fat contents encourage the buildup of fat, especially around the abdomen, which is a major sign of obesity. Overweight gain in childhood not only has an impact on one's immediate health but also lays the groundwork for many chronic illnesses in the future.

Children who are obese are more likely to develop several serious health conditions. Type 2 diabetes is a major risk factor that is becoming more prevalent in younger populations. Excess body fat, especially visceral fat, causes insulin resistance, which impairs the body's ability to control blood sugar levels. Type 2 diabetes and high blood glucose levels can eventually result from this insulin resistance. Moreover, dyslipidemia (abnormal lipid levels) and hypertension (high blood pressure), which both raise the risk of cardiovascular diseases like heart attacks and strokes, are associated with obesity.

These conditions can cause health problems that last a lifetime if they manifest early. In addition to metabolic diseases, childhood obesity raises the risk of musculoskeletal issues. Being overweight puts extra strain on developing bones and joints, which can result in problems like arthritis, joint pain, and decreased mobility. Additionally, obesity is linked to obstructive sleep apnea, a disorder that causes breathing to stop and start repeatedly while you sleep, resulting in poor sleep quality and additional health issues.

There are also significant psychosocial effects of obesity. Children who struggle with being overweight are frequently the targets of stigma and bullying, which can result in psychological problems like anxiety, depression, and low self-esteem. These mental health issues can exacerbate the physical health issues brought on by obesity, perpetuating a vicious cycle of ill health and a lower standard of living.

Reducing the use of UPFs and promoting healthier eating practices are two of the many strategies needed to combat childhood obesity.

Policies, parents, and schools all have important roles to play in encouraging a healthy diet. Children can choose healthier diets if educational programs highlight the advantages of whole foods and the dangers of UPFs. Furthermore, healthier eating environments can be promoted by regulatory actions like restricting the marketing of UPFs to children and enhancing food labeling. The rise in childhood obesity and related diseases is largely attributed to the consumption of ultra-processed foods.

The risks of obesity for one's health, such as type 2 diabetes, cardiovascular disease, musculoskeletal disorders, and mental health difficulties, underscore the pressing need for all-encompassing approaches to lower UVF intake and encourage better eating practices in kids. By tackling this problem comprehensively, we can enhance general health outcomes and lay the groundwork for a healthier generation of people in the future.

7.3.2. Behavioral Issues and Academic Performance

There is growing evidence connecting children's and adolescents' consumption of UPFs to behavioral problems and poor academic performance. Fast food, sugary snacks, and highly processed convenience meals are examples of UPFs. These foods are low in vital nutrients and high in added sugars, unhealthy fats, and artificial additives. This combination may have major effects on behavior as well as cognitive function, which may ultimately compromise a child's performance in an educational environment.

An increase in hyperactivity and attention deficits is one of the main behavioral issues linked to high UPF intake. Many UPFs have a high sugar content, which can cause blood glucose levels to fluctuate and cause energy spikes and crashes. These abrupt shifts can cause irritability, mood swings, and difficulty focusing. Research has indicated a connection between high sugar consumption and symptoms that mimic attention deficit hyperactivity disorder (ADHD), including hyperactivity, impulsivity, and inattention. These behavioral problems may have an immediate effect on a child's capacity to participate in class activities and complete assigned readings.

Furthermore, there is a link between UPF consumption and a higher risk of emotional and psychological problems like depression and anxiety. Diets heavy in processed foods and low in vitamins, antioxidants, and omega-3 fatty acids—nutrients critical to brain health—can impair mood regulation and cognitive performance. For example, low levels of essential fatty acids can affect neurotransmitter function and lead to mood disorders. Essential fatty acids are critical for brain health. Children who are depressed and anxious can have trouble focusing, be less motivated, and have a harder time processing information, all of which can negatively impact their academic performance.

Notable is also the effect of UPFs on cognitive function. Diets deficient in nutrients and high in ultraviolet photochemicals (UPFs) can cause deficiencies in important vitamins and minerals for brain development. For instance, low levels of certain nutrients, like iron, zinc, and B vitamins, have been linked to memory problems, sluggish cognitive processing, and poorer academic performance. A diet high in unprocessed plant foods (UPFs) can impair a child's capacity to carry out intricate cognitive functions, remember knowledge, and solve problems.

A lack of academic progress, low test scores, and bad grades are some manifestations of this cognitive impairment. Moreover, UPF consumption frequently replaces healthier food choices that enhance cognitive function and offer long-lasting energy. Because stable glucose levels are necessary for the brain to function at its best, meals heavy in processed sugars and fats can cause problems with sustained attention and focus.

Lower classroom performance and academic achievement may result from the ensuing cognitive fatigue and diminished mental acuity. To address these problems, a multifaceted strategy that promotes nutrient-rich, balanced diets and reduces UPF consumption is needed. Parents and schools can collaborate to give kids access to wholesome foods and teach them good eating habits.

Furthermore, policy initiatives like limiting the promotion of UPFs to minors and improving food labels can encourage the consumption of healthier foods. By concentrating on these tactics, UPFs' negative effects on behavior and academic performance can be lessened, improving children's and adolescents' overall outcomes.

7.3.3. Long-term Effects of Early-Life Exposure

Public health research is beginning to focus more on the long-term effects of early exposure to ultra-processed foods (UPFs), emphasizing the significant implications for mental and physical health across the lifespan. Ultra-processed foods are low in vital nutrients and high in unhealthy fats, added sugars, and artificial additives. These foods can have long-term negative effects on a person's health, even after childhood.

The chance of acquiring chronic illnesses is one of the most important long-term effects of early-life exposure to UPFs. Regular UPF consumption increases a child's risk of obesity, which can last into adulthood. The high calorie density and inadequate nutritional value of such foods lead to excessive weight gain and fat accumulation. Obesity frequently increases the chance of later in life developing diseases like type 2 diabetes, cardiovascular disorders, and specific types of cancer. Early obesity can cause metabolic changes, such as insulin resistance and inflammation, that can have long-lasting effects on health and increase an individual's susceptibility to chronic illnesses as they age.

Concerningly, UPFs have an effect on cardiovascular health. Eating unprocessed plant foods (UPFs) is linked to increased consumption of added sugars and unhealthy fats, which can cause dyslipidemia, hypertension, and endothelial dysfunction. Heart disease and stroke are conditions for which these are risk factors. Research indicates that early exposure to these dietary patterns may lay the groundwork for the onset of atherosclerosis and other cardiovascular issues in later life, resulting in a progressive decline in heart health. Beyond obesity, early-life UPF consumption has an impact on metabolic health. Consuming a lot of sugar is associated with a higher chance of developing metabolic syndrome, which is a collection of disorders that includes high blood pressure, elevated blood sugar, and abnormal cholesterol levels. Early UPF exposure can cause metabolic disruptions that can impair lipid metabolism and glucose regulation over time, raising the risk of type 2 diabetes and cardiovascular diseases. Moreover, UPFs have an impact on cognitive function and mental health.

An increased chance of mood disorders like depression and anxiety later in life has been linked to early exposure to diets low in nutrients. The inflammatory properties of UPFs coupled with a diet deficient in vital nutrients can cause brain damage and cognitive decline. These consequences may show up as issues with memory, learning, and emotional control, which can affect an adult's overall quality of life and mental health.

Long-term impacts may also result from early UPF exposure's behavioral and educational implications. Poor dietary habits as a child can cause them to prefer UPFs as adults, continuing their unhealthy eating habits. This dietary choice may have an impact on long-term health consequences and feed the vicious cycle of inadequate nutrition and chronic illness.

It is imperative to address early dietary habits through policy and education to mitigate these long-term effects. Healthy eating habits can be established from an early age by encouraging the consumption of whole, minimally processed foods and limiting the availability and marketing of UPFs. Promoting healthier eating habits and enhancing long-term health outcomes requires the implementation of public health campaigns, educational initiatives, and welcoming food environments. By concentrating on these tactics, we can lessen the likelihood of chronic diseases linked to early exposure to highly processed foods and break the cycle of poor nutrition (Baron-Cohen et al., 1999).

REFERENCES

1. Andersen, S., Thompson, A., Krenzel, E., & Teicher, M. (2002). Pubertal changes in gonadal hormones do not underlie adolescent dopamine receptor overproduction. *Psychoneuroendocrinology, 27*, 683–691.
2. Babalola, S. (2004). Perceived peer behavior and the timing of sexual debut in Rwanda: A survival analysis of youth data. *Journal of Youth and Adolescence, 33*, 353–363.
3. Baird, A., Fugelsang, J., & Bennett, C. (2005). What were you thinking? An fMRI study of adolescent decision making. Poster presented at the 12th Annual Cognitive Neuroscience Society (CNS) Meeting, New York.
4. Baird, A., Gruber, S., Fein, D., Maas, L., Steingard, R., Renshaw, P., et al. (1999). Functional magnetic resonance imaging of facial affect recognition in children and adolescents. *Journal of the American Academy of Child and Adolescent Psychiatry, 38*, 195–199.
5. Baron-Cohen, S., Tager-Flusberg, T., & Cohen, D. (1999). *Understanding Other Minds: Perspectives from Developmental Cognitive Neuroscience*. New York: Oxford University Press.
6. Bearman, P., Jones, J., & Udry, J. R. (1997). The National Longitudinal Study of Adolescent Health: Research Design. Chapel Hill, NC: Carolina Population Center.
7. Bechara, A. (2005). Decision making, impulse control and loss of willpower to resist drugs: A neurocognitive perspective. *Nature Neuroscience, 8*, 1458–1463.
8. Bechara, A., Damasio, A., Damasio, H., & Anderson, S. (1994). Insensitivity to future consequences following damage to human prefrontal cortex. *Cognition, 50*, 7–15.
9. Berg, W., & Byrd, D. (2002). The Tower of London spatial problem-solving task: Enhancing clinical and research implementation. *Journal of Experimental and Clinical Neuropsychology, 25*, 586–604.
10. Berridge, K. C. (2003). Comparing the emotional brain of humans and other animals. In R. J. Davidson, H. H. Goldsmith, & K. Scherer (Eds.), *Handbook of Affective Sciences* (pp. 25–51). Oxford: Oxford University Press.
11. Beyth-Marom, R., Austin, L., Fischhoff, B., Palmgren, C., & Jacobs-Quadrel, M. (1993). Perceived consequences of risky behaviors: Adults and adolescents. *Developmental Psychology, 29*, 549–563.
12. Bjork, J., Knutson, B., Fong, G., Caggiano, D., Bennett, S., & Hommer, D. (2004). Incentive-elicited brain activation in adolescents: Similarities and

differences from young adults. *Journal of Neuroscience, 24*, 1793–1802.

13. Blum, R., & Nelson-Mmari, K. (2004). The health of young people in a global context. *Journal of Adolescent Health, 35*, 402–418.
14. Brooks-Gunn, J., & Furstenberg, F. F., Jr. (1989). Adolescent sexual behavior. *American Psychologist, 44*, 249–257.
15. Brown, B. (2004). Adolescents' relationships with peers. In R. Lerner & L. Steinberg (Eds.), *Handbook of Adolescent Psychology* (Vol. 2, pp. 363–394). New York: Wiley.
16. Brown, T., Lugar, H., Coalson, R., Miezin, F., Petersen, S., & Schlaggar, B. (2005). Developmental changes in human cerebral functional organization for word generation. *Cerebral Cortex, 15*, 275–290.
17. Buchanan, C., Eccles, J., & Becker, J. (1992). Are adolescents the victims of raging hormones? Evidence for activational effects of hormones on moods and behavior at adolescence. *Psychological Bulletin, 111*, 62–107.
18. Cameron, J. (2004). Interrelationships between hormones, behavior, and affect during adolescence: Understanding hormonal, physical, and brain changes occurring in association with pubertal activation of the reproductive axis. *Annals of the New York Academy of Sciences, 1021*, 110–123.
19. Casey, B. J., Giedd, J., & Thomas, K. (2000). Structural and functional brain development and its relation to cognitive development. *Biological Psychology, 54*, 241–257.
20. Casey, B. J., Tottenham, N., Liston, C., & Durston, S. (2005). Imaging the developing brain: What have we learned about cognitive development? *Trends in Cognitive Science, 9*, 104–110.
21. Cauffman, E., Steinberg, L., & Piquero, A. (2005). Psychological, neuropsychological, and psychophysiological correlates of serious antisocial behavior in adolescence: The role of self-control. *Criminology, 43*, 133–176.
22. Centers for Disease Control and Prevention. (2006). Youth risk behavior surveillance—United States, 2005. *Morbidity & Mortality Weekly Report, 55*(SS5), 1–108.
23. Chambers, R., Taylor, J., & Potenza, M. (2003). Developmental neurocircuitry of motivation in adolescence: A critical period of addiction vulnerability. *American Journal of Psychiatry, 160*, 1041–1052.
24. Chassin, L., Hussong, A., Barrera, M., Jr, Molina, B., Trim, R., & Ritter, J. (2004). Adolescent substance use. In R. Lerner & L. Steinberg (Eds.), *Handbook of Adolescent Psychology* (Vol. 2, pp. 665–696). New York: Wiley.

25. Chibbar, R., Toma, J., Mitchell, B., & Miller, F. (1990). Regulation of neural oxytocin gene expression by gonadal steroids in pubertal rats. *Molecular Endocrinology, 4*, 2030–2038.
26. Cicchetti, D., & Dawson, G. (2002). Editorial: Multiple levels of analysis. *Development and Psychopathology, 14*, 417–420.
27. Collins, W. A., & Steinberg, L. (2006). Adolescent development in interpersonal context. In N. Eisenberg, W. Damon, & R. Lerner (Eds.), *Handbook of Child Psychology* (pp. 1003–1067). New York: Wiley.
28. Conklin, H., Luciana, M., Hooper, C., & Yarger, R. (2007). Working memory performance in typically developing children and adolescents: Behavioral evidence of protracted frontal lobe development. *Developmental Neuropsychology, 31*, 103–128.
29. Crone, E. A., & van der Molen, M. W. (2004). Developmental changes in real-life decision making: Performance on a gambling task previously shown to depend on the ventromedial prefrontal cortex. *Developmental Neuropsychology, 25*, 251–279.
30. Crone, E. A., Donohue, S., Honomichl, R., Wendelken, C., & Bunge, S. A. (2006). Brain regions mediating flexible rule use during development. *Journal of Neuroscience, 26*, 11239–11247.
31. Dahl, R. E. (2004). Adolescent brain development: A period of vulnerabilities and opportunities. *Annals of the New York Academy of Sciences, 1021*, 1–22.
32. Darling, N., & Steinberg, L. (1993). Parenting style as context: An integrative model. *Psychological Bulletin, 113*(3), 487–496.
33. Diamond, J. (1992). *The Third Chimpanzee: The Evolution and Future of the Human Animal*. New York: HarperCollins.
34. DiBlasio, F., & Benda, B. (1992). Gender differences in theories of adolescent sexual activity. *Sex Roles, 27*, 221–240.
35. DiIorio, C., Dudley, W. N., Kelly, M., Soet, J., Mbwara, J., & Sharpe Potter, J. (2001). Social cognitive correlates of sexual experience and condom use among 13- through 15-year-old adolescents. *Journal of Adolescent Health, 29*, 208–216.
36. Drevets, W. C., & Raichle, M. E. (1998). Reciprocal suppression of regional cerebral blood flow during emotional versus higher cognitive processes: Implications for interactions between emotion and cognition. *Cognition and Emotion, 12*, 353–385.
37. Dumont, N., Andersen, S., Thompson, A., & Teicher, M. (2004). Transient dopamine synthesis modulation in prefrontal cortex: *In vitro* studies.

Developmental Brain Research, 150, 163–166.

38. Durston, S., Davidson, M., Tottenham, N., Galvan, A., Spicer, J., Fossella, J., & Casey, B. J. (2006). A shift from diffuse to focal cortical activity with development. *Developmental Science, 9*, 1–20.
39. Durston, S., Thomas, K. M., Yang, Y., Ulug, A. M., Zimmerman, R. D., & Casey, B. J. (2002). A neural basis for the development of inhibitory control. *Developmental Science, 5*, 9–16.
40. East, P., Felice, M., & Morgan, M. (1993). Sisters' and girlfriends' sexual and childbearing behavior: Effects on early adolescent girls' sexual outcomes. *Journal of Marriage and the Family, 55*, 953–963.
41. Eluvathingal, T., Hasan, K., Kramer, L., Fletcher, J., & Ewing-Cobbs, L. (2007). Quantitative diffusion tensor tractography of association and projection fibers in normally developing children and adolescents. *Cerebral Cortex.*
42. Ennett, S. T., Tobler, N. S., Ringwalt, C. L., & Flewelling, R. L. (1994). How effective is drug abuse resistance education? A meta-analysis of Project DARE outcome evaluations. *American Journal of Public Health, 84*, 1394–1401.
43. Ernst, M., Nelson, E. E., Jazbec, S., McClure, E., Monk, C., Blair, R. J. R., Leibenluft, E., Blair, J., & Pine, D. S. (2005). Amygdala and nucleus accumbens activation in response to receipt and omission of gains in adults and adolescents. *Neuroimage, 25*, 1279–1291.
44. Greenberger, E., & Steinberg, L. (1986). *When Teenagers Work: The Psychological and Social Costs of Adolescent Employment*. Basic Books.
45. Mortimer, R. K., et al. (1994). Genetic and physical maps of *Saccharomyces cerevisiae* (Edition 12). In 1994 *Yeast Genetics and Molecular Biology Meeting Program and Abstracts* (pp. 374–377). Bethesda, MD: The Genetics Society of America.

Chapter 8

Digestive Health and Gastrointestinal Issues

Contents

INTRODUCTION

Ultra-processed foods have become a mainstay of modern diets due to their high levels of added sugars, unhealthy fats, and artificial additives, as well as their extensive industrial processing. These foods, which include packaged meals, sugary snacks, and soft drinks, frequently lack fiber and other nutrients that are vital for preserving digestive health. Their consumption has been linked to a variety of gastrointestinal problems because of their high additive content and low fiber content, which can upset the gut microbiota's normal balance and hinder digestion.

A diet high in whole foods is essential for digestive health because they supply the right nutrients and maintain beneficial gut flora. On the other hand, eating a lot of processed food can cause disruptions in the gut flora, which can aggravate conditions like inflammatory bowel syndrome, bloating, and constipation. These foods' high content of unhealthy fats and refined sugars can cause inflammation and damage to the gut lining, which can cause symptoms like indigestion and abdominal pain. It can also raise the risk of developing gastrointestinal disorders like inflammatory bowel disease (IBD) and irritable bowel syndrome (IBS).

Additionally, foods that have undergone extreme processing frequently have high salt and preservative content, which can aggravate gastrointestinal problems by impairing gut motility and fluid balance. These elements work together to affect not only digestion but also the health of the gut as a whole, which may have long-term effects. Focusing on a diet high in whole, unprocessed foods is crucial to reducing these problems because they promote digestive health and lessen the negative effects of consuming ultra-processed foods.

8.1. EFFECTS ON DIGESTIVE HEALTH AND GASTROINTESTINAL FUNCTION

The consumption of ultra-processed foods, which are characterized by their intensive industrial processing and the presence of multiple additives, preservatives, and artificial ingredients, has become a more prevalent part of the modern diet. These foods, which include packaged snacks, fast food, ready-to-eat meals, and sugary drinks, frequently have high concentrations of unhealthy fats, refined sugars, and little dietary fiber. This essay investigates how eating a lot of processed food affects gut health and gastrointestinal function. It also looks at the mechanisms by which these foods affect the gut and discusses the wider health implications.

8.1.1. Characteristics of Ultra-Processed Foods

Foods that have undergone extensive processing usually have a long shelf life, are convenient, and taste good. They frequently lack fiber and important nutrients while being high in calories, fats, sugars, and sodium. These foods are produced through industrial processing techniques that use additives like colorants, flavor enhancers, emulsifiers, and preservatives, all of which can have different physiological effects. Frequent consumption of highly processed foods has been connected to several health problems, such as obesity, metabolic disorders, and cardiovascular diseases. It is also linked to poor dietary patterns.

8.1.2. Impact on Gut Microbiota

The influence of ultra-processed foods on gut microbiota is one of the main ways that these foods affect digestive health. An intricate community of microorganisms living in the gastrointestinal tract, the gut microbiota is vital to immunological response, digestion, and general health. Foods that have undergone extensive processing, especially those that are heavy in sugars and bad fats, can upset the gut microbiota's balance and cause dysbiosis. An imbalance in the microbial community known as dysbiosis is linked to several gastrointestinal problems, including inflammation, a decrease in microbial diversity, and the growth of pathogenic bacteria.

8.1.3. Effects on Digestion

Ultra-processed foods can also impact the digestive process, which in turn can impact digestion. These foods can affect the effectiveness and well-being of the digestive system because they frequently have high concentrations of unhealthy fats, refined sugars, and artificial additives. Many ultra-processed foods contain refined sugars and unhealthy fats that can interfere with normal digestive processes. Consuming too much sugar can cause harmful bacteria and yeast to proliferate in the gut, which can cause bloating, gas, and discomfort in the digestive system. High-fat diets have also been linked to altered gut motility and an increased risk of gastrointestinal disorders like dyspepsia and pancreatitis, especially those that are high in trans and saturated fats.

Additionally, artificial additives and preservatives have the potential to irritate the lining of the stomach, resulting in symptoms like diarrhea, nausea, and abdominal pain. For instance, alterations in the gut microbiota and gastrointestinal symptoms have been connected to certain food colorings and artificial sweeteners. Furthermore, because fiber is essential for regular bowel movements and the prevention of constipation, the low fiber content of ultra-processed foods may compromise the digestive system's ability to operate at its best.

8.1.4. Gastrointestinal Disorders and Health Implications

Ultra-processed food consumption is linked to a higher risk of various gastrointestinal disorders. For example, research indicates that a high-ultra-processed-food diet is associated with an increased risk of inflammatory bowel disease (IBD), which includes ulcerative colitis and Crohn's disease. These foods' low fiber content, unhealthy fat content, and high additive content can cause long-term inflammation and harm to the lining of the stomach.

Furthermore, eating a lot of ultra-processed food has been associated with a higher incidence of irritable bowel syndrome (IBS), a functional gastrointestinal disorder that causes bloating, altered bowel habits, and abdominal pain. Ultra-processed foods can alter digestive processes and disturb the gut microbiota, which can worsen IBS symptoms and lower the quality of life for those who experience them. Consuming a lot of highly processed foods has also been linked to an increased risk of colorectal cancer. The World Health Organization (WHO) has categorized processed meats in particular as Group 1 carcinogens because of their link to an increased risk of cancer. This elevated risk is attributed to the presence of carcinogenic compounds like nitrates and nitrites, as well as the effects of low-fiber diets.

8.2. ROLE OF DIETARY FIBER AND GUT MICROBIOTA

Dietary fiber, an essential part of foods derived from plants, is vital for preserving and enhancing digestive health. Dietary fiber comes in soluble and insoluble forms and is not broken down by human enzymes, so it passes through the digestive system mostly undigested. Because of its special qualities, it has several physiological advantages, especially for the digestive system. The various functions, advantages, and effects of dietary fiber on digestive health are explored in this essay, along with their implications for general health.

8.2.1. Dietary Fiber

Depending on the nation, food, and pharmaceutical industry definition and classification, the term "fiber" can mean different things. Whatever definition it may have, dietary fiber generally refers to carbohydrates that the host's gut microbiota can ferment and produce health benefits for. The indigestible leftovers also have a laxative or bulking effect (Mandaliya et al., 2018). Since fibers are traditionally categorized based on their physicochemical solubility, undigestible, and unabsorbable carbohydrate polymers that have been shown to have positive health effects are included in the European Food Safety

Authority's (EFSA) list of dietary fiber. Dietary fiber was defined by the US Codex Alimentarius Commission in 2009 as an edible carbohydrate polymer with three or more monomeric units that are resistant to endogenous digestive enzymes and unabsorbable in the small intestine. This definition was largely adopted by other countries. These digestible fibers are further divided into three categories: (i) naturally occurring edible carbohydrate polymers found in foods like fruits, vegetables, legumes, and cereals; (ii) edible carbohydrate polymers extracted from food raw materials using physical, enzymatic, and chemical methods and shown to have physiological benefits; and (iii) synthetically produced edible carbohydrate polymers. Because dietary fibers come from a variety of sources, their solubility alone may not be a reliable indicator of their functional qualities because most foods are complex mixtures and cooking can alter the fibers' availability. Several factors determine the current classification, such as the main food source, chemical structure, water solubility, viscosity, and fermentability. Dietary fibers are classified as soluble or insoluble based on the monomeric units and non-carbohydrate components they contain. While wheat bran, oats, and barley contain more insoluble dietary fibers of cellulose or hemicellulose, fruits, and vegetables are rich sources of soluble fibers (pectin, inulin) (*see* Figure 8.1). Soluble fibers are fermented by gut colon bacteria, and the metabolites have advantageous metabolic effects in comparison to poorly or non-fermentable insoluble fibers that primarily serve as bulking or laxatives in the colon. The viscosity of soluble fibers, like oligosaccharides, affects not only their solubility and fermentability but also how well ingested nutritional components are absorbed in the stomach.

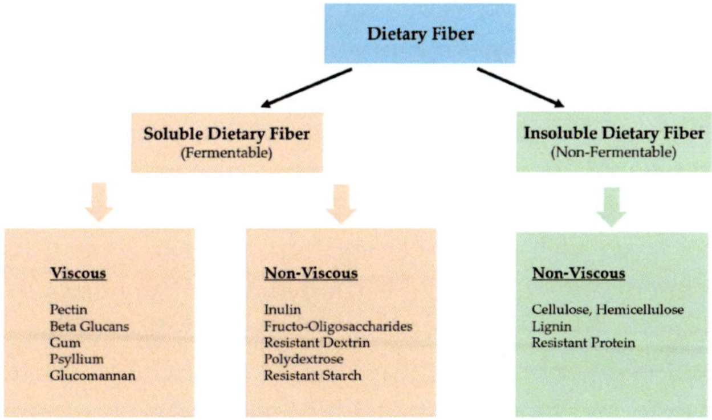

Figure 8.1. *Classification of dietary fibers based on their physicochemical characteristics.*

Source: Ranganathan N, Anteyi E. The Role of Dietary Fiber and Gut Microbiome Modulation in Progression of Chronic Kidney Disease. Toxins. 2022; 14(3):183. https://doi.org/10.3390/toxins14030183.

The protective effects of fiber against disease have been partially linked to its alteration of the gut microbiota, which regulates satiety and body weight, lipid, and bile acid metabolism, cancer risk, and cardiovascular disease risk (Anderson et al., 2009). Additional effects of dietary fiber on gut microbiota include the regulation of inflammation, as demonstrated by a decrease in the C-reactive protein (CRP) marker in an experimental model of chronic kidney disease (CKD) mice fed a high-fat diet after consuming fiber. By altering gut-derived neuropeptides that regulate gluconeogenesis in the brain and intestines, dietary fiber controls glucose and energy homeostasis via the hypothalamic pathway. Furthermore, consuming fiber increases satiating hormones and delays gastric emptying, which results in a feeling of fullness that is mediated by the intestinal L cells' secretion of the gut hormone incretin. Insulin secretion and glucose homeostasis are regulated by these gut incretin hormones. Increased dietary fiber consumption has been linked in other studies to a lower risk of colorectal cancer. Dietary fiber's capacity to change gut microbiota in the general population by preventing and treating a variety of disease conditions has been linked to all these positive effects. The problem with dietary fiber is that there isn't a nutritional database to describe various food sources and estimate the types of fiber that are required for therapeutic interventions that target the gut microbiota.

The general public is advised to consume 20–35 g of fiber per day, or 14 g/1000 kcal, as part of a healthy diet. A mean daily fiber intake of 17 g is recommended by the USA guideline, but only 5% of the population can meet this requirement. The recommendation did not go into detail about the precise kinds or amounts of various fiber-containing foods that are necessary for an adequate intake, beyond the overall amount of fiber. This disparity has led some researchers to propose that components of carbohydrates that are accessible to the microbiota and that the colonic microbiota can metabolize to obtain health benefits should be included in any recommendation for dietary fiber. Carbohydrates that are resistant to being broken down and absorbed by the enzymes of the host, whether they come from plant or animal tissue sources, are known as microbiota-accessible carbohydrates. They can also come from mucus secretions in the host's intestine. The production of short-chain fatty acids, altered colon microbial composition and metabolite production, enhanced colonic transit time, and small intestinal lipid and glucose absorption are some of the mechanisms that have been proposed to explain how dietary fiber promotes overall health. Production of short-chain fatty acids (SCFAs) has been shown to decrease inflammation, and modify glucose and lipid metabolism, which inhibits the development of cancer, influences cardiovascular and metabolic disorders, and may even slow the onset and course of chronic kidney disease (CKD).

8.2.2. The Composition of Healthy Gut Microbiota

Living microorganisms that coexist in a symbiotic relationship with the host and are primarily bacteria, including viruses, archaea, fungi, and unicellular eukaryotes, make up the healthy gut microbiota. Approximately one trillion of these microorganisms, comprising thousands of distinct species, are found in the human gut. They encode approximately three million genes, in contrast to the 23,000 genes found in the human genome. Gut microbiota are categorized taxonomically based on phyla, classes, orders, families, genera, and species. The microbiota of a healthy colon is composed of five phyla: 90% of the species are Firmicutes and Bacteroidetes, with Actinobacteria, Verrucomicrobia, and a minor amount of Proteobacteria. The Bacteroides and Prevotella genera are the most common in the Bacteroidetes phyla, while the Bifidobacterium genus represents the actinobacterium phyla. The Phyla Filmicutes contain over 200 genera, including Clostridium, Lactobacillus, Enterococcus, and Ruminococcus (Figure 8.2). Everybody has a different microbiome in terms of composition, diversity, and function, which is influenced by a range of factors including location, age, sex, race, lifestyle, use of antibiotics, and diet. The primary factor influencing changes in the composition of the gut microbiome is the type of food consumed; some of these changes can be seen as soon as a week after a person switches from a diet high in plant-based protein to one high in animal protein.

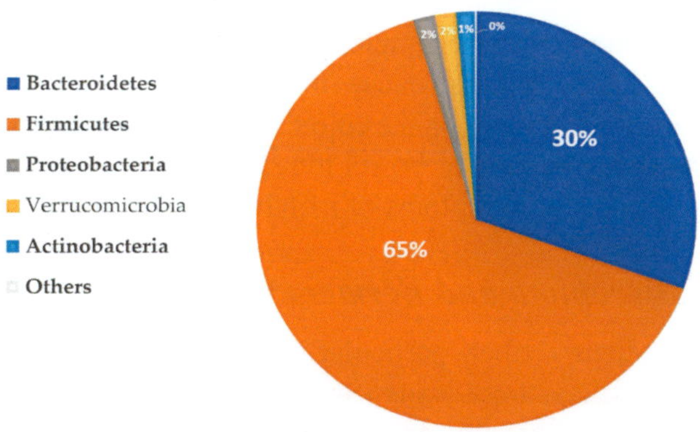

Figure 8.2. *The healthy human gut microbiome phyla classification.*

Source: Ranganathan N, Anteyi E. The Role of Dietary Fiber and Gut Microbiome Modulation in Progression of Chronic Kidney Disease. Toxins. *2022; 14(3):183. https://doi.org/10.3390/toxins14030183.*

Dietary fiber breakdown results in the production of SCFA and vitamins, the metabolism of conjugated bile acids, and immune system regulation. An overabundance of bacteria called Bacteroidetes and Actinobacterium produce trimethylamine (TMA), a precursor to the uremic toxin trimethyl amine-N-oxide (TMAO), which has been linked to chronic systemic inflammation and cardiovascular disease (CVD). This is caused by a diet high in animal protein and low in dietary fiber. The production of antimicrobial substances, development of intestinal mucosa and endothelial integrity, immune modulation, prevention of cancer, and protection against pathogenic organisms through competitive exclusion are some of the additional health benefits that the gut microbiome confers upon its host (Ma et al., 2006).

8.2.3. Benefits of Dietary Fiber for Digestive Health

8.2.3.1. Improved Bowel Regularity

It is widely acknowledged that dietary fiber helps to maintain regular bowel movements. Constipation is less common when insoluble fiber helps stool pass through the intestines more quickly and with greater bulk. Increased stool bulk can facilitate defecation and lessen straining, which is especially advantageous for people with irregular bowel movements or chronic constipation.

8.2.3.2. Prevention and Management of Digestive Disorders

Dietary fiber is important for both managing and preventing several digestive disorders. A high-fiber diet, for example, has been linked to a decreased risk of diverticulitis, a condition that can cause inflammation or infection of intestine pouches. Fiber relieves pressure on the colon and keeps the digestive tract clean. Additionally, by restoring regular bowel movements and reducing discomfort, fiber may help manage the symptoms of irritable bowel syndrome (IBS).

8.2.3.3. Gut Microbiota and Fermentation

Gut bacteria use dietary fiber, particularly soluble fiber, as a substrate for fermentation. Short-chain fatty acids (SCFAs), like butyrate, propionate, and acetate, are produced during this process and have several advantages for gut health. SCFAs lower inflammation, give colon cells energy, and support the integrity of the gut lining. In addition to improving overall digestive health, dietary fiber lowers the risk of gastrointestinal disorders by promoting a diverse and healthy gut microbiota.

8.2.3.4. Reduction in Colon Cancer Risk

A high-fiber diet has been linked to a lower risk of colon cancer, according to research. SCFAs, which offer protection against colon cancer, are produced when gut bacteria ferment fiber. Furthermore, fiber makes stools bulkier, which facilitates and speeds up the digestive tract's elimination of possible carcinogens. The incidence of colorectal cancer and dietary fiber intake are inversely correlated in epidemiological studies, indicating that fiber may have a protective effect against cancer.

8.2.3.5. Impact on Digestive Transit Time

The amount of dietary fiber has a big impact on how quickly food passes through the digestive system. The gel-like form that soluble fiber produces during digestion permits nutrients to be absorbed gradually. This delayed transit time has the potential to improve blood sugar control and extend satiety, both of which can support weight management. On the other hand, insoluble fiber expedites food and waste passage through the intestines, minimizing the amount of time that potentially harmful substances come into contact with the gut lining and lowering the likelihood of experiencing gastrointestinal distress.

8.2.3.6. Hydration and Fiber

Drinking enough water is crucial when following a high-fiber diet. Fiber helps to form soft, bulky stools that are easy to pass by absorbing water. Fiber can cause hard, dry stools if not enough water is consumed; this will exacerbate constipation rather than help it. Therefore, to fully benefit from dietary fiber's digestive properties, it's critical to increase water consumption in addition to it.

8.2.3.7. Fiber and Nutrient Absorption

Although fiber is well known for improving digestion, it can also affect how well nutrients are absorbed. Soluble fiber facilitates more effective nutrient absorption in the small intestine by delaying digestion. However, by binding to and preventing the uptake of certain minerals, such as calcium, iron, and zinc, an excessive intake of fiber can obstruct their absorption. To guarantee the best possible nutrient absorption, it is crucial to balance your intake of fiber.

8.2.3.8. Dietary Recommendations and Sources

Age and gender-specific recommendations for daily dietary fiber intake differ. The recommended daily intake for adults is typically 38 grams for men and 25 grams for women. But the majority of people don't follow these guidelines. Achieving an appropriate fiber intake can be facilitated by increasing the consumption of foods high in fiber, such as fruits, vegetables, whole grains, legumes, nuts, and seeds. It is recommended that fiber be added to the diet gradually to minimize potential gastrointestinal discomfort and to give the digestive system time to adjust.

8.2.3.9. Potential Drawbacks and Considerations

Although eating too much fiber can have many positive effects on the digestive system, it can also cause gas, bloating, and cramping. These are usually transient symptoms that can be lessened by consuming more fiber gradually and drinking enough water. To prevent aggravating their symptoms, people with specific digestive disorders, such as gastroparesis or severe IBS, may need to carefully control their fiber intake under medical supervision (Ranganathan et al., 2022).

8.3. COMMON GASTROINTESTINAL PROBLEMS ASSOCIATED WITH ULTRA-PROCESSED FOODS

Many gastrointestinal (GI) disorders have been related to the ultra-processed foods that are a common part of the modern diet. Ultra-processed foods are generally low in fiber and vital nutrients and high in artificial additives, unhealthy fats, added sugars, and salt. These foods, which include sugar-filled drinks, packaged snacks, instant noodles, and prepared meals, are frequently tasty and convenient, but they can have a negative effect on digestive health. This essay examines the typical gastrointestinal issues linked to eating a lot of highly processed food.

1. **Constipation**: One of the most prevalent gastrointestinal issues associated with a diet heavy in highly processed foods is constipation. Dietary fiber, which is necessary for regular bowel movements, is frequently lacking in these foods. Fiber gives the stool more volume and facilitates its easier passage through the digestive system. Constipation can result from hard, difficult-to-pass stool, which can be brought on by insufficient fiber. Moreover, since hydration is essential for softening stool, the low water content in many processed foods can make the problem worse.
2. **Bloating and Gas**: Gas and bloating are common complaints from people who eat a lot of ultra-processed food. Artificial sweeteners like sorbitol and mannitol, which are poorly absorbed in the intestines and have the potential to ferment and produce gas, are frequently found in these foods. Furthermore, a lot of processed foods have a lot of sodium, which can lead to bloating and water retention. The absence of fiber in these foods also deprives beneficial gut bacteria of the nutrition they require, upsetting the microbiota's balance and causing an increase in gas production.

3. **Irritable Bowel Syndrome (IBS)**: The symptoms of Irritable Bowel Syndrome (IBS), a chronic illness, include bloating, diarrhea, constipation, and abdominal pain. Ultra-processed food consumption can cause or worsen symptoms of IBS. These foods frequently have high concentrations of FODMAPs, or fermentable oligosaccharides, disaccharides, monosaccharides, and polyols, which are hard-to-digest forms of carbohydrates. One of the main symptoms of IBS is gas, bloating, and discomfort, which can be brought on by the fermentation of FODMAPs in the gut.
4. **Gastroesophageal Reflux Disease (GERD)**: Heartburn and irritation are caused by stomach acid frequently flowing back into the esophagus, a condition known as gastroesophageal reflux disease (GERD). Foods with high processing levels can aggravate GERD in many ways. These foods frequently contain a lot of fat, which can relax the LES and let acid leak out of the stomach. Furthermore, acidic or caffeine-containing processed foods and drinks, like sodas and some snack foods, can exacerbate the symptoms of GERD. Reflux can also result from the high salt content in many processed foods, which increases the production of stomach acid.
5. **Inflammatory Bowel Disease (IBD)**: Conditions like Crohn's disease and ulcerative colitis, which cause persistent inflammation of the digestive tract, are included in the category of inflammatory bowel disease (IBD). Even though the precise cause of IBD is unknown, diet is a key component of disease management. Ultra-processed foods, with their high levels of unhealthy fats, sugars, and additives, can worsen gut inflammation. These food ingredients have the potential to affect gut microbiota, raise intestinal permeability, and stimulate inflammatory processes—all of which have the potential to precipitate or exacerbate IBD symptoms.
6. **Diverticular Disease**: The lining of the digestive system develops tiny protuberances or pockets called diverticula as a result of diverticular disease. These may become infected or inflammatory, which can result in diverticulitis, a condition that produces fever, intense abdominal pain, and altered bowel habits. Diverticular disease has been linked to a diet high in processed foods and low in fiber. Insufficient fiber intake may result in elevated colon pressure during bowel movements, potentially facilitating the development of diverticula.
7. **Dysbiosis**: An imbalance in the gut microbiota, or the community of microorganisms that reside in the digestive tract, is referred to as

dysbiosis. Foods with high processing levels may have a detrimental effect on the variety and makeup of the gut microbiota. These foods frequently include artificial additives, emulsifiers, and preservatives that may upset the natural balance of gut flora. Immune system performance, nutrition absorption, and digestion all depend on a healthy gut microbiota. Dysbiosis can result in some gastrointestinal issues, such as inflammation, poor digestion, and an increased risk of infections.

8. **Leaky Gut Syndrome**: Increased intestinal permeability, another name for leaky gut syndrome, is a condition in which the small intestine's lining is compromised, allowing bacteria, toxins, and undigested food particles to enter the bloodstream. This may set off an immunological reaction and exacerbate many health conditions, such as allergies, autoimmune diseases, and gastrointestinal disorders. Ultra-processed foods can cause inflammation and damage to the gut barrier, which can lead to leaky gut. The integrity of the intestinal lining can be weakened by additives found in processed foods, such as thickeners and emulsifiers.

9. **Food Intolerances and Allergies**: An increase in food allergies and intolerances has been connected to the consumption of highly processed foods. These foods frequently include additives, preservatives, and artificial ingredients that may cause allergic reactions in those who are sensitive to them. Artificial sweeteners, flavorings, and colorings are common offenders. These substances have the potential to sensitize the immune system over time, which can result in allergic reactions or intolerances that cause GI symptoms like nausea, diarrhea, and abdominal pain.

10. **Obesity and Metabolic Syndrome**: Obesity and metabolic syndrome have major effects on digestive health even though they are not directly related to gastrointestinal issues. Ultra-processed foods can lead to weight gain and the development of obesity because they are frequently high in calories, unhealthy fats, and sugars. Several gastrointestinal disorders, such as gallstones, fatty liver disease, and GERD, are associated with obesity. By encouraging inflammation and raising the risk of GI disorders, metabolic syndrome—which is characterized by a number of conditions including high blood pressure, high blood sugar, and abnormal cholesterol levels—can also have a detrimental effect on digestive health.

8.3.1. Mechanisms Linking Ultra-Processed Foods to GI Problems

Understanding the mechanisms through which ultra-processed foods contribute to GI problems is essential for addressing and mitigating these issues. Several key mechanisms are involved:

1. **Alteration of Gut Microbiota**: Consistency of the gut microbiota is essential for digestive health maintenance. Ultra-processed foods have the potential to upset the diversity and balance of gut flora. When fiber is deficient and synthetic additives, preservatives, and emulsifiers are present, good bacteria can be harmed while bad bacteria can proliferate. Dysbiosis, the term for this imbalance, can cause a number of gastrointestinal issues, such as inflammation, poor digestion, and an increased risk of infections.
2. **Increased Inflammation**: There are components in a lot of highly processed foods that aggravate gut inflammation. Certain additives, unhealthy fats, and refined sugars can all cause inflammatory reactions. Prolonged inflammation within the digestive tract can harm the lining of the gut, resulting in ailments such as leaky gut syndrome and playing a role in the emergence of inflammatory gastrointestinal disorders like IBD.
3. **Impaired Gut Barrier Function**: Maintaining the integrity of the gut barrier is crucial in preventing the bloodstream from being contaminated. Ultra-processed foods can cause inflammation and upset the gut microbiota's balance, which can weaken the gut barrier. Processed food additives like thickeners and emulsifiers can further weaken the gut barrier and cause intestinal permeability, or leaky gut syndrome.
4. **Disruption of Digestive Processes**: Ultra-processed foods contain artificial ingredients and additives that can disrupt regular digestive processes. For instance, synthetic sweeteners such as sorbitol and mannitol can result in fermentation and gas and bloating because they are not well absorbed in the intestines. Foods high in fat can relax the LES and slow down stomach emptying, which can exacerbate GERD symptoms. Furthermore, these foods' low fiber content can cause constipation by affecting bowel movements.
5. **Nutrient Deficiencies**: Ultra-processed foods frequently lack fiber and other vital nutrients that are needed to keep the digestive tract healthy. An inadequate intake of these nutrients can cause the immune system to deteriorate and the digestive system to malfunction, leaving the gut more vulnerable to inflammation and infections.

8.3.2. Mitigating the Impact of Ultra-Processed Foods on GI Health

To mitigate the negative impact of ultra-processed foods on digestive health, it is essential to adopt a diet rich in whole, unprocessed foods. Here are some strategies:

1. **Increase Fiber Intake**: Eating a diet rich in fruits, vegetables, whole grains, legumes, and other fiber-rich foods can help lower the risk of GI disorders, promote healthy gut bacteria, and prevent constipation. Fiber gives the stool more volume, encourages regular bowel movements, and feeds the good bacteria in the gut.
2. **Stay Hydrated**: Adequate hydration is crucial for maintaining digestive health. Drinking plenty of water throughout the day helps soften stool, prevent constipation, and support overall digestive function. Avoiding excessive consumption of sugary and caffeinated beverages, which can dehydrate the body, is also important.
3. **Limit Processed Foods**: Reducing the intake of ultra-processed foods and replacing them with whole, unprocessed alternatives can significantly improve digestive health. Focus on consuming fresh fruits and vegetables, lean proteins, whole grains, and healthy fats. Preparing meals at home using whole ingredients can help control the quality and nutritional content of your diet.
4. **Probiotics and Prebiotics:** Incorporating probiotics and prebiotics into your diet can support a healthy gut microbiota. Probiotics are beneficial bacteria found in fermented foods like yogurt, kefir, sauerkraut, and kimchi. Prebiotics are types of fiber that feed beneficial gut bacteria and are found in foods like garlic, onions, bananas, and asparagus.
5. **Mindful Eating**: Practicing mindful eating can improve digestion and reduce the risk of GI problems. Eating slowly, chewing food thoroughly, and paying attention to hunger and fullness cues can help prevent overeating and support the digestive process. Mindful eating also promotes a better connection with the body's natural signals and reduces stress, which can positively impact digestive health.
6. **Stress Management**: Chronic stress can negatively affect digestive health by disrupting gut motility, altering gut microbiota, and increasing inflammation. Engaging in stress-reducing activities like meditation, yoga, deep breathing exercises, and regular physical activity can help manage stress levels and support digestive health (Ranganathan et al., 2022).

8.4. STRATEGIES FOR MAINTAINING DIGESTIVE HEALTH

Sustaining digestive health is essential for overall health and can be accomplished in a number of ways. Eating a balanced diet high in fiber is a fundamental strategy that promotes the growth of good gut bacteria and helps regulate bowel movements. Whole grains, legumes, fruits, and vegetables are examples of foods high in fiber. It's also critical to stay hydrated because water promotes healthy digestion and helps ward off constipation. Drinking lots of water is advised throughout the day, and adding herbal teas with digestive properties, like ginger or peppermint, can help even more. Eating mindfully, which involves chewing food well, eating slowly, and paying attention to signals of hunger and fullness, is another important tactic. This facilitates better food processing by the digestive system and helps avoid overeating. Furthermore, consistent exercise is very important for digestive health. Exercise encourages regular bowel movements and lowers the risk of digestive disorders because it stimulates the intestines.

8.4.1. Ways to Improve Your Digestive Health

Your digestive system is constantly producing energy for each cell in your body and effectively eliminating waste. Disregarding your digestive system can have negative impacts on all aspects of your life. Your digestive health is significantly influenced by your diet and other lifestyle decisions, which are frequently areas where modest adjustments can have a big influence. Here are some ways that lifestyle choices can improve your digestive health.

1. **Eat Fruits and Vegetables**: Fruits and vegetables are rich in fiber and beneficial nutrients that help maintain your general health and digestive system. Fruits and vegetables are rich in antioxidants that can fight cancer, and fiber helps reduce constipation.
2. **Choose Whole Grains and Nuts**: The dietary fiber, iron, antioxidants, and other beneficial nutrients that are lost when grains are refined to produce processed foods like white flour, white bread, crackers, and pastries are still present in whole grains. Constipation can be avoided with fiber, and certain whole grains help maintain healthy gut flora.
3. **Limit red and Processed Meats**: Processed meats typically contain harmful levels of fat, sodium, and nitrates, which have been connected to colon cancer. Red meat has been connected to heart disease, cancer, and digestive problems.

4. **Limit Added Sugars**: Inflammation and an imbalance of beneficial bacteria can result from consuming too much sugar. Many products, particularly baked goods, ice cream, and desserts, contain added sugars. Even unexpectedly, they are added to a lot of processed foods, such as soups, salad dressings, and pasta sauces.
5. **Cook Healthier**: Eating healthily involves not just the foods you choose to eat but also the way you prepare them. Although they are delicious treats, fried, and grilled foods have a drawback. Fried foods are high in unhealthy fats, and grilling at a high temperature can produce carcinogens, which are chemicals that cause cancer.
6. **Eat Foods with Probiotics**: Probiotics are good bacteria that support healthy digestion by nourishing your gut and controlling harmful bacteria.
7. **Stay Hydrated**: Constipation could be a sign that you are not drinking enough. Water consumption promotes healthy bowel function and helps avoid constipation. Your body needs different amounts of water depending on your size, activity level, and overall health. Adults require four to six glasses of water daily on average.
8. **Exercise Regularly**: Frequent exercise lowers your risk of nausea, bloating, and constipation by accelerating the passage of food through your digestive system. Frequent exercise can also help with digestive health, which includes reducing inflammatory bowel disease and constipation symptoms.
9. **Maintain a Healthy Weight**: Keeping a healthy weight lowers your risk of gallstones, GERD, certain types of digestive cancer, and other conditions. Maintaining a healthy weight can be facilitated by adhering to dietary recommendations and engaging in regular exercise.
10. **Don't smoke**: You may be shocked to learn that smoking is harmful to your digestive system in addition to your lungs. It has been demonstrated that smoking greatly increases your risk of developing acid reflux and that symptoms can be alleviated by quitting. Additionally, smoking has been linked in studies to stomach ulcers, an increased risk of gastrointestinal cancers, and ulcerative colitis.
11. **Get enough Sleep**: Every bodily system, including the digestive system, is impacted by sleep deprivation. Lack of sleep can cause intestinal inflammation, which can result in gastrointestinal problems.
12. **Reduce Stress**: Prolonged stress is bad for your health in general and is associated with ulcers, heartburn, IBS, constipation, and diarrhea. Digestion can be improved by practicing stress management techniques (Ranganathan et al., 2022).

REFERENCES

1. Anderson, J. W., Baird, P., Davis, R. H., Ferreri, S., Knudtson, M., Koraym, A., & Williams, C. L. (2009). Health benefits of dietary fiber. *Nutrition Reviews, 67*(4), 188–205.
2. Bermudez-Brito, M., Plaza-Diaz, J., Munoz-Quezada, S., Gomez-Llorente, C., & Gil, A. (2012). Probiotic mechanisms of action. *Annals of Nutrition and Metabolism, 61*(2), 160–174.
3. Canani, R. B., Cirillo, P., Roggero, P., et al. (2006). Therapy with gastric acidity inhibitors increases the risk of acute gastroenteritis and community-acquired pneumonia in children. *Pediatric Neonatology, 117*(5), e817–e820.
4. Dial, S., Delaney, J. A., Barkun, A. N., & Suissa, S. (2005). Use of gastric acid-suppressive agents and the risk of community-acquired Clostridium difficile-associated disease. *JAMA, 294*(23), 2989–2995.
5. Farhadi, A., Banan, A., Fields, J., & Keshavarzian, A. (2003). Intestinal barrier: An interface between health and disease. *Journal of Gastroenterology and Hepatology, 18*(5), 479–497.
6. Funaki, Y., Tokudome, K., Izawa, S., et al. (2013). Comparison of the effect of a single dose of omeprazole or lansoprazole on intragastric pH in Japanese participants: A two-way crossover study. *Journal of the Chinese Medical Association, 76*(3), 131–134.
7. Gau, J. T., Yang, Y. X., Chen, R., & Kao, T. C. (2012). Uses of proton pump inhibitors and hypomagnesemia. *Pharmacoepidemiology and Drug Safety, 21*(5), 553–559.
8. Gray, S. L., LaCroix, A. Z., Larson, J., et al. (2010). Proton pump inhibitor use, hip fracture, and change in bone mineral density in postmenopausal women: Results from the Women's Health Initiative. *Archives of Internal Medicine, 170*(9), 765–771.
9. Gutierrez-Castrellon, P., Lopez-Velazquez, G., Diaz-Garcia, L., et al. (2014). Diarrhea in preschool children and *Lactobacillus reuteri*: A randomized controlled trial. *Pediatrics, 133*(4), e904–e909.
10. Hegar, B., Hutapea, E. I., Advani, N., & Vandenplas, Y. (2013). A double-blind placebo-controlled randomized trial on probiotics in small bowel bacterial overgrowth in children treated with omeprazole. *Journal of Pediatrics (Rio J), 89*(4), 381–387.
11. Laheij, R. J., Sturkenboom, M. C., Hassing, R. J., Dieleman, J., Stricker, B. H., & Jansen, J. B. (2004). Risk of community-acquired pneumonia and use of gastric acid-suppressive drugs. *JAMA, 292*(16), 1955–1960.

12. Lam, J. R., Schneider, J. L., Zhao, W., & Corley, D. A. (2013). Proton pump inhibitor and histamine 2 receptor antagonist use and vitamin B12 deficiency. *Nature Medicine, 310*(22), 2435–2442.
13. Leyer, G. J., Li, S., Mubasher, M. E., Reifer, C., & Ouwehand, A. C. (2009). Probiotic effects on cold and influenza-like symptom incidence and duration in children. *Pediatrics, 124*(2), e172–e179.
14. Lovegrove, A., Edwards, C. H., De Noni, I., Patel, H., El, S. N., Grassby, T., & Shewry, P. R. (2017). Role of polysaccharides in food, digestion, and health. *Critical Reviews in Food Science and Nutrition, 57*(2), 237–253.
15. Ma, Y., Griffith, J. A., Chasan-Taber, L., Olendzki, B. C., Jackson, E., Stanek, E. J., III, & Ockene, I. S. (2006). Association between dietary fiber and serum C-reactive protein. *American Journal of Clinical Nutrition, 83*(4), 760–766.
16. Macdonald, L. E., Brett, J., Kelton, D., Majowicz, S. E., Snedeker, K., & Sargeant, J. M. (2011). A systematic review and meta-analysis of the effects of pasteurization on milk vitamins, and evidence for raw milk consumption and other health-related outcomes. *Journal of Food Protection, 74*(11), 1814–1832.
17. Mandaliya, D., Patel, S., & Seshadri, S. (2018). Fiber in our diet and its role in health and disease. In *Functional Food and Human Health* (pp. 247–255). Springer.
18. Orel, R., & Turk, H. (2010). Re: Might the use of acid-suppressive medications predispose to the development of eosinophilic esophagitis? *American Journal of Gastroenterology, 105*(2), 468; author reply 469.
19. Ranganathan, N., & Anteyi, E. (2022). The role of dietary fiber and gut microbiome modulation in progression of chronic kidney disease. *Toxins, 14*(3), 183.
20. Richter, C., Tanaka, T., & Yada, R. Y. (1998). Mechanism of activation of the gastric aspartic proteinases: Pepsinogen, progastricsin, and prochymosin. *Biochemical Journal, 335*(Pt 3), 481–490.
21. Valuck, R. J., & Ruscin, J. M. (2004). A case-control study on adverse effects: H2 blocker or proton pump inhibitor use and risk of vitamin B12 deficiency in older adults. *Journal of Clinical Epidemiology, 57*(4), 422–428.
22. Wilhelm, S. M., Rjater, R. G., & Kale-Pradhan, P. B. (2013). Perils and pitfalls of long-term effects of proton pump inhibitors. *Expert Review of Clinical Pharmacology, 6*(4), 443–451.

Chapter 9

Impacts of Food Additives in Ultra-Processed Food on Human Health

Contents

INTRODUCTION

Food additives are substances added to food to enhance its flavor, appearance, shelf-life, or texture. While they play a crucial role in the food industry, their inclusion in ultra-processed foods (UPFs) raises significant health concerns. UPFs, characterized by their extensive industrial processing and multiple ingredients, often contain a high number of additives, including preservatives, colorants, flavor enhancers, and emulsifiers.

One of the primary concerns regarding food additives is their potential to contribute to chronic health conditions. For instance, some preservatives like sodium nitrite, commonly found in processed meats, have been linked to an increased risk of cancer. Artificial colorants, such as those found in many snack foods and beverages, have been associated with behavioral issues in children, including hyperactivity and attention deficits. Additionally, flavor enhancers like monosodium glutamate (MSG) can cause adverse reactions in sensitive individuals, leading to symptoms such as headaches, nausea, and chest pain.

Emulsifiers, which are used to improve texture and extend shelf-life, can disrupt gut health by altering the composition of the gut microbiota. This disruption can lead to inflammation, which is a known risk factor for a range of diseases, including inflammatory bowel disease and metabolic syndrome. Furthermore, the cumulative effect of consuming multiple additives from various sources can exacerbate these health risks.

Despite regulatory agencies setting acceptable daily intake levels for many additives, the long-term effects of their consumption, especially in combination, remain poorly understood. The widespread use of food additives in UPFs thus necessitates further research to fully elucidate their impact on human health. Consumers are advised to be vigilant and to reduce their intake of UPFs, opting instead for whole, minimally processed foods that are naturally free from these potentially harmful substances. This approach can help mitigate the health risks associated with food additives and promote overall well-being.

9.1. FOOD ADDITIVES

Any substance added to food is considered a food additive. According to legal definitions, the term encompasses any substance whose intended use could reasonably be expected to result in its incorporation into, or alteration of, any food item, either directly or indirectly. Any material used in the preparation, processing, handling, packing, shipping, or storing of food is covered by this definition (Kunkel as well as Barbara, 2004). Substances added to food to preserve flavor or improve its appearance are also known as food additives. Certain additives, like pickling (with vinegar), salting (as with bacon), preserving sweets, or using sulfur dioxide (as in some wines), have been used for centuries. Many more additives, both natural and artificial, have been added since the second half of the 20th century with the introduction of processed foods. "Any substance, the intended use of which results or may reasonably be expected to result, directly or indirectly, in its becoming a component or otherwise affecting the characteristics of any food" is the definition of food additives as given by the US Food and Drug Administration (FDA). Stated differently, an additive is any material that is incorporated into food. Indirect additives are ones to which food is exposed while it is being processed, packaged, or stored, whereas direct additives are purposefully added to food. A substance is called a direct additive if it is added to food with a specific intention. For instance, aspartame, a low-calorie sweetener found in yogurt, drinks, puddings, and other foods, is regarded as a direct additive. Food ingredient labels identify many direct additives. Food additives that are incorporated into food in trace amounts as a result of handling, storage, or packaging are known as indirect food additives. For example, during food storage, trace amounts of packaging materials could end up in food.

Food additives may be classified as:

1. Antimicrobial agents, which prevent spoilage of food by mold or micro-organisms. These include not only vinegar and salt, but also compounds such as calcium propionate and sorbic acid, which are used in products such as baked goods, salad dressings, cheeses, margarines, and pickled foods.
2. Antioxidants, which prevent rancidity in foods containing fats and damage to foods caused by oxygen. Examples of antioxidants include vitamin C, vitamin E, BHA (butylated hydroxyl anisole), BHT (butylated hydroxytolene), and propyl gallate.
3. Artificial colors, which are intended to make food more appealing and to provide certain foods with a color that humans associate with a particular flavor (e.g., red for cherry, green for lime).
4. Artificial flavors and flavor enhancers are the largest class of additives; their function is to make food taste better or to give it a specific taste. Examples are salt, sugar, and vanilla, which are used to complement the flavor of certain foods. Synthetic flavoring agents, such as benzaldehyde for cherry or almond flavor, may be used to simulate natural flavors. Flavor enhancers, such as monosodium glutamate (MSG) intensify the flavor of other compounds in a food.
5. Bleaching agents, such as peroxides, are used to whiten. foods such as wheat flour and cheese.

6. Chelating agents, which are used to prevent discoloration, flavor changes, and rancidity that might occur during the processing of foods. Examples are citric acid, malic acid, and tartaric acid.
7. Nutrient additives, including vitamins and minerals, are added to foods during enrichment or fortification. For example, milk is fortified with vitamin D, and rice is enriched with thiamin, riboflavin, and niacin.
8. Thickening and stabilizing agents, which function to alter the texture of a food. Examples include the emulsifier lecithin, which, keeps oil and vinegar blended in salad dressings, and carrageen, which is used as a thickener in ice creams and low-calorie jellies.

9.1.1. Classification of Food Additives

Food additives can be divided into several groups, although there is some overlap between them.

1. ***Acids***: Food acids are added to make flavors "sharper," and also act as preservatives and antioxidants. Common food acids include vinegar, citric acid, tartaric acid, malic acid, fumaric acid, and lactic acid.
2. ***Acidity regulators***: Acidity regulators are used to change or otherwise control the acidity and alkalinity of foods.
3. ***Anticaking agents***: Anticaking agents keep powders such as milk powder from caking or sticking.
4. ***Antifoaming agents***: Antifoaming agents reduce or prevent foaming in foods.
5. ***Antioxidants***: Antioxidants such as vitamin C act as preservatives by inhibiting the effects of oxygen on food, and can be beneficial to health.
6. ***Bulking agents***: Bulking agents such as starch are additives that increase the bulk of a food without affecting its nutritional value.
7. ***Food coloring***: Colorings are added to food to replace colors lost during preparation, or to make food look more attractive and appealing (McCann, et al., 2007).
8. ***Color retention agents***: In contrast to colorings, color retention agents are used to preserve a food's existing color.
9. ***Emulsifiers***: Emulsifiers allow water and oil to remain mixed together in an emulsion, as in mayonnaise, ice cream, and homogenized milk.
10. ***Flavors***: Flavors are additives that give food a particular taste

or smell, and may be derived from natural ingredients or created artificially.

11. ***Flavor enhancers:*** Flavor enhancers enhance a food's existing flavors. They may be extracted from natural sources (through distillation, solvent extraction, and maceration, among other methods) or created artificially.
12. ***Flour treatment agents:*** Flour treatment agents are added to flour to improve its color or its use in baking.
13. ***Glazing agents:*** Glazing agents provide a shiny appearance or protective coating to foods.
14. ***Humectants:*** Humectants prevent foods from drying out.
15. ***Tracer gas:*** Tracer gas allows for package integrity testing preventing foods from being exposed to the atmosphere, thus guaranteeing shelf life.
16. ***Stabilizers:*** Stabilizers, thickeners, and gelling agents, like agar or pectin (used in jam for example) give foods a firmer texture. While they are not true emulsifiers, they help to stabilize emulsions.
17. ***Sweeteners:*** Sweeteners are added to foods for flavoring. Sweeteners other than sugar are added to keep the food energy (calories) low, or because they have beneficial effects for diabetes mellitus, tooth decay and diarrhea.
18. ***Thickeners:*** Thickeners are substances which, when added to the mixture, increase its viscosity without substantially modifying its other properties (Czarra, 2009).

Instead of cooking their own food these days, the majority of people prefer to consume it already prepared from the market. These foods have preservatives and additives of some kind to preserve their flavor and quality and prevent bacteria and yeast from ruining them. There are over 3000 preservatives and additives on the market that function as antimicrobials and antioxidants. Among additives, salt, and sugar are the most frequently used. Aluminum silicate, amino acid compounds, ammonium carbonates, sodium nitrate, propyl gallate, butylated hydroxyl toluene (BHT), butylated hydroxyanisole (BHA), monosodium glutamate, white sugar, potassium bromate, potassium sorbate, sodium benzoate, etc., are a few of the frequently used food additives and preservatives. The foods are also given some artificial coloring to give them a visually appealing appearance. Erythrosine (red), cantaxanthin (orange), amaranth (Azoic red), tartrazine (Azoic yellow), and annatto bixine (yellow orange) are these coloring agents. To preserve the food's quality and flavor during long-term storage, additives, and preservatives must be used. Foods containing too much water may encourage the growth of bacteria, fungi, and yeasts. The use of preservatives and additives stops food from spoiling because they stop bacteria and fungi from growing. Preservatives and additives keep food consistent and of high quality. Additionally, they preserve the food's wholesomeness and palatability, boost or preserve its nutritional value, regulate pH levels appropriately, add color and leavening, and improve flavor.

9.1.2. Chemistry of Food Additives

Any food that needs to be kept for a longer period of time needs to be properly processed. The food is given certain materials and chemicals, referred to as additives, during this processing. To keep the food from spoiling, we also need to add some preservatives. Foods that have direct additives added to them on purpose do so for specific reasons. Food is infused with indirect additives while it is being processed, packaged, and stored (Richard and Totowa, 1995). Whether one produces chemical food additives or relies on food preservative suppliers to supply additives and preservatives for baked goods or beverages, food additives require the proper machinery, dispensing, and storing equipment. When it comes to supplies, the food preservatives industry takes into account the necessity of using food additives in food products. Stated differently, food additive supplies and equipment should contribute to the preservation of food products' freshness, consistency, texture, taste, color, and nutrient levels. Food preservation presses, industrial mixers, homogenizers, pasteurizers, and dispensers are among the supplies for food additives. Other typical equipment used for food preservatives includes vacuum tumblers, storage, and processing tanks, and conveyors for marinating or curing meats. Chemical food additives that come in syrup, crystal, or powder form, such as potassium sorbate, sodium nitrate, and corn starch, are typically ordered from food preservative suppliers.

9.1.3. Effect of Additives in Food

Due to evidence linking certain contemporary synthetic preservatives to respiratory and other health issues, these preservatives have gained controversy. Certain research indicates that artificial coloring agents and synthetic preservatives exacerbate the symptoms of ADD and ADHD in those who are impacted (Gustafsson et al., 2003). Previous research yielded conflicting results, most likely because clinical techniques for assessing deviant behavior were insufficient. Clinical tests were not as reliable as parental reports in detecting the presence of additives. When artificial ingredients, including preservatives, were removed from school food programs, several significant studies have shown that academic performance increased and disciplinary issues decreased in sizable populations of non-ADD students (Schoenthaler and Doraz (1986).

In those who are vulnerable, allergic preservatives in food or medication can result in anaphylactic shock, which is frequently fatal in a matter of minutes if emergency care is not received. It is almost a given that very few people actually understand what it is that they eat and that could be dangerous. Additives that pose a threat or are dangerous primarily come from two sources. The first group consists of those added during the processing process. These consist of flavorings, sweeteners, texture agents, colorings, preservatives, and processing agents. These can be identified with a little knowledge and by paying attention to the manufacturer's information, and details about them need to be included on the label. Food without any additives should ideally be chosen, especially if it will be fed to youngsters. These additives cause far too many of our children to become sensitized, which can result in allergies and disorders like Attention Deficit Disorder and Hyperactivity (Schoenthaler and Doraz, 1986).

9.1.4. Advantages of Food Additives

The modern food supply is significantly impacted by food additives. They make a wide range of foods available all year round to our expanding urban population. Additionally, they enable a variety of convenience foods without the hassle of regular shopping. Food additives serve many beneficial purposes in foods that are frequently overlooked. Additives help keep food wholesome and appealing while it is being transported to markets, sometimes thousands of miles away from where it is grown or manufactured, as the majority of people no longer live on farms. In addition to increasing a food's nutritional content, additives can enhance a food's flavor, texture, consistency, or color (Houghton, 2002).

Additives are used in foods for five main reasons:

1. To maintain product consistency. Emulsifiers give products a consistent texture and prevent them from separating. Stabilizers and thickeners give a smooth uniform texture. Anti-caking agents help substances such as salt to flow freely.
2. To improve or maintain nutritional value. Vitamins and minerals are added to many common foods such as milk, flour, cereal, and margarine to make up for those likely to be lacking in a person's diet or lost in processing. Such fortification and enrichment has helped reduce malnutrition among the U.S. population. All products containing added nutrients must be appropriately labeled.
3. To maintain palatability and wholesomeness. Preservatives retard product spoilage caused by mold, air, bacteria, fungi or yeast. Bacterial contamination can cause foodborne illness, including life-

threatening botulism. Antioxidants are preservatives that prevent fats and oils in baked goods and other foods from becoming rancid or developing an off-flavor. They also prevent cut fresh fruits such as apples from turning brown when exposed to air.

4. To provide leavening or control acidity/alkalinity. Leavening agents that release acids when heated can react with baking soda to help cakes, biscuits, and other baked goods to rise during baking. Other additives help to modify the acidity and alkalinity of foods for proper flavor, taste, and color.
5. To enhance flavor or impact the desired color. Many spices, and natural and synthetic flavors enhance the taste of foods. Colors, likewise, enhance the appearance of certain foods to meet consumer expectations (Houghton, 2002).

Preservatives and additives are used to preserve the wholesomeness and consistency of products, as well as to improve or maintain their nutritional value, leaven, regulate pH, improve flavor, and add color.

9.1.5. Disadvantages of Food Additives

While preservatives and additives are necessary for food storage, they can also cause some health issues. In certain individuals who are sensitive to particular chemicals, they can result in various allergies and disorders like hyperactivity and attention deficit disorder (Louis and Botulism, 1991).

Additive-containing foods can aggravate eczema and cause hay fever, asthma, and other reactions like rashes, vomiting, headaches, tight chests, and hives. Some of the known dangers of food additives and preservatives are as follows:

1. Benzoates can trigger allergies such as skin rashes and asthma as well as believed to be causing brain damage.
2. Bromates destroy the nutrients in the foods. It can give rise to nausea and diarrhea.
3. Butylates are responsible for high blood cholesterol levels as well as impaired liver and kidney function.
4. Caffeine is a colorant and flavorant that has diuretic, and stimulant properties. It can cause nervousness, heart palpitations and occasionally heart defects.
5. Saccharin causes toxic reactions and allergic responses, affecting the skin, gastrointestinal tract and heart. It may also cause tumors and bladder cancer.

6. Red Dye 40 is suspected to cause certain birth defects and possibly cancer.
7. Mono and di-glycerides can cause birth defects, genetic changes and cancer.
8. Caramel is a famous flavoring and coloring agent that can cause vitamin B6 deficiencies. It can cause certain genetic defects and even cancer.
9. Sodium chloride can lead to high blood pressure, kidney failure, stroke, and heart attack.

Foods containing additives and preservatives should be avoided in order to reduce the chance of health issues resulting from them. You need to review the ingredients of the canned food before buying it. Organic foods are free of artificial additives, so you should purchase them. Eat as many freshly made foods as you can instead of processed or canned ones (Louis and Botulism, 1991).

9.2. NATURAL ALTERNATIVES TO CHEMICAL ADDITIVES IN ULTRA-PROCESSED FOODS

One of the primary goals of the food industry is to find natural substitutes for chemical additives. The food industry is going through a revolution when it comes to the ingredients and methods they use in food today since health and wellness are top priorities. Customers are calling for foods that are more natural, less processed, and additive-free.

Foods labeled as ultra-processed are notorious for having a lot of chemical additives. Are there any natural alternatives to these additives, which are used to enhance the food's organoleptic properties or safety?

9.2.1. Chemical Additives Are Used in Ultra-Processed Foods

Chemical additives are substances that are added to food to enhance its flavor, texture, color, or shelf life. They are governed by Regulation (EC) No 1333/2008 of the European Parliament and of the Council of December 16, 2008, which enumerates all additives approved for use in the European Union along with their usage restrictions under the corresponding E number.

Studies involving excessive additive consumption in ultra-processed foods (like sausages, ready meals, pastries, etc.) have raised concerns about these additives at the moment.) may be connected to several health risks, including allergies, metabolic diseases, and negative effects on digestive health.

Customers frequently find it challenging to make educated food decisions due to opaque product labels. Because of this, more, and more customers are searching for natural products without artificial additives.

9.2.2. Natural Alternatives to Chemical Additives Are There on the Market

The market for ingredients offers a vast array of natural substitutes for chemical additives. Ultra-processed foods frequently require additives to help preserve color, impart a particular flavor, or keep the food fresher for longer than it would otherwise; however, natural ingredients can serve the same purpose as artificial additives.

Herbal and spice extracts, fruit or vegetable juices, and other natural extracts that can safely and healthily serve as preservatives, colorings, and flavor enhancers are the most well-known alternatives to additives. They are a healthy, natural, and eco-friendly choice. But other naturally occurring substances, like ferments or lactic acid bacteria, work better as preservatives.

9.2.3. Starters as an Alternative to Chemical Additives

Starters are lactic acid bacteria colonies that provide a great substitute for synthetic additives.

It has long been understood that starters carry out this preservative role by allowing the product to mature. This typically involves an organoleptic alteration in the food as a result of acidification, which decreases pH and alters the matrix's organoleptic properties. They boost the food's nutritional content and flavor in addition to preserving it better. Manufacturers are familiar with these starters since they have historically been used to mature cheeses and sausages.

However, some ferments preserve food through other mechanisms, like competition, without causing acidification in the final product.

9.2.4. Benefits of Opting for Natural Alternatives to Ultra-Processed Foods

There are many advantages to selecting foods high in natural ingredients and low in artificial additives. First of all, by eschewing artificial additives, you lower your exposure to dangerous chemicals and encourage a more sustainable and natural way of living in addition to advocating a healthier and more nutrient-dense diet. However, the use of these natural substitutes in highly processed foods, like ferments, permits a reduction in food processing, leading to a product that is more "homemade." Along with the removal of chemical additives, the food will have a more favorable impression on the consumer.

9.3. IMPACT OF FOOD ADDITIVES IN ULTRA-PROCESSED FOOD ON HUMAN HEALTH

A 2020 INIC consumer survey revealed the food safety habits, trends, and opinions of American consumers. Of those who try to avoid certain food ingredients, 77% cite artificial ingredients, sugar, and salt as the most common ones. 43% of those surveyed stated that avoiding particular food ingredients would improve their long-term health due to those ingredients' negative effects on health. This indicates that, when it comes to nutrition content, almost one in four consumers are worried. Food additives are artificial or natural materials that are added to food during processing to enhance or preserve the food's freshness, safety, texture, appearance, or flavor. Additives have long been used by humans, and contemporary food processing has introduced several new additives to food. Based on their characteristics and uses, FAO, and WHO have combined to classify food additives into three primary categories: flavoring agents, enzyme preparations, and other additives. There are many uses for other additives, such as e.g., Food coloring, food preservation, etc. There are severe limits on the amount that can be consumed each day because some of them are unhealthy for humans.

9.3.1. Ultra-Processed Food and NOVA System

Foods that have been processed to alter their qualities are referred to as ultra-processed foods. g. add color, add salt, and so forth. According to NOVA, there are four categories into which all foods fall: Group 1 comprises unprocessed or very minimally ultra-processed foods; unprocessed foods are defined as any food that comes directly from nature, including fruits, leaves, roots, eggs, milk, and fresh animal meat. Foods that have undergone minimal processing involve removing undesired or unfit components by methods like freezing, drying, grinding, roasting, and so on. to prolong the shelf life of food. Group 1 foods include vacuum-packed seafood and frozen steak, while Group 2 foods are processed culinary ingredients. Group 3 is processed food; this category includes foods from Groups one or two, such as cheese, fresh bread, and canned fish, as well as products made with additional salt, oil, sugar, or foods from those groups. Group 1 foods are the source of salt, oil, butter, and sugar, which are considered processed culinary ingredients. Usually, there are two or three ingredients in these dishes. They go well with other foods or can be eaten on their own. Food in this group has undergone processing to increase stability without sacrificing quality; Group 4 is Ultra-processed foods, which are defined as foods whose primary nutritional content consists of additives and components derived from food. Group 4 is still intact with Groups 1 and 2. Ultra-processed foods like beverages, pre-made frozen dishes, and fried foods frequently contain Group 1 foods like salt, oil, sugar, and chemical additives like artificial sweeteners, preservatives, and antioxidants (Monteiro et al., 2018).

9.3.2. Recent Studies about Diseases and Food Additives

A UK study found that, as Figure 9.1 illustrates, the average percentage of UPF consumption was 22.9% of the total diet. They came to the conclusion that rising UPF consumption is directly correlated with rising cancer risk. Higher UPF consumption is particularly linked to a higher death rate from ovarian cancer in women (Gunter et al., 2023). Because UPFs have the highest additive content, consumers may unintentionally consume a high dose of additives, increasing their risk of illness.

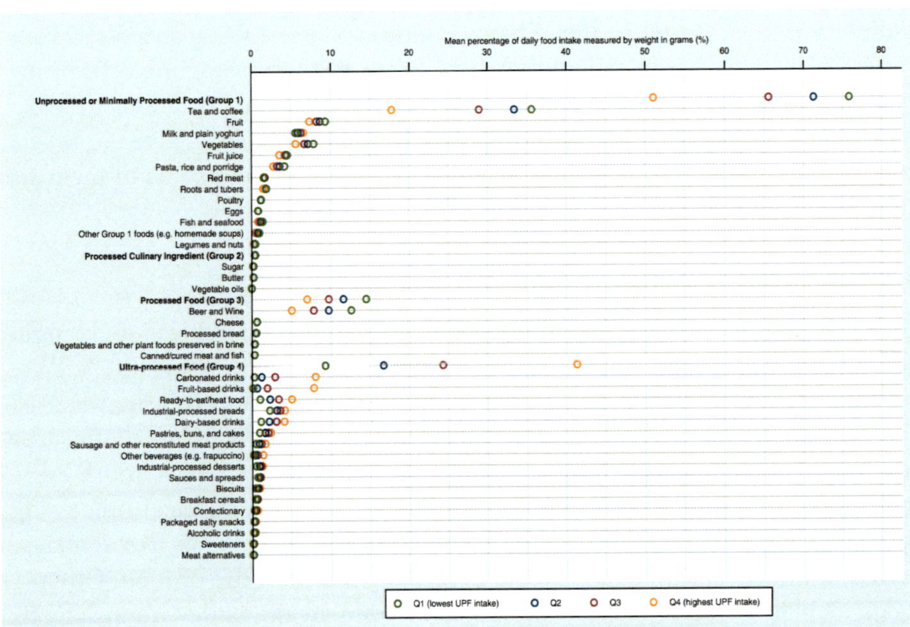

Figure 9.1. *The total diet of ultra-processed food consumption (Gunter et al., 2023).*

Source: Liang, X. (2023).Impact of food additives in ultra-processed food on human health. Theoretical and Natural Science,6,437-444.

To increase awareness about nutrition and wellbeing, many news outlets have shared a variety of articles on social media about the negative effects of highly processed foods on wellbeing. Of these, the harm caused by food additives has been discussed extensively. Examples include the crushing of human cells by anti-cancer agents, the effect of color on children's development, and the link between food additives and cancer. Therefore, consumers need to be aware of the harm that dietary additives do to health and the need to regulate daily added substance intake. This article will provide an overview of the harm that substances added to food can do to the human body from the perspective of atomic structure and chemical properties, as well as the connection to an increased risk of cancer.

9.3.3. Common Food Additives

Food manufacturers have added a variety of chemicals to food to enhance its flavor and quality. Food additives can be used to modify the food's texture, flavor, and other characteristics to satisfy customers. These three additives are frequently used when processing food.

9.3.3.1. Antioxidants

Antioxidants are molecules that provide electrons to counteract free radicals within the body, thereby reducing lipid and protein damage. Antioxidants that prevent cancer include carotenoids, vitamin C, vitamin E, and selenium. Free radicals are molecules with unpaired electrons that react violently with other substances. They are primarily produced by radiation, chemicals, tobacco use, and consuming an excessive amount of sugar. These unrestricted radicals have the potential to cause "oxidative stress" and kill human cells. Dietary antioxidants, endogenous antioxidants, exogenous antioxidants, synthetic antioxidants, and natural antioxidants are the various types of antioxidants that have been identified (Kebede et al., 2019).

According to the information in Antioxidants, they are compounds that provide radicals with electrons. A higher antioxidant diet lowers the rate of death from cancer and cardiovascular diseases, per a recent research report (Parohan et al., 2019). This is because antioxidants donate electrons to lower active radicals, which are safe for the body and reduce the amount of highly reactive radicals. Antioxidants are therefore frequently regarded as vital nutrients in the diet. Vitamin C, also known as L-ascorbic acid, is a common antioxidant found in food and serves as an essential cofactor for several enzymatic reactions. e.g., the biosynthesis of carnitine, the post-translational hydroxylation of collagen, the neurotransmitter dopamine's conversion to norepinephrine, etc. Vitamin C is an important antioxidant that breaks down chains when it dissolves in water. Ascorbate's ability to scavenge superoxide, hydrogen peroxide, hypochlorous acid, aqueous peroxyl radicals, and other chemicals has been demonstrated.

Ascorbate undergoes two reductions during the antioxidant process of those free radicals, first forming semi-dehydroascorbic and then forming dehydro-L-ascorbate, as illustrated in Figure 9.2 (Young et al., 2001). With a minimal chance of harm, vitamin C lowers the amount of free radicals in the human body. Numerous scientists hypothesize that antioxidants could be incorporated into cancer therapies. Antioxidants may be used to treat cancer, even though more clinical research is required before using vitamin C to treat CVDs (Morelli et al., 2020).

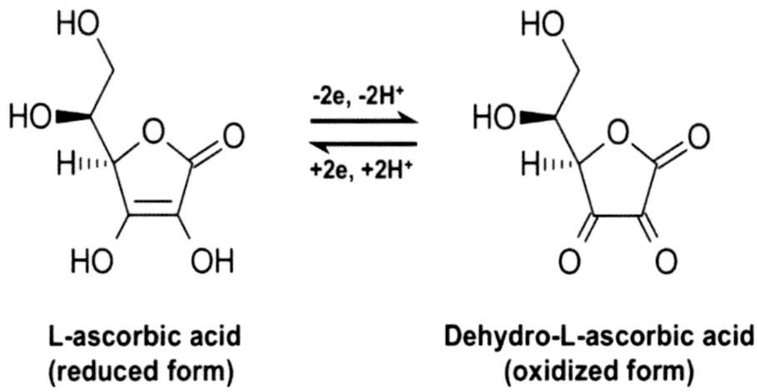

Figure 9.2. *Reduced (left) and oxidized (right) forms of vitamin C by Stanislaw Gackowski (Parohan et al., 2019).*

Source: Liang, X. (2023).Impact of food additives in ultra-processed food on human health. Theoretical and Natural Science,6,437-444.

9.3.3.2. Food Preservatives

Food additives known as preservatives or antimicrobials shield food from enzymatic, microbiological, or chemical reactions that cause natural food spoilage over extended periods of storage. Food preservation can be divided into two main categories: chemical and natural preservatives. Natural preservatives in Class I include salts, alcohol, sugar, honey, and vegetable oil are a few instances of organic preservatives. The use of them in food is less restricted because they are found naturally and pose less of a threat to health. Class II chemical preservatives are defined as artificially created in a laboratory chemically synthesized preservatives. For example, nitrites, sulfur dioxide, acetates, and benzoates. They might have an impact on the body's equilibrium. Preservatives are therefore subject to stringent regulatory measurement requirements set by agencies such as the FDA and the EU. Chemical preservatives are further divided into categories such as antimicrobial, antioxidant, and anti-enzyme. There are restrictions on the dosage of some preservatives (Gupta et al., 2021).

Sodium benzoate is a colorless powder that is synthesized in a laboratory by neutralizing sodium hydroxide and benzoic acid, as illustrated in Figure 9.3.

A wide variety of plants, including berries, plums, prunes, and spices, naturally contain benzoic acid. Benzoic acid is produced commercially in the food industry by reacting oxygen and toluene, which uses inexpensive raw materials but produces a high yield. Because of this, benzoic acid works well as a food preservative in fruit juices, carbonated drinks, and acidic foods. Additionally, it could be used to process jams and sauces. When food is exposed to low pH levels, sodium benzoate prevents the growth of mold and other microorganisms (Dey et al., 2022). The entire molecule and pH are related to its antimicrobial activity. Benzoate has a pKa of 4.2, which makes it potentially harmful to bacterial growth. Benzoate can sometimes effectively inhibit the growth of mold and yeast (Shahmohammadi et al., 2016).

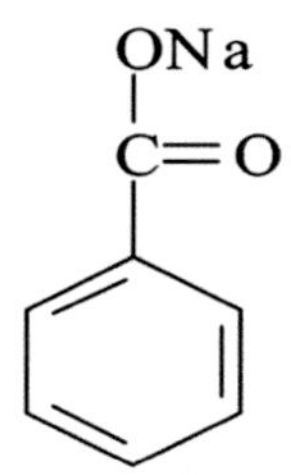

Figure 9.3. *Sodium benzoate (World Health Organization, (2000)).*

Source: Liang, X. (2023).Impact of food additives in ultra-processed food on human health. Theoretical and Natural Science,6,437-444.

9.3.3.3. Coloring Agents

A color additive is any chemical that can add color to food, including dyes and pigments. The following forms of coloring agents are available: powder, granular, Lake Colors, FD, and C Colors, and D & C Colors (Julie Barrows et al., 2017). They are very useful in many aspects of our daily lives, such as e.g., Bakery, drinks, candies, dairy products, meat, and savory foods, pharmaceuticals, seafood, and pet food. They are regarded as necessary components in highly processed food because colorings make food look more appetizing by making it seem enticing and tempting. There are additional uses for color agents, such as e.g., enhancing the original color, guarding against nutrient loss or damage from light exposure, giving food a distinctive appearance, enhancing flavor, and hiding color variations in food. Yellow No. is among the food colorings that have received approval from the FDA and EFSA. Yellow No. 5 (Tartrazine). Figure 9.4, Red No., displays Figure 6 (Sunset Yellow). Red No. 3 (Erythrosine). Blue No. 40 (Allura Red) in Figure 9.5. Blue No. 1 and Brilliant Blue. Figure 9.6 (Indigo Carmine) (Dey et al., 2022).

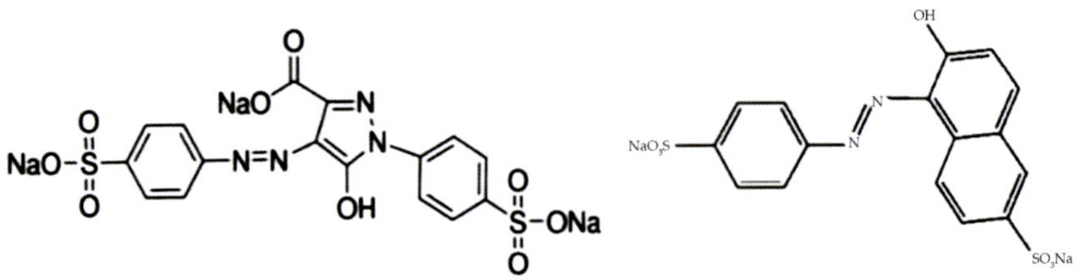

Figure 9.4. *Tartrazine (left), Sunset Yellow (right) (Subhashish Dey et al., 2022).*

Source: Liang, X. (2023).Impact of food additives in ultra-processed food on human health. Theoretical and Natural Science,6,437-444.

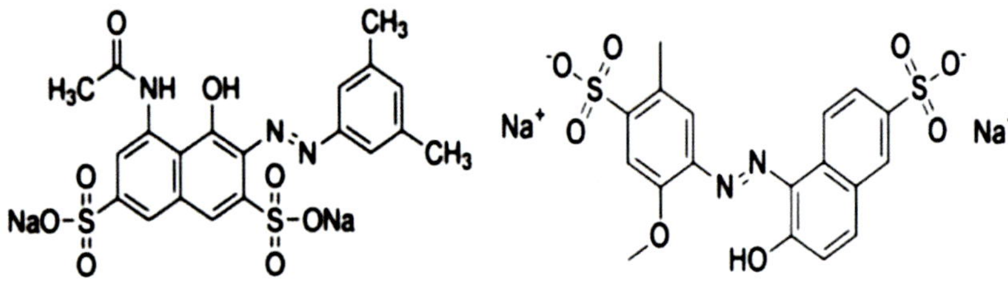

Figure 9.5. *Erythrosine (left), Allura Red (right) (Subhashish Dey et al., 2022).*

Source: Liang, X. (2023).Impact of food additives in ultra-processed food on human health. Theoretical and Natural Science,6,437-444.

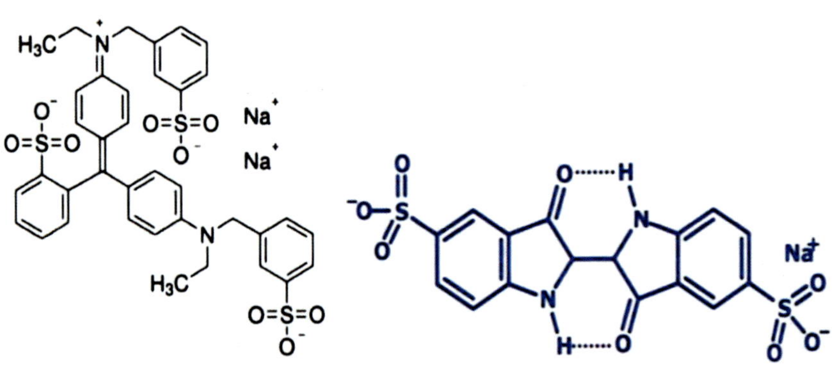

Figure 9.6. *Brilliant Blue (left), Indigo Carmine (right) (Subhashish Dey et al., 2022).*

Source: Liang, X. (2023).Impact of food additives in ultra-processed food on human health. Theoretical and Natural Science,6,437-444.

9.3.4. Threats and Disease

9.3.4.1. Antioxidants' Mechanism of the Reaction and Impact on Health

The short-term oxidative stretch may occur in tissues damaged by trauma, infection, heat shock, hyperoxia, toxins, or excessive exercise. These injured tissues release chemicals that produce more radicals, activate phagocytes, release free radicals, release copper particles, or disrupt the electron transport chains of oxidative phosphorylation, all of which increase ROS levels. The onset, progression, and spread of cancer as well as the adverse effects of radiation and chemotherapy have all been linked to the imbalance between ROS and the antioxidant defense system. Age-related macular degeneration, Parkinson's disease, and neurodegenerative infections have all been linked to ROS production and progression. There have been two main suggested modes of action for cancer prevention agents. The primary antioxidants may act as a chain-breaking component, providing electrons to the system's free radicals, including lipid radicals. Chain-breaking antioxidants function by taking out free radicals and donating hydrogen atoms. The second method involves removing the RNS initiator and ROS by quenching the chain initiator catalyst. Reductants and metal chelators that shield other antioxidants *in vivo* are the most common types of antioxidants that function as prophylactics. These reactive species have the power to upset homeostasis and harm vital biological components such as proteins, lipids, carbohydrates, and deoxyribonucleic acid (DNA) (Lobo along with others, 2010).

Antioxidants are therefore used in the treatment and prevention of cancer. Nevertheless, some research indicates that they might exacerbate or enlarge the tumor. A 2020 analysis revealed a link between a diet high in beta-carotene or vitamin E and an increased risk of lung cancer. However, the opinion is only partially supported by research, and there is no concrete evidence linking beta-carotene consumption to lung cancer. There is little to no difference in the risk of lung cancer between high dosages of vitamins A, C, D, and E, calcium, selenium, and a combination of cancer-prevention drugs. Use vitamin E as an example. It appears that there are small contrasts. While vitamin E does have some side effects, it has no discernible impact on lung cancer mortality or risk (Gaber El-Saber Batiha et al., 2021).

9.3.4.2. Preservatives and Threat to Health

According to a diagram, although sodium benzoate is thought to be a safe substance, prolonged exposure can disrupt the respiratory system, eyes, and skin, while too much time spent in contact or repeated exposure can exacerbate skin sensitization. Tall amounts result in altered production of gastrointestinal bodily fluids, ulcers, and the release of prostaglandins and histamines. According to a 2017 study, sodium benzoate raised blood pressure, which ultimately caused the rats' blood vessels to burst (Eberechukwu et al., 2017). Using sodium benzoate can have several unfavorable effects, such as destruction of the hepatocyte cell film and crista setbacks in mitochondria, attachment to the outermost shell of the vacuole mitochondria within the cytoplasm, and abnormalities in the liver and kidney (Cortés-Jofré et al., 2020). The research helped demonstrate how benzene forms in natural products and soft drinks when benzoic and ascorbic corrosives react without the use of metal catalysts (Shahmohammadi et al., 2016). In 2015, another study examined the carcinogenic potential of eating processed meat. There were also easily accessible results for 15 more cancer genres. There was a correlation between the consumption of manufactured meat and stomach cancer as well as between the consumption of red meat and pancreatic and prostate cancer throughout the prospective study and demographically case analysis. Based on extensive data and ongoing associations between colorectal cancer and rising prepared meat consumption across a range of demographics, almost all of the studies found sufficient evidence to support the conclusion that mechanically delivered meat admissions pose a cancer risk to humans. However, there is still uncertainty surrounding the use of red meat because certain studies did not reveal a clear correlation, and extrapolating data from other dietary habits and risks is difficult. In summary, more investigation is needed to demonstrate that eating prepared meat increases the risk of cancer (Bouvard et al., 2015).

9.3.4.3. Food Coloring and Impact on Health

Artificial food coloring raises the possibility of children developing ADHD and attention deficit hyperactivity disorder, according to Harvard Health Publishing (Claire McCarthy, 2021). In the production of food, synthetic colorings were widely used to enhance the appearance of the finished goods. Yellow No. was used by Kraft Macaroni and Cheese. 5 and Yellow Number 1. 6 to add more appeal and attractiveness to the macaroni. Because customers wouldn't buy food with artificial ingredients, the company switched from using artificial colors to natural ones (Harrington, 2015). Subsequent studies on artificial coloring revealed that kids with a higher intake of food coloring had higher hyperactivity scores. However, recent research indicates that while some kids exhibit greater hyperactivity, others show no discernible change. Additionally, the ratio is 1/2 (Kannarek, 2011). The idea that food coloring is the direct cause of children's hyperactivity is not supported by enough research. It is true that consuming too much food coloring is unhealthy, so one must limit their intake.

In any case, the real ingredients, like sugar and carbs, rather than the chemical additions, are what cause them to be harmful to our bodies. The three main types of fixing substances found in food packages are typically sugar, fats, and starch. Eating sugar makes people happy because it causes the brain to produce an affront, which enhances positive emotions and encourages people to continue eating. Consequently, the elevated blood sugar level in the circulatory system causes high blood pressure, which in turn causes heart diseases like heart attack and blockage. Furthermore, a high-fat diet that causes weight loss can harm heart tissue. Therefore, the primary cause of heart disease is ultra-processed food.

9.4. INGREDIENTS TO AVOID IN PROCESSED FOOD

As a registered dietitian, minimally processed foods and some processed foods are acceptable, but ultra-processed foods (UPFs) make me uneasy. No, a single Twinkie or Big Mac won't be the end of you. However, it is unhealthful if your diet consists primarily of foods with added sugar and other dubious nutrients. First, let's address the obvious: processed foods are thought to be a contributing factor to the obesity epidemic because they are often high in empty calories.

However, obesity is merely one of the health issues linked to a diet heavy in processed foods. Ultra-processed foods have been linked to accelerated cognitive decline. A study presented at the Alzheimer's Association International Conference 2022 found that individuals who consume high quantities of ultra-processed foods experience a faster rate of cognitive decline. Over eight years, researchers examined 10,775 participants and discovered that consuming high levels of ultra-processed foods (20% or more of daily intake) was associated with a 28% faster decline in global cognitive scores, which include executive function, memory, and verbal fluency. In addition, a high intake of these UPFs has been connected to a host of other issues, including inflammation, cancer, poor dental health, microbiome imbalances, behavioral issues, sleep disturbances, disruption of the endocrine system, heart problems, and more.

UPFs do tend to be high in calories, carbohydrates, and fat, but the additives also seem to be a cause for concern. Now let's examine the connections between particular ingredients and health issues. The general consensus on additives has been that most substances are safe in moderation, which is why eating a varied diet and avoiding processed foods is crucial. The 10 ingredients on this list are acceptable to have occasionally; however, if they are a regular ingredient in your bread, you might want to think about replacing them with less processed alternatives.

9.4.1. Hydrogenated and Trans Fat

In the US food system, vegetable, and seed oils—derived from soybean, canola, corn, cottonseed, and other sources—are widely used in everything from highly processed foods to home cooking and culinary applications. Vegetable oils contain polyunsaturated fats, including omega-6, which are unstable. Therefore, to make these oils more stable, the food industry hydrogenates them, changing their chemical structure. Artificial trans fats, a type of fat harmful to health, are developed with the help of this process.

Despite FDA regulations prohibiting the use of artificial trans fats, foods like margarines, shortenings, biscuits, pie crusts, shelf-stable pastries, tortillas, crackers, and more frequently contain hydrogenated oils. Inflammation and ill health can result from consuming too many omega-6 fats from vegetable oils and not enough omega-3 fats.

9.4.2. High Fructose Corn Syrup (HFCS)

A common sweetener in highly processed foods is HFCS, which is derived from corn. Because HFCS is widely used in so many packaged foods, it is simple for consumers to consume too much of it. Similar to sucrose, or table sugar, HFCS is also made up of glucose and fructose, the sweetest type of sugar; the difference is that HFCS has a slightly higher fructose ratio than table sugar, which is 50/50. This is significant because fructose metabolizes differently from glucose and because eating too much fructose from processed foods can harm the liver. In the end, HFCS should be avoided as an additional source of added sugar in the diet.

High fructose corn syrup consumption was linked in a dose-dependent manner to worsened glucose parameters and increased liver fat in a clinical study involving 85 participants. This finding suggests that high fructose corn syrup consumption may play a role in the development of nonalcoholic fatty liver disease and type 2 diabetes.

9.4.3. Artificial Colors

The jury is still out because the research is contradictory. Although the consumption of artificial colors has been connected to behavioral issues like ADHD, the cause of these issues has not been proven, although some research indicates that these food colorings may exacerbate the problems. In the end, the American Academy of Pediatrics' clinical guidelines and practice parameters, as well as the National Institute for Health and Care Excellence's in the UK, do not consistently advise against removing artificial food coloring from diets in order to treat ADHD.

Allergy responses to dyes and the possibility of contaminants present in dyes are additional worries. As processed foods frequently contain other ingredients that should be avoided regardless, avoiding foods containing artificial colors will likely benefit you in the long run if you want to err on the side of caution. Colors can be added to food by using turmeric, beets, blueberries, cherries, and other naturally occurring foods with vibrant colors.

9.4.4. Added Sugar

It's common knowledge by now that consuming large amounts of sugar is unhealthy. But we should be mindful of more than just the amount of sugar we put in our coffee and baked goods. It is noteworthy that a significant factor contributing to excessive sugar consumption is packaged foods and beverages. Yogurt is one example of a generally healthy food that may have too much sugar. Make sure to check the nutrition facts breakdown to see how much is specifically accounted for by added sugar.

9.4.5. Artificial Sweeteners

Well, we all know that sugar consumption should be reduced, but artificial sweeteners also come with "baggage"—some of it very serious. People who consumed large amounts of artificial sweeteners had an overall higher risk of cancer compared to non-consumers in a large cohort study involving over 100,000 people. Aspartame and acesulfame-K in particular have been linked to a higher risk of cancer. Furthermore, synthetic sweeteners might be a factor in dysbiosis, or an unbalanced state of gut health. Watch out for saccharin and sucralose in addition to aspartame and acesulfame-K.

If you're not sure which of these is in the packet you prefer from your neighborhood café, do a quick Google search to find out which sweetener that particular brand uses. If you'd like something without sugar, xylitol, and stevia are natural sugar substitutes that you can use without risk.

9.4.6. Nitrites/Nitrates

In addition to being naturally occurring in soil and water (and consequently in food, including vegetables!), nitrates, and nitrites are also added as preservatives to processed meats to stop the meat from browning and to stop bacteria from growing. After analyzing hundreds of studies, the World Health Organization concluded that processed meats, like hot dogs, raise our risk of colorectal cancer. But nobody wants to eat bad meat, so what's the deal? In fact, processed meats have been identified by the WHO as carcinogens—substances that encourage the development of cancer. Higher nitrite intake was associated with an increased risk of stomach, bladder, and glioma cancer. The risk of thyroid cancer increased with increased nitrate intake. Food nitrates and nitrites have been linked to a lower risk of prostate and breast cancer, according to another study.

As usual, there's more to this story; if nitrites/nitrates are present in vegetables as well as other foods grown in soil, then why aren't vegetables included in the WHO's list of carcinogens and part of the problem? Vegetable nitrites and nitrates have a lower tendency to combine to form related substances known as nitrosamines. Because processed meats are cooked at high temperatures and contain a higher concentration of amino acids from protein, the nitrites produce advanced-glycated end products and nitrosamines, which can cause cancer. The main lessons to be learned from this would be to choose off-the-frame deli meats rather than those that are packaged in yellow and red plastic bags and to think twice before lighting the hot dog grill, especially if you do it frequently.

9.4.7. Monosodium Glutamate (MSG)

This flavor enhancer, whose umami (savory) taste is frequently linked to its use in Chinese cuisine, is also present in the well-known brand of triangle tortilla chips with a nacho flavor. It is called monosodium glutamate and is listed under ingredients on the bag. The FDA has classified MSG as generally recognized as safe (GRAS), but some researchers have connected prolonged use of the spice to problems with the liver, fat tissue, central nervous system, and other organ systems. Additionally, some people should avoid MSG because they are particularly sensitive to it.

That being said, not all glutamate is harmful. Tomato, cheese, and seaweed are among the foods that naturally contain glutamate, an amino acid. When MSG is added to food, it is present in larger quantities than when it is present naturally.

9.4.8. Synthetic Preservatives

Natural food preservatives include extract from rosemary. However, I've put the synthetic ones on my "avoid" list. Synthetic compounds with antioxidant qualities like TBHQ and BHA/BHT are used, for example, to help keep oils and fats from oxidizing. Although BHT is recognized as safe by the FDA and other international agencies when consumed at levels permitted in foods, there have been reports of its potential to interfere with sleep, alter serotonin and norepinephrine levels, and cause endocrine disruption. Selecting foods with a shorter shelf life is preferable.

9.4.9. Artificial Flavors

Artificial flavors are frequently used to improve the flavor and aroma of ultra-processed foods. Diacetyl, an artificial butter flavoring, is one such. The use of this ingredient in the making of microwaveable popcorn is the source of the cautionary tale known as "popcorn lung." Lung disease was developed by popcorn factory workers as a result of high diacetyl inhalation. Alzheimer's in laboratory animals has also been connected to the flavor, which is similar to butter. Six of the seven artificial flavorings that the FDA removed from the list of food additives in 2018 did so because they were shown to cause cancer in laboratory animals. The good news is that manufacturers typically list artificial flavoring right on the label, so if you're a picky consumer, it shouldn't be too difficult to stay away from these additives.

9.4.10. Excessive Sodium

Sodium is a vital mineral that affects many bodily processes, including heart health, neurological system function, cellular function, and electrolyte balance. Excessive amounts of good things can also be harmful, just as excessive amounts of bad things are bad. Large amounts of sodium, a component of table salt, are used in the manufacturing of processed foods not only to add flavor but also as a preservative. The average daily intake of sodium among Americans is 3,400 mg, which is almost 50% higher than the 2,300 mg upper limit advised by federal guidelines for individuals 14 years of age and above. A significant portion of this excess sodium is derived from packaged and prepared foods.

9.4.11. Cut Out Processed Foods in a Healthy Diet

If you're sold on reducing the processed foods from your diet, it's pretty easy to make small changes to how you shop and prepare food. Start with these tips:

- ***Discerning deli shopping***: Choose "off the frame" sliced meats instead of those that have been cured.
- ***Be a baker***: Make your own tortilla shells and baked goods.
- ***Read labels***: Read the nutritional facts *and* ingredient list. Keep a close eye on added sugars and sodium in particular, and compare brands rather than just tossing your favorite in the shopping cart.
- ***Play the numbers game***: Choose packaged goods that have 3–5 ingredients—or less!
- ***Make a lifestyle change***: Follow the Mediterranean or flexitarian diet, both of which will naturally steer you towards unprocessed foods, with their emphasis on fruits and vegetables and whole grains.
- ***Stay in balance***: Make sure you are getting enough omega-3, as this nu-

trient helps balance out the more harmful omega-6 fatty acids found in processed foods.

- ***Opt for EVOO***: Choose extra-virgin olive oil and avocado oil over other vegetable oils.
- ***Be colorful***: Eat foods that provide color naturally, like brightly colored fruits and vegetables, which are rich in flavonoids—and flavor!
- ***Pass on the packets***: Avoid those sweeteners you see at restaurants and cafes.

REFERENCES

1. Abdullaev, F. I. (2002). Cancer chemopreventive and tumoricidal properties of saffron (*Crocus sativus* L.). *Experimental Biology and Medicine, 227*(1), 20–25. PMID 11788779.
2. Adefegha, S. A., Okeke, B. M., & Oboh, G. (2021). Antioxidant properties of eugenol, butylated hydroxylanisole, and butylated hydroxyl toluene with key biomolecules relevant to Alzheimer's diseases—*In vitro*. *Journal of Food Biochemistry.* https://pubmed.ncbi.nlm.nih.gov/32458455/ (accessed on 21 October 2024).
3. Alzheimer's Association. (2022). Highlights from the Alzheimer's Association International Conference 2022. Retrieved from: https://aaic.alz.org/releases_2022/overview.asp (accessed on 21 October 2024).
4. American Society of Clinical Oncology. (2019). Food and cancer risk. *Cancer.Net.*
5. Anon, M. I. (1991). Food irradiation–A technique for preserving and improving the safety of food. *World Health Organization, Geneva.*
6. Arnold, L. E., Lofthouse, N., & Hurt, E. (2012). Artificial food colors and attention-deficit/hyperactivity symptoms: Conclusions to dye for. *Neurotherapeutics, 9*(3), 590–601. https://www.ncbi.nlm.nih.gov/pmc/articles/PMC3441937/ (accessed on 21 October 2024).
7. Barrows, J. N., Lipman, A. L., & Bailey, C. J. (2017). Color additives history. *Food and Drug Administration.*
8. Batiha, G. E. S., Hussein, D. E., Algammal, A. M., George, T. T., Jeandet, P., Al-Snafi, A. E., Tiwari, A., Pagnossa, J. P., Lima, C. M., Thorat, N. D., Zahoor, M., El-Esawi, M., Dey, A., Alghamdi, S., Hetta, H. F., & Cruz-Martins, N. (2021). Application of natural antimicrobials in food preservation: Recent views. *Food Control, 126*, 108066. https://doi.org/10.1016/j.foodcont.2021.108066.
9. Behall, K. M. (2000). Whole food. *Journal of the American College of Nutrition, 19*(1), 61–70. PMID 10682877.
10. Bouvard, V., Loomis, D., Guyton, K. Z., Grosse, Y., El Ghissassi, F., Benbrahim-Tallaa, L., Guha, N., Mattock, H., & Straif, K. (2015). Carcinogenicity of consumption of red and processed meat. *The Lancet Oncology, 16*(16), P1599–1600. https://doi.org/10.1016/S1470-2045(15)00444-1.
11. Brendon, O. N. (2006). Is this what you call junk food? Health effect. Pg. version ID 400238957.
12. Centers for Disease Control and Prevention. (2017). Exposures to flavoring

chemicals. Retrieved from: https://www.cdc.gov/niosh/topics/flavorings/exposure.html (accessed on 21 October 2024).

13. Church, J. J., & Parson, A. L. (1995). Modified atmosphere packaging technology: A review. *Journal of the Science of Food and Agriculture, 67*, 143–152.
14. Cortés-Jofré, M., Rueda, J. R., Asenjo-Lobos, C., Madrid, E., & Bonfill Cosp, X. (2020). Drugs for preventing lung cancer in healthy people. *Cochrane Database of Systematic Reviews, 2020*(3), CD002141.
15. Czarra, F. (2009). *Spices and Condiments Food Additives Spices* (ISBN 9781861894267). pp. 128.
16. Daniel, M. (2007). Reactions to food additives and preservatives.
17. Debras, C., et al. (2022). Artificial sweeteners and cancer risk: Results from the NutriNet-Santé population-based cohort study. *PLOS Medicine, 19*(3). https://pubmed.ncbi.nlm.nih.gov/35324894/ (accessed on 21 October 2024).
18. Deeley, C. M., Gao, M., & Hunter, R. (2006). The development of food in Asia. In *Pacific Radiation Processing* (pp. 11–15).
19. Deeley, C. M., Gao, M., Hunter, R., & Ehlermann, D. A. E. (2006). The development of food irradiation in the Asia Pacific, the Americas, and Europe: Tutorial presented to the International Meeting on Radiation Processing, Kuala Lumpur. Retrieved from: http://www.doubleia.org/index.php?sectionid=43&parentid=13&contentid=494 (accessed on 21 October 2024).
20. Dey, S., & Nagababu, B. H. (2022). Applications of food color and bio-preservatives in the food and their effects on human health. *Food Chemistry Advances, 1,* 100019. https://doi.org/10.1016/j.fca.2022.100019.
21. Eberechukwu, S., Amadikwa, A., & Okechukwu, M. (2017). Effect of oral intake of sodium benzoate on some hematological parameters of Wistar albino rats. *Scientific Research Essays, 2*(1), 6–9.
22. Educhange. (n.d.). The NOVA food classification system. Retrieved from: https://ecuphysicians.ecu.edu/wp-content/pv-uploads/sites/78/2021/07/NOVA-Classification-Reference-Sheet.pdf.
23. Elhkim, M. O., Heraud, F., Bemrah, N., Tanaka, T., & Ogata, A. (2007). New considerations regarding the risk assessment, intolerance reactions and maximum theoretical daily intake in France. *Regulatory Toxicology and Pharmacology, 43*(3), 308–316.
24. Family Survival Planning. (2009). Food storage guidelines. Retrieved from: http://www.family-survival-planning.com/foodstorage-guidelines.html#axzz1jLVAb3Fl (accessed on 21 October 2024).

25. Fennema, O. R. (1985). *Food Chemistry* (2nd ed., revised and expanded). New York: Marcel Dekker, Inc.
26. Food Irradiation. (2000). Position of ADA. *Journal of the American Dietetic Association, 100*, 246–253. Retrieved from: http://www.mindfully.org/Food/Irradiationposition-ADA.htm (accessed on 21 October 2024).
27. Francis, F. J. (2000). Pioneer in food science and quality. In *A Century of Food Science* (pp. 13–14). Chicago: Institute of Food Technologists.
28. Gunter, M. J., Rauber, F., Levy, R. B., Huybrechts, I., Kliemann, N., Millett, C., Vamos, E. P. (2023). Ultra-processed food consumption, cancer risk, and cancer mortality: A large-scale prospective analysis within the UK Biobank. *eClinicalMedicine, 56*, 101840. https://doi.org/10.1016/j.eclinm.2023.101840.
29. Gupta, R., & Yadav, R. K. (2021). Impact of chemical food preservatives on human health. *PalArch's Journal of Archaeology of Egypt / Egyptology, 18*(15), 811–818.
30. Gustafsson, E., Edlund, M., & Hagberg, M. (2003). Effect of food additives. Health beliefs, *34*, 565–570.
31. Harrington, R. (2015). Does artificial food coloring contribute to ADHD in children?
32. Hauther, W. W. (2008). *Irradiation and the Death of Food.* Food and Water Watch Press.
33. Heldman, E. R. (1980). Frozen foods: Challenges for the future. *Food Australia, 45*(8), 374–380.
34. Hoover, D., & Milich, R. (1994). Food additives may affect kid's hyperactivity. *Journal of Abnormal Child Psychology, 22*(5), 501–515.
35. Houghton, M. (2002). *The American Heritage Food Science Dictionary.* Retrieved from: http://www.amazon.com/AmericanHeritage%C2%AE-Student-ScienceDictionary/dp/061818919X (accessed on 21 October 2024).
36. International Food Information Council. (2019). Food safety: Consumer trends, habits, attitudes.
37. Jacques, A., et al. (2019). The impact of sugar consumption on stress-driven, emotional and addictive behaviors. *Neuroscience & Biobehavioral Reviews.* https://www.sciencedirect.com/science/article/pii/S0149763418308613 (accessed on 21 October 2024).
38. Jakle, J. A. (1999). *Fast Food: Roadside Restaurants in the Automobile Age.* Johns Hopkins University Press.

39. John, E. M. (2003). Ionizing radiations sources, biological effects emission and exposures. *83*, 1766.
40. Kaayla, T., Kimura, Y., Tsutsumi, K., & Funikawa, Y. (1993). The effect of various phospholipids on plasma lipoprotein and liver lipids in hypercholesterolemic rats. *Journal of Nutritional Science and Vitaminology, 39*(1), 63–71.
41. Kannarek, R. B. (2011). Artificial food dyes and attention deficit hyperactivity disorder. *Nutrition Reviews, 69*(7), 385–391.
42. Kebede, M., & Admassu, S. (2019). Application of antioxidants in food processing industry: Options to improve the extraction yields and market value of natural products. *Advances in Food Technology and Nutritional Sciences Open Journal, 5*(2), 38–49.
43. Kinderlerer, J. L., & Hatton, P. (1990). Fungal metabolites of sorbic acid. *Food Additives and Contaminants, 7*(5), 657–669.
44. Kunkel, E. M., & Barbara, H. D. (2004). *Gale Nutrition and Well-Being A to Z*. The Gale Group Inc., Macmillan Reference USA.
45. Liang, X. (2023). Impact of food additives in ultra-processed food on human health. *Theoretical and Natural Science, 6*, 437–444.
46. Lobo, V., Patil, A., Phatak, A., & Chandra, N. (2010). Free radicals, antioxidants, and functional foods: Impact on human health. *Pharmacognosy Reviews, 4*(8), 118–126.
47. Lopez, A. (1987). *A Complete Course in Canning* (Books I, II, III). The Canning Trade Inc.
48. Louis, S. T., & Botulism, M. E. (1991). *Complete Guide to Home Canning.* Epidemiology and Control (2nd ed.). Washington, D.C.: U.S. Government Printing Office.
49. McCann, D., Barrett, A., Cooper, A., Crumpler, D., Dalen, L., Grimshaw, K., Kitchin, E., Lok, K., Porteous, L. E., Prince, E., Sonuga-Barke, E. J., Warner, J. O., & Stevenson, J. (2007). Food additives and hyperactive behavior in 3-year-old and 8/9-year-old children in the community: A randomised, double-blinded, placebo-controlled trial. *The Lancet, 370*(9598), 1560–1567. https://doi.org/10.1016/S0140-6736(07)61306-3.
50. McCarthy, C. M. (2021). Common food additives and chemicals harmful to children. *Harvard Health Publishing.*
51. McGee, H. (2004). *On Food and Cooking: The Science and Lore of the Kitchen* (Rev. upd. ed.). Scribner.
52. Monteiro, C. A., Cannon, G., Moubarac, J. C., Levy, R. B., Louzada, M. L. da, & Jaime, P. C. (2018). The UN decade of nutrition, the NOVA

food classification and the trouble with ultra-processing. *Public Health Nutrition, 21*(1), 5–17.

53. More, S., et al. (2012). The butter flavorant, diacetyl, exacerbates β-amyloid cytotoxicity. *Chemical Research in Toxicology.* https://doi.org/10.1021/tx3001016.
54. Morelli, M. B., Gambardella, J., Castellanos, V., Trimarco, V., & Santulli, G. (2020). Vitamin C and cardiovascular disease: An update. *Antioxidants, 9*(12), 1227. https://doi.org/10.3390/antiox9121227.
55. Niaz, K., Zaplatic, E., & Spoor, J. (2018). Extensive use of monosodium glutamate: A threat to public health? *EXCLI Journal, 17*, 593–605. https://doi.org/10.17179/excli2018-934.
56. Parish, J. (2009). Use of stored food: Saving your stored food by the use of oxygen absorber.
57. Parohan, M., Anjom-Shoae, J., Nasiri, M., et al. (2019). Dietary total antioxidant capacity and mortality from all causes, cardiovascular disease and cancer: A systematic review and dose–response meta-analysis of prospective cohort studies. *European Journal of Nutrition, 58*, 2175–2189. https://doi.org/10.1007/s00394-018-1741-5.
58. Potter, N. N., & Hotchkiss, J. H. (1995). *Food Science* (5th ed.). New York: Chapman & Hall.
59. Pressinger, R. W. (1997). Environmental circumstances that can damage the developing brain. *Graduate Student Research Project Conducted at the University of South Florida, Journal of Pediatrics, 92*(1), 64–67.
60. Richard, R., & Totowa, N. J. (1995). The history of food preservation. *Technology in Food Production, 152*.
61. Riddervold, A. (2008). High pressure food preservation. In *Food Conservation* (pp. 12–16). ISBN 9780907325406.
62. Said Abasse, K., et al. (2022). Association between dietary nitrate, nitrite intake, and site-specific cancer risk: A systematic review and meta-analysis. *Nutrients.* https://doi.org/10.3390/nu14020225.
63. Saulo, A. A. (2005). Sugars and sweeteners in foods. In *Food Safety and Technology* (FST16). College of Tropical Agriculture and Human Resources, University of Hawai'i at Manoa.
64. Schoenthaler, S. J., & Doraz, W. E. (1986). The impact of a low food additives and sucrose diet on academic performance in 803 New York City public schools. *International Journal of Biological and Social Resources, 8*(2), 185–195.
65. Shahmohammadi, M., Javadi, M., & Nassiri-Asl, M. (2016). An overview

on the effects of sodium benzoate as a preservative in food products. *Biotechnology and Health Sciences, 3*(3), e35084.

66. Sigala, D. M., et al. (2022). The dose-response effects of consuming high fructose corn syrup-sweetened beverages on hepatic lipid content and insulin sensitivity in young adults. *Nutrients.* https://doi.org/10.3390/nu14040776.
67. Talwar, J. P. (2003). *Fast Food, Fast Food Track: Immigrants, Big Business and the American Dream*. Westview Press.
68. U.S. Food and Drug Administration. (1993). *Everything Added to Food in the United States.* Boca Raton, FL: C.K. Smoley (c/o CRC Press, Inc.).
69. U.S. Food and Drug Administration. (2018). FDA removes 7 synthetic flavoring substances from food additives list. Retrieved from: https://www.fda.gov/food/cfsan-constituent-updates/fda-removes-7-synthetic-flavoring-substances-food-additives-list (accessed on 21 October 2024).
70. Young, I. S., & Woodside, J. V. (2001). Antioxidants in health and disease. *Journal of Clinical Pathology, 54*, 176–186.

Chapter 10

Socioeconomic and Cultural Factors

Contents

INTRODUCTION

Access to healthful food, nutrition, and food security can all be impacted by socioeconomic factors. Food security can be impacted by variables such as household size, income, education, and access to health care and insurance. By raising family income and job prospects, education can also increase food security. Nutritious status can be impacted by racial, environmental, and neighborhood factors. A poor diet can raise the risk of undernutrition, weight gain, and obesity as well as infections like dental caries. Chronic illnesses like diabetes, high blood pressure, high cholesterol, and cardiovascular disease can also be exacerbated by these conditions. Access to healthful food can be impacted by variables such as geography and income. Since health food is more expensive, it may be more difficult for low-income households to obtain it. Food deserts are areas of extreme poverty where it is difficult for locals to obtain nutritious food, which is linked to poor eating habits and health problems.

The food consumption patterns in this region are significantly influenced by socio-cultural factors, including but not limited to religion, beliefs, food preferences, gender discrimination, education, and the employment status of women. The mass media, particularly food commercials on television, has a significant impact on changing people's eating habits.

10.1. THE FACTORS THAT INFLUENCE OUR FOOD CHOICES

There is a need for a deeper comprehension of the factors influencing food choice given the importance of changing the dietary habits of the population. This section looks at the main factors influencing food choice, highlighting those that can be changed, and it also covers some effective interventions.

The key driver for eating is of course hunger but what we choose to eat is not determined solely by physiological or nutritional needs. Some of the other factors that influence food choice include:

1. Biological determinants such as hunger, appetite, and taste.
2. Economic determinants such as cost, income, availability.
3. Physical determinants such as access, education, skills (e.g., cooking) and time.
4. Social determinants such as culture, family, peers, and meal patterns.
5. Psychological determinants such as mood, stress, and guilt.
6. Attitudes, beliefs, and knowledge about food.

The above list is by no means exhaustive, but it does demonstrate the complexity of food choice. Factors influencing food choices also differ based on a person's stage of life, and the importance of a given factor varies among individuals and groups. As a result, not every population group will benefit from the same kind of intervention to change eating habits. Instead, treatments must be targeted at various population segments while taking into account the various variables affecting their dietary preferences.

10.1.1. Biological Determinants of Food Choice

Biological determinants of food choice are physiological needs and sensory mechanisms that influence what people eat.

10.1.1.1. Hunger and Satiety

Our basic determinants of food choice are provided by our physiological needs. Humans respond to feelings of hunger and satiety (satisfaction of appetite, state of no hunger between two eating occasions) and require energy and nutrients to survive. The balance between food intake, appetite stimulation, and hunger is regulated by the central nervous system.

Differentially strong satiety signals are produced by the macronutrients i.e., carbohydrates, proteins, and fats. Based on the available data, it appears that protein is the most satiating food, fat has the least ability to satisfy hunger, and carbs have an intermediate effect (Stubbs et al., 1996).

Dietary energy density has been demonstrated to have strong effects on satiety; diets with lower energy densities produce more satiety than diets with higher energy densities. The high energy density of foods high in fat and/or sugar can also result in "passive overconsumption," which is the inadvertent ingestion of excess energy without the conscious consumption of additional bulk.

The amount of food or portion size consumed could be a key indicator of satiety. Because they don't know what constitutes a healthy portion size, many people unintentionally consume too much energy.

10.1.1.2. Palatability

The level of enjoyment one gets from consuming a certain food is directly correlated with its palatability. It is based on the food's sensory qualities, including texture, taste, smell, and appearance. There is no denying the sensory appeal of sweet and high-fat foods. That food is frequently consumed for its pleasurable qualities rather than just as a source of nourishment is therefore not surprising. Numerous studies have looked into the impact of palatability on human appetite and food intake. When food is more palatable, people eat more of it, but it's unclear how this affects appetite in the hours after consumption. Increasing the variety of foods consumed can also increase energy and food intake, changing the short-term energy balance. Effects on the long-term regulation of energy, however, are unknown.

10.1.1.3. Sensory Aspects

"Taste" is frequently cited as having a significant impact on eating habits. Actually, the entire range of sensory stimulation a food produces during ingestion is what is referred to as "taste." This covers the flavor in and of itself as well as the food's aroma, presentation, and texture. In particular, it is believed that these sensory factors affect impulsive food choices.

Food-related behavior is influenced by taste and familiarity from an early age. A preference for sweetness and a distaste for bitterness are thought to be inborn human characteristics. Food aversions and taste preferences are shaped by our attitudes, beliefs, and expectations and emerge from our experiences (Clarke, 1998).

10.1.2. Economic and Physical Determinants of Food Choice

Economic and physical factors are two of many determinants that influence food choices.

1. **Economic Factors**: These include income, food prices, and availability. For example, income fluctuations can change food preferences.
2. **Physical Factors**: These include access to food, education, skills, and time. Access to food can be impacted by income, food prices, markets, and food distribution within the household.

10.1.2.1. Cost and Accessibility

Without a doubt, one of the main factors influencing food choice is price. The determination of whether a cost is prohibitive is contingent upon an individual's income and socio-economic standing. Low-income populations are more likely to eat diets that are out of balance and to consume less fruit and vegetables in particular. Money does not, however, always translate into a higher-quality diet; rather, there should be a wider selection of foods to choose from.

Another significant physical factor influencing food choice is accessibility to stores, which depends on resources like transportation and location. When it comes to supermarkets on the outskirts versus towns and cities, healthy food is typically more expensive. Nevertheless, increasing access by itself does not lead to a rise in the purchase of extra fruits and vegetables, which are still thought to be excessively costly (Dibsdall et al., 2003).

10.1.2.2. Education and Knowledge

Research suggests that an adult's dietary habits may be influenced by their educational attainment. On the other hand, there is little correlation between sound eating practices and nutrition knowledge. This is due to the fact that when people are unsure of how to put their health knowledge to use, it does not result in direct action. Moreover, conflicting or mistrusted information about nutrition is propagated from a range of sources, which deters people from making a change in their eating habits. Therefore, it's critical to spread accurate and consistent messages via a variety of media, food packaging, and, of course, medical professionals.

10.1.3. Social Determinants of Food Choice

Food choice is influenced by a large number of factors, including social and cultural factors.

10.1.3.1. Influence of Social Class

People's eating habits are shaped and limited by primarily social and cultural factors. Population studies demonstrate that there are observable disparities in food and nutrient intakes between social classes. Unhealthy eating habits can lead to under- (micronutrient deficiency) and over-nutrition (energy overconsumption resulting in overweight and obesity); these issues affect various societal groups and call for varying degrees of skill and intervention strategies.

10.1.3.2. Cultural Influences

Cultural influences can result in variations in the customs surrounding the preparation and regular consumption of particular foods, as well as in some situations, dietary restrictions like the avoidance of milk and meat. Cultural influences can, however, be altered. For example, people frequently take up specific eating customs from the local culture when they relocate to a new nation.

10.1.3.3. Social Context

Social influences on food intake are the effects that one or more people have on other people's eating habits. These effects can be direct, such as when someone purchases food, indirect, such as when someone learns from the behavior of a peer, and subconscious, such as when someone transfers beliefs. Social factors influence food choices even when eating alone because social interactions shape attitudes and habits. However, it can be challenging to quantify the social influences on food intake because people can have a variety of effects on other people's eating habits and because individuals aren't always aware of the social influences on their own eating habits.

Social support can influence eating decisions and healthy dietary modifications in a positive way. For instance, it has been discovered that adult consumption of fruits and vegetables is highly predicted by social support. Social support can improve health promotion by making people feel more competent and self-sufficient and by giving them a sense of belonging to a group.

Families are widely acknowledged to play a significant role in dietary decisions. Studies demonstrate that food choices are shaped at home. A person may benefit from adopting dietary strategies that are acceptable to family and friends while also having an impact on the eating habits of others, as they can serve as a source of encouragement for making and maintaining dietary changes.

10.1.3.4. Social Setting

While eating at home is still the norm for most food consumption, eating out is becoming more and more common. g. in restaurants, workplaces, and schools. The setting in which food is consumed can influence the selection of foods, especially with regard to what is available. Consumption of healthy foods is increased when they are readily available both at home and when traveling. However, in many workplaces and educational settings, there is limited access to healthy food options. This is particularly true for those with irregular hours or with particular requirements, e.g., vegetarian. Given that most adult men and women work, it is crucial to look into how work affects decisions about what to eat and other health-related behaviors.

10.1.3.5. Meal Patterns

Every day, people have a wide variety of eating occasions, each with its own set of reasons for eating. The majority of research looks into the variables that affect eating habits, but it could be helpful to look into the variables that affect eating preferences during certain mealtimes.

Snacking's impact on health has been the subject of much discussion. Research indicates that eating snacks can affect nutrient and energy intakes but not always body mass index. But when snack foods are readily available, people who are normal weight or overweight may have different coping mechanisms and compensatory mechanisms for subsequent meals. Furthermore, the makeup of snacks may play a significant role in a person's capacity to modify intake in order to satisfy energy requirements.

It can be difficult for many health professionals to guide young adults in making healthy snack choices. Introducing healthy snack options gradually could be a more positive approach in the home than outright banning unhealthy snacks. Furthermore, it is imperative to increase the accessibility of healthy food options beyond one's residence.

10.1.4. Psychological Factors

10.1.4.1. Stress

Stress on the psychological level is a prevalent aspect of contemporary life and can alter behaviors that impact health, like exercising, smoking, or eating habits.

The complexity of the relationship between stress and food choice is largely due to the range of stressors that can arise. Depending on the person, the stressor, and the situation, stress can have varying effects on food intake. In general, when under stress, some people eat more and some people eat less than usual.

Stress-related physiological changes (reduced appetite brought on by stress-related processes), motivational differences (reduced concern about weight control), and practical changes in meal preparation, eating opportunities, and food availability are the mechanisms suggested to explain changes in eating and food choice. Additionally, research indicates that prolonged or frequent work-related stress may lead to unfavorable dietary changes, which may raise the risk of weight gain and, in turn, cardiovascular disease.

10.1.4.2. Mood

Hippocrates was the first to propose that food has therapeutic properties, but it wasn't until the Middle Ages that food was thought of as a means of influencing mood and temperament. It is now widely acknowledged that our moods are strongly influenced by the food we eat and that our moods in turn affect the food we choose.

It's interesting to note that attitudes toward specific foods seem to have an impact on how food affects mood. Many people struggle with their ambivalent relationship with food—wanting to enjoy it but being aware of weight gain. Dieters, those who exercise extreme self-control, and some women claim to feel guilty when they don't eat what they believe is right. Furthermore, attempting to limit one's consumption of a certain food can intensify one's craving for that food, resulting in what are known as food cravings.

Compared to men, women report food cravings more frequently. It seems that depression affects how strong these cravings are. Cravings are also more frequently reported during the premenstrual phase when there is an increase in overall food intake and a corresponding shift in basal metabolic rate. Hence, stress, and mood can affect eating habits and potentially the immediate and long-term effects of dietary interventions.

10.1.5. Eating Disorders

Unlike many other biological processes, eating behavior is frequently controlled by complex cognitive processes. Dieting is one of the most popular strategies used to exercise cognitive control over food consumption. Many people engage in methods to reach their ideal body mass index because they express a desire to reduce weight or enhance their physical appearance. But when exercise and/or dieting are overdone, issues can occur. Eating disorders typically have a complex etiology involving biological, psychological, familial, and sociocultural factors. Eating disorders are frequently linked to low self-esteem, a distorted self-image, generalized anxiety, obsessions, stress, and unhappiness.

Stabilizing one's weight and individual psychotherapy are usually necessary for the treatment of an eating disorder. Although prevention is harder to define, some ideas include avoiding child abuse, avoiding exaggerating diet and health problems, expressing affection without becoming overly controlling, avoiding setting unrealistic expectations, rewarding modest achievements in the here and now, and promoting independence and sociability.

10.1.6. Consumer Attitudes, Beliefs, Knowledge, and Optimistic Bias

10.1.6.1. Consumer Attitudes and Beliefs

Our knowledge of consumer attitudes in the domains of nutrition and food safety is inadequately researched. The development and execution of healthy eating programs would benefit from a deeper comprehension of the public's perceptions of their diets.

According to the Pan-European Survey of Consumer Attitudes toward Food, Nutrition, and Health, "quality/freshness" (74%), "price" (43%), "taste" (38%), "trying to eat healthy" (32%), and "what my family wants to eat" (29%) are the top five factors influencing food choice in 15 European member states. These are average numbers derived from a grouping of results from 15 European member states, which varied greatly between nations. According to reports, the following factors influence food choices in the USA: taste, affordability, nutrition, ease of preparation, and weight concerns.

Females, older participants, and participants with higher levels of education in the Pan-European study thought that "health aspects" were especially significant. Males chose "taste" and "habit" as the primary factors influencing their food choices more often than females. "Price" appeared to be the most significant factor for retired and jobless subjects. The perceived factors influencing food choices for these groups should be taken into account in interventions aimed at them.

Beliefs and attitudes are malleable; during the past 50 years, our attitude toward dietary fat has evolved, leading to a reduction in the total amount of fat consumed and a shift in the proportion of unsaturated to saturated fat.

10.1.6.2. Optimistic Bias

For health-related reasons, Europeans don't think they need to change their eating habits all that much—71% of those surveyed said their diets are already sufficiently healthy. Subjects from Australia, America, and England have reported feeling very satisfied with their current diets.

The absence of dietary modifications appears to indicate a high degree of optimistic bias, a condition in which people think they are less vulnerable to a risk than other people. This misplaced optimism is also evident in research that demonstrates how people undervalue the possibility that they eat a diet high in fat in comparison to others and how some consumers who consume little in the way of fruits and vegetables think of themselves as "high consumers."

If individuals think their diets are already healthy, it might not be fair to expect them to change, or to value nutrition and healthful eating above all else when making food choices. Even though these consumers are more likely to recognize that their diet needs to be improved than those who don't, they still fall well short of the widely acknowledged nutrition goals for public health. It is improbable that dietary guidelines will inspire any more motivation in these groups. Future interventions may therefore need to raise public awareness that their own diet may not be entirely sufficient in terms of, say, fat or fruit and vegetable consumption. Research suggests that individuals who maintain a healthy diet may be more inclined to make dietary changes if their attitudes towards the changes become more positive and they are convinced of the benefits of the changes. Therefore, the first step in implementing a dietary change is to perceive a need for change.

10.1.7. Barriers to Dietary and Lifestyle Change

10.1.7.1. Focus on Cost

Food choices are significantly influenced by household income and food costs, particularly for consumers with low incomes. Food waste makes people reluctant to try "new" foods for fear of being rejected by their families. Furthermore, purchasing, and making meals with basic ingredients can be hampered by ignorance and a decline in cooking abilities. One possible solution is to provide affordable education on how to increase fruit and vegetable consumption so that no additional costs in terms of money or effort are incurred. The promotion of fruit and vegetable dishes as affordable options by producers, retailers, public health authorities, and governments may also have a beneficial impact on dietary changes.

10.1.7.2. Time Constraints

Time constraints are a common excuse given by the young and well-educated for not adhering to dietary recommendations. Convenience foods are preferred over meals made with basic ingredients by people who cook for themselves or live alone. Fruit and vegetable sales have shifted from loose to prepackaged, prepared, and ready-to-cook items in response to this demand. People are willing to pay a higher price for these products due to their convenience, even though they are more expensive than loose goods. One way to improve the quality of these groups' diets is to develop a wider variety of palatable, easily-transportable foods with good nutritional profiles.

10.1.8. Models for Changing Behavior

10.1.8.1. Health Behavioral Models

When developing health promotion strategies, it can be helpful to understand how people make decisions about their health. This is where theory-based models related to social psychology and their influence come into play. These models aid in the explanation of behavior in general and health decision-making in particular. Additionally, they've been used to forecast the chance of dietary behavior changes. Only a few are covered in this section.

10.1.8.2. The Health Belief Model (HBM) and the Protection Motivation Theory

The HBM, which was first put forth by Rosenstock and then altered by Becker7, has been used to forecast preventive health behaviors like screening, immunization uptake, and following medical advice. According to the model, before contemplating a behavior change, a person must feel personally threatened by a disease or illness and perform a cost-benefit analysis. This model also implies that individuals require a cue of some kind to act to alter their behavior or make decisions about their health.

10.1.8.3. The Theory of Reasoned Action (TRA) and the Theory of Planned Behavior (TPB)

A particular behavior's intention can be predicted and explained with the aid of the Theory of Reasoned Action or its extension, the Theory of Planned Behavior5. These models are predicated on the idea that behavioral intention serves as the most reliable indicator of behavior. The model proposes that an individual's behavioral intention is jointly derived from three components:

1. attitudes,
2. perception of social pressure to perform the behavior, and
3. perceived control over the behavior.

TPB/TRA can be utilized in dietary research to compare the impact of factors on individuals and sample groups and to gain insight into the factors that influence food selection. It has been successful for the TRA to explain behaviors like the consumption of fat, salt, and milk. The UK's attitudes and beliefs regarding starchy foods were also explained by the TPB model.

10.1.8.4. Stage Classification for Health-Related Behavior

According to Prochaska and colleagues' Stages of Change model, there are five distinct stages through which changes in behavior related to health can occur. These are planning, acting, maintaining, and reflecting before acting. According to the model, people should react most favorably to interventions that are matched to their stage of change if various factors influence transitions at different stages. Among the models that have been discussed, the Stages of Change model has been shown to be more effective when used to modify behavior than when used to explain existing behavior. This is most likely because the model provides practitioners with guidance on practical interventions that they can learn. Furthermore, messages customized to an individual's degree of change readiness can be tested on sizable random samples.

It has been proposed that a stage model would be more suitable for more straightforward, discrete behaviors like consuming low-fat milk or eating five servings of fruit and vegetables each day (food-based goals) as opposed to more intricate dietary adjustments like low-fat eating (nutrient-based goal).

The entire spectrum of food-choice behaviors is currently not adequately explained or predicted by a single theory or model. All models should be seen as tools for deciphering the variables affecting personal choices and actions. Though numerous behavior change models exist, their application in nutrition interventions has been comparatively limited; the most widely used model is the Stages of Change model. The best test of this model has not yet been carried out, though, to see if stage-matched dietary interventions work better than standardized methods.

10.1.9. Changing Food Behavior: Successful Interventions

Making dietary changes means breaking long-standing habits, which makes it difficult to do. Diverse environments, including schools, workplaces, grocery stores, primary care offices, and community-based research projects, have been utilized to determine what is most effective for specific populations. These targeted interventions have shown some degree of success, even though the results of such trials are difficult to generalize to other contexts or the general public. This suggests that different strategies are needed for different populations or dietary components.

Given that supermarkets are where most people purchase the majority of their food, interventions there are quite popular. Information can be given through point-of-purchase interventions, shop tours, and screening. While nutrition awareness and knowledge are successfully increased by such interventions, it is currently unknown how effective they will be in changing behavior in a meaningful and long-term way.

Because schools can reach students, parents, and staff, they are another obvious setting for interventions. Children's consumption of fruits and vegetables has increased as a result of the use of tuck shops, multimedia, the internet, and growing, preparing, and cooking their own food. Additionally, the nutritional profile of school dinners was enhanced by subtly altering recipes to reduce fat, sodium, and energy content without lowering student participation in the school lunch program.

Interventions in the workplace can also target individuals who are at risk and reach a large number of people. In worksite canteens, increasing the availability and attractiveness of fruits and vegetables proved successful, and lower prices for healthier snacks in vending machines increased sales. Therefore, there is a greater chance of success when nutrition education is combined with workplace modifications, especially when interactive activities are used and maintained over an extended period. In the primary care context, addressing multiple dietary factors at once, such as decreasing dietary fat and increasing fruit and vegetables, has shown to be successful. In these types of situations, behavioral counseling appears to be most successful when combined with nutrition counseling; however, it is currently unknown how much it would cost to train primary care physicians in behavior counseling. In public health and community contexts, behavioral, and educational tactics have also been employed, and it has been demonstrated that these approaches enhance fruit and vegetable consumption.

10.2. MOVING TOWARDS A HEALTHIER DIETARY PATTERN FREE OF ULTRA-PROCESSED FOODS

Ultra-processed food (UPF) consumption has surged globally in recent years; in several countries, like the USA and the UK, it accounts for nearly 60% of total dietary intake; in Mediterranean countries, like Spain and Italy, it ranges from 17 to 24%. Thus, this rise has been directly linked to changes in dietary habits around the world as well as an increase in the prevalence of chronic illnesses and obesity. Because of this, people are becoming more curious about its effects on health and whether consuming large amounts of it indicates an unhealthy diet. High UPF consumption is currently well-documented to be linked to a worse cardiometabolic risk profile, as well as an increased risk of cancer, frailty, overweight, and obesity, altered renal function, depression, and all-cause mortality. Indeed, independent of previously identified risk factors, such as a global assessment of overall diet quality as reflected by adherence to a Mediterranean diet, the Moli-Sani study found that high consumption of UPFs is associated with a 58% increased risk of CVD mortality, a 52% higher risk of dying from ischemic heart disease/cerebrovascular causes, and a 26% higher risk of all-cause mortality. Reported outcomes for secondary CVD prevention were comparable.

In fact, the ubiquity and innovation of the food industry, coupled with the intense advertising and aggressive marketing (lower price for super-size servings) of this kind of inexpensive, highly palatable processed food, targeted primarily at children and adolescents, may help to partially explain the observed dietary changes. First of all, this kind of highly processed food technology took over in high-income nations; however, it is currently spreading quickly to lower- and upper-middle-income nations. Additionally, socioeconomic factors should be emphasized as one of the primary explanations for the emergence and rise in the consumption of ultra-processed foods. The desire to spend less time preparing and processing food emerged in urbanized societies as the number of women entering the workforce increased. This, combined with globalization and factors related to the food environment, has encouraged the production and consumption of "ready-to-eat" and "ready-to-heat" products.

Five or more ingredients are used in the formulation of the finished food products, according to the general definition of UPF. Usually inexpensive, industrial sources of fat, carbohydrates, or various additives used to flavor, aromatize, and enhance the palatability of food are these ingredients. These additives are synthetic mixes made mostly (if not entirely) of whole foods. Depending on the area, there is a large range of UPFs, which include pastries (cakes, sweets, and industrial bread), instant soups, ice creams, carbonated soft drinks, fatty or salty snacks, margarine, processed fruits and vegetables, baked goods, etc. UPFs are the main sources of added sugars, trans, and saturated fats, sodium, and other dietary components, although they typically have poor nutritional quality (low levels of fiber, proteins, and micronutrients).

It is crucial to determine which factors, including sociodemographic and behavioral ones, are linked to UPF consumption as well as the primary dietary sources of UPFs to develop and implement food policies that will mitigate the detrimental effects of UPF consumption and its impact on health. An unhealthy nutritional profile and an increased risk of chronic disease are directly linked to unhealthy dietary patterns, which are characterized by a high intake of energy-dense foods and a low intake of fruits and vegetables, as well as socioeconomically disadvantaged conditions (lower educational attainment or socioeconomic position, or living in disadvantaged areas). In this regard, Calixto Andrade et al. investigated the relationship between UPF intake and socioeconomic attributes and diet quality in a cross-sectional nationally representative survey of French adults (2642 participants, ages 18 to 74) called the Étude Nationale Nutrition Santé. The authors first discussed dietary disparities among adult French citizens based on UPF consumption and the negative impact of UPF consumption on the overall quality of diets. UPF intake accounted for nearly one-third of adult French energy intake. Furthermore, a poorer dietary profile (higher energy density, higher intake of total and saturated fat, free sugar, and carbohydrates) was linked to higher UPF consumption, with young people consuming the most UPFs. Marchese et al. and Magalhães et al. reported similar findings, demonstrating how diet quality and sociodemographic factors (age, gender, educational attainment, marital status, smoking status, living in underprivileged areas or with a lower socioeconomic position) affect UPF consumption. The relationship between age and UPF consumption is well supported by data. Adolescents and younger adults have the highest percentage of energy derived from UPF intake, whereas older adults ($\geq$45 years of age) have the lowest percentage of energy from UPF. The youngest generation consumes a lot of UPF, which can be explained by their openness to trying new foods and eating habits, their belief that eating healthily is expensive, and the challenge of meal planning and/or grocery shopping during long work or study hours. In addition, compared to older people, their eating habits are primarily erratic and they typically eat more snacks, fast food, and foods high in energy. Yet, research by Calixto Andrade et al. revealed that UPF consumption is inversely correlated with age. Similarly, Marchese et al. discovered that older people (≥71 years old) consumed more UPF than adults (51–70 years old). The authors contended that the variations seen with other populations under study may be explained by a lack of access to convenience goods or a lack of motivation.

Calixto Andrade et al. claim that because the French people resist "westernization" and have preserved their traditional culinary culture, France has a lower dietary contribution of UPF to total energy intake (31.1% of daily energy intake) than other countries. Globalization, technological advancement, industrialization of food systems, and a lack of health policies encouraging better eating practices all contribute to a greater supply of UPFs (more production and a greater range of products) and a rise in UPF consumption.

Second, Calixto Andrade et al. discovered that people who consumed the most UPFs were those who lived in cities or had less education (only completed primary school); these correlations have been noted in both developed and developing nations in the past. These results are consistent with data from Marron-Ponce et al., who discovered that UPF consumption exhibited a direct correlation with living in an urban area—such as northern Mexico—and an inverse correlation with age. Sociodemographic factors were also found to be directly associated with higher UPF intake. Colombia has also reported similar findings.

Third, a different study examined the impact of educational attainment on UPF consumption, finding that those with lower educational attainment consumed more UPFs. Consumers are readily misled by marketing tactics, low-cost UPFs, and a lack of nutritional education, all of which encourage UPF consumption. These results are in line with earlier research that found a negative correlation between educational attainment and UPF consumption. However, research from Brazil, Colombia, and Mexico revealed that the wealthiest people with the highest levels of education also consumed more UPF. On the other hand, people with lower incomes and educational levels in the USA, the UK, or France consumed more UPFs.

To determine the most effective health policies and programs to reduce UPF consumption and, consequently, improve diet quality, future research should take into account additional socioeconomic and sociodemographic factors as well as psychosocial factors like employment status, family income level, number of children in the household, the head of the household's educational attainment and occupation, sex, ethnicity, or toxic habits (including smoking status and alcohol intake). Given the extensive body of research linking UPF consumption to several detrimental health effects, practical public policies restricting UPF consumption are required. Therefore, educating people about healthy eating environments and minimizing the promotion of obesogenic food advertising should target people from all sociodemographic and socioeconomic backgrounds to encourage them to adopt healthier lifestyle habits. To discourage the consumption of these foods, some countries have recently begun to implement various measures, such as taxes on UPF foods, limiting the locations that can sell them, or enforcing regulations regarding advertisements, particularly those directed toward children or adolescents.

10.3. STRATEGIES FOR PROMOTING HEALTHIER EATING HABITS

Daily food and exercise decisions have an impact on both the short- and long-term health of an individual. Physical activity can lower the risk of heart disease, high blood pressure, and other conditions, and it is directly linked to weight-related outcomes. Diabetes, osteoporosis, some cancers, and being overweight or obese are all related to healthy eating. These illnesses and ailments affect the person and his or her quality of life.

They are also linked to rising healthcare expenses that burden the public sector and private industry. Childhood offers the chance to lay a strong foundation that can result in wholesome eating habits that last a lifetime. Promoting a healthier and more productive society requires preventing childhood obesity. Furthermore, childhood, and adolescence are the earliest stages of the development of many chronic diseases linked to diet.

The United States' food and agricultural policies, population growth, technological advancements, and changes in the country's economy, society, and way of life have all contributed to significant changes in the country's food system and eating environments in recent decades. Food is now easily accessible and available all day long in a variety of settings. Energy-dense foods make up a big portion of the food supply in the United States today. These foods include refined grains, foods high in fats and/or sugars, and foods low in nutrients. A lot of these foods are frequently offered for comparatively cheap prices in ever-larger portion sizes. Additionally, Americans are consuming more calories and eating out more frequently than in the past when they are not at home. In addition, there are fewer meals shared by families. Furthermore, compared to a few decades ago, the school food environment has drastically changed. High-calorie, low-nutrition foods are now frequently offered and promoted throughout the school day in many schools. The amount of food that is marketed to kids through various media platforms, including digital media, has also significantly increased. Finally, access to healthier foods varies by income and race/ethnicity, partly due to the decline of grocery stores and the rise of fast-food restaurants in lower-class urban areas.

The current obesity epidemic is largely a result of these environmental changes that have affected Americans' eating habits in terms of what, where, and how much they eat. According to the Institute of Medicine's (IOM) Preventing Childhood Obesity: Health in the Balance report, preventing childhood obesity should be a national health priority and public health at its broadest and most inclusive. Public health population-based strategies, such as alterations to the environment and policies that impact a large number of people, should be used in obesity prevention initiatives if they are to be successful. Many stakeholders, including those in the public and private sectors, must collaborate to bring about change to solve the issue.

10.3.1. What Is Meant by Healthy Eating and Healthy Foods?

The committee looked to the 2005 Dietary Guidelines for Americans to develop working definitions for healthy eating and foods. These recommendations, which are updated every five years and are based on the most recent scientific research, offer guidance on how to choose a healthy diet, keep a healthy weight, and get enough exercise. There are 16 major dietary and food-related recommendations in the guidelines (*see* Table 10.1 for a snapshot). The 2005 guidelines advise limiting intake of added sugars, alcohol, cholesterol, saturated, and trans fats, and a range of nutrient-dense foods and beverages both within and between the basic food groups for all healthy Americans two years of age and older. Foods high in nutrients are those that have a high nutrient content for a relatively small calorie count.

Examples of these foods include fiber, vitamins, and minerals. Foods with low nutritional density provide calories but little or no vitamins, minerals, or other nutrients that are good for you. Achieving energy balance—the equilibrium between calories consumed and calories burned through physical activity and bodily processes—while still meeting nutrient needs becomes more challenging when the more low-nutrient density foods and beverages that are high in fats and sugars are consumed. Unhealthy weight gain can result from an imbalance in energy intake. Choosing foods following the recommendations (i.e., fruits, vegetables, whole grains, and dairy products with low or no fat) and controlling portion sizes is the most effective strategy to keep a healthy weight while fulfilling dietary requirements.

Table 10.1. Comparison of Nutritional/Lifestyle Components Among Different Healthy Diet Options

Dietary Component	**Recommended Servings**				
	Mediterranean	**DASH[a]**	**MIND**	**Healthy Nordic**	**Traditional Asian**
Fruits	1–2/meal	4–5 servings/day	Berries: ≥ 2 servings/week	Fruits, berries, vegetables, and potatoes: ≥ 500 g/day	Daily
Vegetables	≥ 2 servings/meal	4–5 servings/day	Green leafy: ≥ 6 servings/week Other: ≥ 1 serving/day		Daily
Whole grains	1–2 servings/meal	7–8 servings/day	≥ 3 servings/day	Bread: 4–6 slices/day Cereal: 1.5 servings/day Pasta: 3 servings/week β-glucan-rich foods: 3 g/d	Daily
Dairy	Low-fat: 2 servings/day	Low- or non-fat: 2–3 servings/day	Cheese: < 1 serving/week Butter: < 1 Tbsp/day	Low-fat milk: ≤ 5 dL/day Cheese: for cooking [b]	Yogurt: daily to weekly
Nuts, seeds, and legumes	Olives/nuts/seeds: 1–2 servings/day Legumes: ≥ 2 servings/week	4–5 servings/week	Nuts: ≥ 5 servings/week Beans: > 3 servings/week	Nuts (mostly almonds): 15 g/day	Daily

Dietary Component	**Recommended Servings**				
	Mediterranean	**DASH**[a]	**MIND**	**Healthy Nordic**	**Traditional Asian**
Beef, pork, ham, lamb, veal, poultry	Red meat: < 2 servings/week Processed meat: ≤ 1 servings/week White meat: 2 servings/week	Lean protein: ≤ 2 servings/ day	Red meat: < 4 servings/week	Meat: ≤ 500 g/ week	Red meat: infrequent
			Poultry: ≥ 2 servings/week	Poultry: ≤ 300 g/ week	Poultry: Daily to weekly
Fish/seafood	≥ 2 servings/ week		≥ 1 serving/ week	3–5 servings/ week	2 servings/ week
Fats, oils, and salad dressing	Olive oil: 1–2 servings/meal	2–3 servings /day	Olive oil as primary oil	5 g/bread slice 0.5 dL/day as dressing	Healthy cooking oils: daily to weekly
Sweets	≤ 2 servings/ week	≤ 5 servings/ week	Pastries & sweets: < 5 servings/week	On weekends	Infrequent
Other	Eggs: 2–4 servings/week Potatoes: ≤ 3 servings// week	Sodium < 2,300 mg/day	Fried or fast food: < 1 serving/week	Eggs: Stay within daily recommended cholesterol intake Fruit/vegetable juice: 4 dL/week	Eggs: daily to weekly
Alcohol	Wine: in moderation	Women: ≤ 1 drink/day Men: ≤ 2 drinks/day	1 glass/day	Habitual amount	In moderation

[a] Recommendations shown here are based on a 2000 calorie per day eating plan.
[b] Contribution of total fat and quality of fat from cheese to stay within the recommended daily intake.

Source: Cena H, Calder PC. Defining a Healthy Diet: Evidence for the Role of Contemporary Dietary Patterns in Health and Disease. Nutrients. 2020; 12(2):334. https://doi.org/10.3390/nu12020334.

Here, "healthy eating" is defined in accordance with the Dietary Guidelines as consuming the kinds and quantities of foods, nutrients, and calories that are suggested. Whole grains, lean meats, dairy products that are low in fat or fat-free, and fruits and vegetables with little to no added sugar, fat, or salt are examples of healthy foods. Nutrients that support general health and wellness, like fiber, vitamins, and minerals, are abundant in healthy foods. These working definitions are adapted to address dietary excesses and deficiencies linked to children's and youth's current eating patterns, as well as the issue of childhood obesity.

There's no general agreement on what constitutes a calorie-dense, nutrient-poor food, nor on what foods belong in this category. The term "discretionary" calories was first used in the 2005 Dietary Guidelines for Americans. These are extra calories that are consumed during the day in the form of added fats and/or sugars after a sufficient amount of healthful foods have been consumed to meet calorie and nutritional needs. Particularly for individuals who are not physically active, the majority of discretionary calorie allowances are relatively low, ranging from 100 to 300 calories. The words "calorie-dense, nutrient-poor foods" in this report describe foods and drinks that are high in calories, high in fat and/or sugar, and low in vitamins and minerals (*see* table). These foods, which include candy, chips, and beverages with added sugar, can cause children to consume too many calories and gain weight unintentionally. Additionally, some micronutrient intake may decrease as a result of these foods taking the place of more nutrient-dense foods.

10.3.2. Creating A Healthy Eating Environment

Even though eating is ultimately a personal decision, children, families, and community members make decisions based on their local food environments. "Individual behavior change can occur only in a supportive environment with accessible and affordable healthy food choices and opportunities for regular physical activity," according to former US Surgeon General David Satcher in the 2001 Surgeon General's Call to Action to Prevent and Decrease Overweight and Obesity. Individuals can take charge of making better food choices by implementing effective obesity prevention policies and practices that address environmental changes. One reason why individual-based nutrition and physical activity programs have not been successful is that the environments in which they have been implemented have not encouraged people to make healthy decisions (Booth et al., 2001). It must be simple to choose the healthy option as well.

A healthy eating environment includes things like lower-income community members being aware of and participating in federal nutrition programs, like the Supplemental Nutrition Assistance Program (SNAP, formerly known as the Food Stamp Program); women feeling supported and at ease in breastfeeding; and an abundance of water fountains in public areas. Families should also have access to supermarkets or other locations where they can afford healthy foods, like fruits and vegetables. This section goes over how local governments can establish these fundamental elements of a setting that encourages healthy eating.

10.3.3. Disparities in Access to Healthy Foods

In certain neighborhoods, making the healthier option is frequently more difficult. Obesity and other preventable diet-related diseases are disproportionately common in lower-income and minority neighborhoods and communities, and these disparities are exacerbated by unequal access to reasonably priced, wholesome, and nutritious food. In many lower-class urban and rural areas, access to nutritious food is frequently lacking. This is especially concerning because it can have a detrimental effect on dietary intake and obesity among vulnerable populations.

Comparatively speaking, higher-, and middle-class, predominately white neighborhoods are more likely to have supermarkets and grocery stores that sell healthful foods. Smaller stores with fewer products and fewer healthy options, like fruits and vegetables, low-fat or non-fat dairy products, and whole grains, are common in lower-class neighborhoods. There is frequently a dearth of public transportation to supermarkets. According to research, locals with better access to supermarkets and less access to convenience stores typically maintain healthier diets and experience lower rates of obesity. The profusion of fast-food restaurants that offer high-calorie, high-fat meals at comparatively low prices may make the poor food environment in lower-income areas worse. Fast food joints and calorie-dense foods are more prevalent in minority and lower-class neighborhoods (Larson et al., 2009).

There is a paucity of research assessing methods for mitigating disparities through increased availability of reasonably priced, healthful foods and decreased availability of high-calorie, low-nutrient foods. Reduced access to calorie-dense foods at fast food restaurants and fast-food chains, increased availability of fruits, vegetables, and whole grain products, and the attraction of supermarkets to underprivileged neighborhoods are just a few of the ideas and strategies that have been put forth. Local governments should think about prioritizing areas and communities that have a relative excess of unhealthy foods or lack access to healthy foods to combat childhood obesity. These communities typically have greater rates of obesity and diseases associated with obesity (Morland and Evenson, 2009).

10.3.4. The Role of Local Governments in Promoting Healthy Eating

Local governments are in a good position to improve the food environments in their communities and can have a significant and direct impact on the health and well-being of people. Providing healthier food options in lower-income areas, utilizing zoning laws to alter local food environments, mandating menu labeling in restaurants, acting as a catalyst for community change by introducing healthier food options at government facilities, building the infrastructure required for localized implementation and evaluation of obesity prevention policies and programs, and educating the public about the significance of healthy eating and obesity prevention are just a few examples of the strategic leadership that local governments can provide. Local governments have the power and obligation to create opportunities for modifying the local food landscape in a way that positively impacts people's eating habits by making healthy eating the default option.

To decrease access to unhealthy foods and increase access to healthy foods, local governments must play a part. As this section illustrates, there are currently more strategies available that increase access to healthy foods than those that could limit it. Though they might have a greater effect on lowering obesity than expanding access to healthful foods, local governments should also concentrate on policies that limit access to unhealthy foods. For instance, limiting children's access to sugar-sweetened beverages in after-school programs may have a greater effect on lowering their intake of excess calories and weight gain than establishing a community garden or holding a once-weekly farmers' market for five months. Ultimately, even though certain programs might not directly influence the adoption of healthy eating habits, they might fortify relationships between various and significant community stakeholders. Wide coalitions or organizational strengths that are formed as a result of these kinds of activities can help to effectively advocate for later, more impactful, but more challenging-to-implement initiatives.

The healthy eating strategies and action steps recommended by the committee for local governments' consideration are organized around three goals:

1. Improve access to and consumption of healthy, safe, and affordable foods.
2. Reduce access to and consumption of calorie-dense, nutrient-poor foods.
3. Raise awareness about the importance of healthy eating to prevent childhood obesity.

As mentioned in the introduction, some action steps are bolded as the most promising. However, this list of goals and the strategies and steps discussed under each are not ranked in order of importance. The best people to determine which plans of action and strategies will work best for their community's needs and circumstances are local leaders, members, and policy makers. The choice will mostly depend on elements including leadership, priorities, resources, and demographics.

10.3.5. Strategies to Encourage Healthy Eating Habits

Eating a healthy diet is crucial for a child's wellbeing. Children who are overweight run the risk of developing long-term health issues. The National Institute of Diabetes and Digestive and Kidney Diseases (NIDDK) provides parents and caregivers with information on how to support children's healthy eating habits through the Weight-control Information Network (WIN).

Developing healthy eating habits (Figure 10.1) in kids can frequently feel like navigating uncharted territory in the hectic world of parenthood.

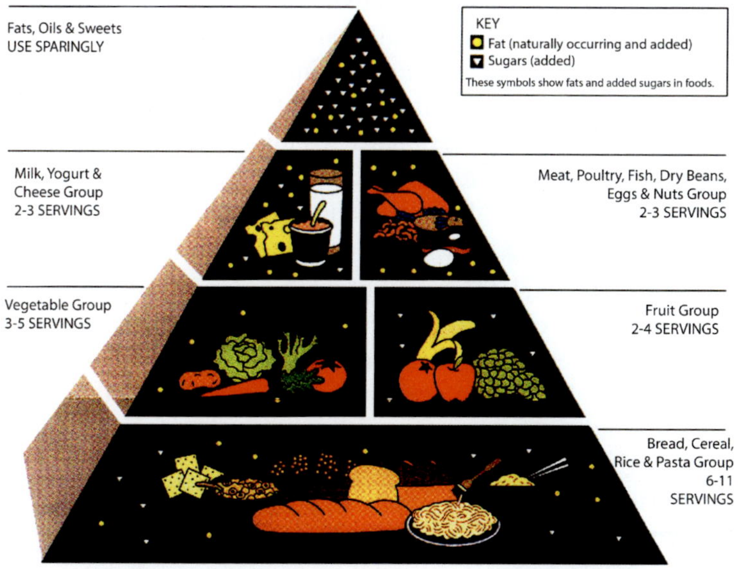

Figure 10.1. *Strategies to encourage healthy eating habits.*

Source: By The food pyramid, from http://www.nal.usda.gov/fnic/Fpyr/pyramid.gif, Public Domain, https://commons.wikimedia.org/w/index.php?curid=680809.

1. **Lead by Example**: Children are keen observers, and they often mimic the behaviors of the adults around them. As parents, caregivers, and role models, it's essential to demonstrate healthy eating habits in our own lives. By embracing a balanced diet rich in fruits, vegetables, whole grains, and lean proteins, we set a powerful example for our children to follow. Make mealtimes a shared experience where the entire family enjoys nutritious foods together, fostering a sense of unity and connection around the dinner table.
2. **Create a Positive Food Environment**: Transforming mealtime into a positive and enjoyable experience can significantly impact children's attitudes towards food. Involve children in meal planning, grocery shopping, and food preparation to empower them and instill a sense of ownership over their food choices. Opt for colorful and visually appealing presentations, and explore creative ways to incorporate a variety of flavors and textures into meals. By making healthy foods accessible, enjoyable, and exciting, you lay the groundwork for a lifelong appreciation of nutritious eating.
3. **Educate and Empower**: Knowledge is a powerful tool in shaping behavior, and educating children about the importance of healthy eating can empower them to make informed choices. Engage children in age-appropriate discussions about the benefits of different food groups, the importance of balanced nutrition, and the impact of food on their overall health and well-being. Encourage critical thinking and exploration by discussing the origins of food, the journey from farm to table, and the environmental implications of food choices.
4. **Encourage Variety and Exploration**: Variety is the spice of life, and exposing children to a diverse range of foods can broaden their palates and expand their culinary horizons. Encourage adventurous eating by introducing new foods regularly and incorporating them into familiar dishes. Create opportunities for exploration and discovery by hosting themed meal nights, taste-testing sessions, or visits to farmers' markets and grocery stores. Celebrate small victories and progress, and approach mealtime with curiosity and excitement.
5. **Foster a Positive Relationship with Food**: Above all, strive to cultivate a positive and balanced approach to food that emphasizes enjoyment, moderation, and flexibility. Avoid labeling foods as "good" or "bad," and instead focus on promoting a sense of mindfulness and self-awareness around eating habits. Encourage children to listen to their bodies, eat intuitively, and honor their

hunger and fullness cues. By fostering a healthy relationship with food based on respect, moderation, and self-care, you empower children to develop lifelong habits that support their overall health and well-being.

10.3.6. Tips for Families to Help Children Eat Healthy

1. Eat breakfast every day. Skipping breakfast can leave your child hungry, tired, and looking for less healthy foods later in the day.
2. Plan healthy meals and eat together as a family. Eating together at mealtimes helps children learn to enjoy a variety of foods.
3. Buy and serve more fruits and vegetables (fresh, frozen, or canned). Let your child choose them at the store.
4. Buy fewer soft drinks and high-fat/high-calorie snack foods like chips, cookies, and candy. These snacks are OK once in a while, but keep healthy snack foods on hand too and offer them to your child more often.
5. Start with small servings and let your child ask for more if he or she is still hungry. It is up to you to provide your child with healthy meals and snacks, but your child should be allowed to choose how much food he or she will eat. One tablespoon per year of age for each component of the meal is a great place to start when considering serving sizes for young children.
6. Offer your child water or low-fat milk more often than fruit juice. Fruit juice is a healthy choice but is high in calories.
7. Eat fast food less often. When you visit a fast food restaurant, try the healthful options offered.
8. Do not get discouraged if your child will not eat a new food the first time it is served. Some kids will need to have a new food served to them 10 times or more before they will eat it.
9. Try not to use food as a reward when encouraging kids to eat. Promising dessert to a child for eating vegetables, for example, sends the message that vegetables are less valuable than dessert.
10. Make healthy choices easy by putting nutritious foods where they are easy to see and keeping high-calorie foods out of sight.

10.3.7. Healthy Snack Ideas

1. Fresh or frozen fruit, or fruit canned in juice or light syrup.
2. Small amounts of dried fruits such as raisins, apple rings, or apricots.

3. Fresh vegetables such as baby carrots, cucumber, squash, zucchini, or tomatoes.
4. Reduced fat cheese or a small amount of peanut butter on whole-wheat crackers.
5. Low-fat yogurt with fruit.
6. Graham crackers, animal crackers, baked pretzels, or low-fat vanilla wafers.

10.3.8. The 5-2-1-0 Message Provides Suggestions for Building Healthy, Active Lives

1. Eat at least **5** fruits and vegetables a day.
2. Keep screen time (like TV, video games, computer) down to **2** hours or less per day.
3. Get **1** hour or more of physical activity every day.
4. Drink **0** sugar-sweetened drinks. Replace soda pop, sports drinks, and even 100% fruit juice with milk or water.

10.3.9. Be Supportive

Offer your child your full support during any process or program that you implement to help with their eating habits. Assist your child with goal-setting and progress monitoring. Give hugs and compliments for accomplishments. Stay upbeat.

Remind your child of his or her importance, love, and specialness. Children's perceptions of themselves are frequently shaped by what they believe their parents and other caregivers to be true. Children require understanding, encouragement, and compassion from kind adults.

Note: Foods that are small, round, sticky, or difficult to chew, like popcorn, nuts, seeds, raisins, whole grapes, hard veggies, hard cheese chunks, and hard vegetables can choke children younger than four years old. For small children, some of these foods can still be prepared; for instance, you can cook and chop vegetables and cut grapes into small pieces. At mealtimes and during snacks, children should always be under supervision (Pagliai et al., 2021).

REFERENCES

1. Abdul, R., et al. (2008). Psychosocial predictors of fruit and vegetable consumption in adults: A review of the literature. *American Journal of Preventive Medicine, 34*(6), 535–543.
2. Ajzen, I. (1988). *Attitudes, Personality and Behavior.* Milton Keynes: Open University Press.
3. Anderson, A. S., et al. (1998). Take Five, a nutrition education intervention to increase fruit and vegetable intakes: Impact on attitudes towards dietary change. *British Journal of Nutrition, 80,* 133–140.
4. Anderson, A., & Cox, D. (2000). Five a day–challenges and achievements. *Nutrition and Food Science, 30*(1), 30–34.
5. Anderson, A., et al. (2003). The development of and evaluation of a novel school-based intervention to increase fruit and vegetable intake in children (Five a Day The Bash Street Way). Report for the FSA, London.
6. Andrade, G. C., Julia, C., Deschamps, V., Srour, B., Hercberg, S., Kesse-Guyot, E., Allès, B., Chazelas, E., Deschasaux, M., Touvier, M., et al. (2021). Consumption of ultra-processed food and its association with sociodemographic characteristics and diet quality in a representative sample of French adults. *Nutrients, 13,* 682.
7. Baraldi, L. G., Steele, E. M., Canella, D. S., & Monteiro, C. A. (2018). Consumption of ultra-processed foods and associated sociodemographic factors in the USA between 2007 and 2012: Evidence from a nationally representative cross-sectional study. *BMJ Open, 8,* e020574.
8. Baranowski, T., et al. (2003). Squire's Quest! Dietary outcome evaluation of a multimedia game. *American Journal of Preventive Medicine, 24,* 52–61.
9. Becker, M. H. (1974). The health belief model and sick role behavior. *Health Education Monographs, 2,* 409–419.
10. Berkman, L. F. (1995). The role of social relations in health promotion. *Psychosomatic Medicine, 57*(3), 245–254.
11. Bielemann, R. M., Santos Motta, J. V., Minten, G. C., Horta, B. L., & Gigante, D. P. (2015). Consumption of ultra-processed foods and their impact on the diet of young adults. *Revista de Saúde Pública, 49,* 28.
12. Bonaccio, M., Costanzo, S., Di Castelnuovo, A., Persichillo, M., Magnacca, S., De Curtis, A., Cerletti, C., Donati, M. B., de Gaetano, G., Iacoviello, L., et al. (2021). Ultra-processed food intake and all-cause and cause-specific mortality in individuals with cardiovascular disease: The Moli-sani Study. *European Heart Journal.*

13. Bonaccio, M., Di Castelnuovo, A., Costanzo, S., De Curtis, A., Persichillo, M., Sofi, F., Cerletti, C., Donati, M. B., de Gaetano, G., Iacoviello, L. (2021). Ultra-processed food consumption is associated with increased risk of all-cause and cardiovascular mortality in the Moli-sani Study. *American Journal of Clinical Nutrition, 113,* 446–455.
14. Booth, S. L., Sallis, J. F., Ritenbaugh, C., Hill, J. O., Birch, L. L., Frank, L. D., Glanz, K., Himmelgreen, D. A., Mudd, M., Popkin, B. M., Rickard, K. A., St. Jeor, S., & Hays, N. P. (2001). Environmental and societal factors affect food choice and physical activity: Rationale, influences, and leverage points. *Nutrition Reviews, 59*(3 II).
15. Clarke, J. E. (1998). Taste and flavour: Their importance in food choice and acceptance. *Proceedings of the Nutrition Society, 57,* 639–643.
16. Cotugna, N., et al. (1992). Nutrition and cancer prevention knowledge, beliefs, attitudes, and practices: The 1987 National Health Interview Survey. *Journal of the American Dietetic Association, 92*(8), 963–968.
17. Cox, D. N., et al. (1998a). Take Five, a nutrition education intervention to increase fruit and vegetable intakes: Impact on consumer choice and nutrient intakes. *British Journal of Nutrition, 80,* 123–131.
18. Cox, D. N., et al. (1998b). UK consumer attitudes, beliefs and barriers to increasing fruit and vegetable consumption. *Public Health Nutrition, 1,* 61–68.
19. Cox, R. H., et al. (1996). Impact of a cancer intervention on diet-related cardiovascular disease risks of white and African-American EFNEP clients. *Journal of Nutrition Education, 28,* 209–218.
20. De Almeida, M. D. V., et al. (1997). Sources used and trusted by nationally-representative adults in the European Union for information on healthy eating. *European Journal of Clinical Nutrition, 51,* S8–15.
21. De Irala-Estevez, J., et al. (2000). A systematic review of socioeconomic differences in food habits in Europe: Consumption of fruit and vegetables. *European Journal of Clinical Nutrition, 54,* 706–714.
22. Devine, C. M., et al. (2003). Sandwiching it in: Spillover of work onto food choices and family roles in low- and moderate-income urban households. *Social Science and Medicine, 56,* 617–630.
23. Dewberry, C., & Ussher, J. M. (1994). Restraint and perception of body weight among British adults. *Journal of Social Psychology, 134*(5), 609–619.
24. Dibsdall, L. A., et al. (2003). Low-income consumers' attitudes and behavior towards access, availability and motivation to eat fruit and vegetables. *Public Health Nutrition, 6*(2), 159–168.

25. Donkin, A. J., et al. (2000). Mapping access to food in a deprived area: The development of price and availability indices. *Public Health Nutrition, 3*(1), 31–38.

26. Drummond, S., Crombie, N., & Kirk, T. (1996). A critique of the effects of snacking on body weight status. *European Journal of Clinical Nutrition, 50*(12), 779–783.

27. Dye, L., & Blundell, J. E. (1997). Menstrual cycle and appetite control: Implications for weight regulation. *Human Reproduction, 12*(6), 1142–1151.

28. Faugier, J., et al. (2001). Barriers to healthy eating in the nursing profession: Part 2. *Nursing Standard, 15*(37), 33–35.

29. Feunekes, G. I. J., et al. (1998). Food choice and fat intake of adolescents and adults: Associations of intakes within social networks. *Preventive Medicine, 27,* 645–656.

30. French, S. A., et al. (2001). Pricing and promotion effects on low-fat vending snack purchases: The CHIPS Study. *American Journal of Public Health, 91,* 112–117.

31. Gatenby, S. (1996). Healthy eating: Consumer attitudes, beliefs and behavior. *Journal of Human Nutrition and Dietetics, 9,* 384–385.

32. Gibney, M. J. (2004). European consumers' attitudes and beliefs about safe and nutritious foods: Concepts, barriers and benefits. In *Proceedings of the International Food Conference: 'Thinking Beyond Tomorrow'* (Dublin, June 2004).

33. Glanz, K., et al. (1998). Psychosocial correlates of healthful diets among male auto workers. *Cancer Epidemiology, Biomarkers and Prevention, 7,* 119–126.

34. Grech, A. L., Rangan, A., & Allman-Farinelli, M. (2017). Dietary energy density in the Australian adult population from national nutrition surveys 1995 to 2012. *Journal of the Academy of Nutrition and Dietetics, 117,* 1887–1899.e2.

35. Hampl, J. S., Heaton, C. L., & Taylor, C. A. (2003). Snacking patterns influence energy and nutrient intakes but not body mass index. *Journal of Human Nutrition and Dietetics, 16*(1), 3–11.

36. Horwath, C. C. (1999). Applying the transtheoretical model to eating behavior change: Challenges and opportunities. *Nutrition Research Reviews, 12,* 281–317.

37. Kearney, M., et al. (1997). Perceived need to alter eating habits among representative samples of adults from all member states of the European Union. *European Journal of Clinical Nutrition, 51,* S30–35.

38. Kearney, M., et al. (2000). Sociodemographic determinants of perceived influences on food choice in a nationally representative sample of Irish adults. *Public Health Nutrition, 3*(2), 219–226.
39. Khandpur, N., Cediel, G., Obando, D. A., Jaime, P. C., & Parra, D. C. (2020). Sociodemographic factors associated with the consumption of ultra-processed foods in Colombia. *Revista de Saúde Pública, 54,* 19.
40. Kristal, A. R., et al. (1999). How can stages of change be best used in dietary interventions? *Journal of the American Dietetic Association, 99,* 679–684.
41. Lappalainen, R., et al. (1997). Difficulties in trying to eat healthier: Descriptive analysis of perceived barriers for healthy eating. *European Journal of Clinical Nutrition, 51,* S36–40.
42. Larson, N. I., Story, M. T., & Nelson, M. C. (2009). Neighborhood environments: Disparities in access to healthy foods in the U.S. *American Journal of Preventive Medicine, 36*(1), 74–81.
43. Lassen, A., et al. (2004). Successful strategies to increase the consumption of fruits and vegetables: Results from the Danish '6 a day' Worksite-Canteen Model Study. *Public Health Nutrition, 7*(2), 263–270.
44. Lowe, C. F., et al. (2004). Effects of a peer modelling and rewards-based intervention to increase fruit and vegetable consumption in children. *European Journal of Clinical Nutrition, 58*(3), 510–522.
45. Mac Evilly, C., & Kelly, C. (2001). Conference report on 'Mood and Food.' *Nutrition Bulletin, 26*(4).
46. Magalhães, V., Severo, M., Correia, D., Torres, D., Costa de Miranda, R., Rauber, F., Levy, R., Rodrigues, S., & Lopes, C. (2021). Associated factors to the consumption of ultra-processed foods and its relation with dietary sources in Portugal. *Journal of Nutritional Science, 10,* E89.
47. Marchese, L., Livingstone, K. M., Woods, J. L., Wingrove, K., & Machado, P. (2021). Ultra-processed food consumption, socio-demographics and diet quality in Australian adults. *Public Health Nutrition,* 1–11. https://doi.org/10.1017/S1368980021003967.
48. Margetts, B. M., et al. (1998). Factors which influence 'healthy' eating patterns: Results from the 1993 Health Education Authority health and lifestyle survey in England. *Public Health Nutrition, 1*(3), 193–198.
49. Marino, M., Puppo, F., Del Bo,' C., Vinelli, V., Riso, P., Porrini, M., & Martini, D. (2021). A systematic review of worldwide consumption of ultra-processed foods: Findings and criticisms. *Nutrients, 13,* 2778. https://doi.org/10.3390/nu13082778.

50. Marrón-Ponce, J. A., Sánchez-Pimienta, T. G., Da Costa Louzada, M. L., & Batis, C. (2018). Energy contribution of NOVA food groups and sociodemographic determinants of ultra-processed food consumption in the Mexican population. *Public Health Nutrition, 21,* 87–93. https://doi.org/10.1017/S1368980017002129.
51. Monteiro, C. A., Moubarac, J. C., Cannon, G., Ng, S. W., & Popkin, B. (2013). Ultra-processed products are becoming dominant in the global food system. *Obesity Reviews, 14,* 21–28. https://doi.org/10.1111/obr.12107.
52. Morland, K. B., & Evenson, K. R. (2009). Obesity prevalence and the local food environment. *Health & Place, 15*(2), 491–495. https://doi.org/10.1016/j.healthplace.2008.10.004.
53. Nestle, M., et al. (1998). Behavioral and social influences on food choice. *Nutrition Reviews, 56*(5), S50-S64. https://doi.org/10.1111/j.1753-4887.1998.tb01743.x.
54. Oliver, G., & Wardle, J. (1999). Perceived effects of stress on food choice. *Physiology & Behavior, 66,* 511–515. https://doi.org/10.1016/S0031-9384(98)00345-1.
55. Pagliai, G., Dinu, M., Madarena, M. P., Bonaccio, M., Iacoviello, L., & Sofi, F. (2021). Consumption of ultra-processed foods and health status: A systematic review and meta-analysis. *British Journal of Nutrition, 125,* 308. https://doi.org/10.1017/S0007114520002688.
56. Paisley, L., et al. (1995). Consumer perceptions of dietary changes for reducing fat intake. *Nutrition Research, 15,* 1755–1766. https://doi.org/10.1016/0271-5317(95)00083-9.
57. Patterson, R. E., et al. (1997). Components of the Working Well Trial intervention associated with the adoption of healthful diets. *American Journal of Preventive Medicine, 13,* 271–276. https://doi.org/10.1016/S0749-3797(97)00017-6.
58. Poobalan, A. S., Aucott, L. S., Clarke, A., & Smith, W. C. S. (2014). Diet behavior among young people in transition to adulthood (18–25 year olds): A mixed method study. *Health Psychology and Behavioral Medicine, 2,* 909. https://doi.org/10.1080/21642850.2014.931232.
59. Popkin, B. M., Barquera, S., Corvalan, C., Hofman, K. J., Monteiro, C., Ng, S. W., Swart, E. C., & Taillie, L. S. (2021). Towards unified and impactful policies to reduce ultra-processed food consumption and promote healthier eating. *The Lancet Diabetes & Endocrinology, 9,* 462–470. https://doi.org/10.1016/S2213-8587(21)00078-4.

60. Prochaska, J. O., DiClemente, C. C., & Norcross, J. C. (1992). In search of how people change: Applications to addictive behaviors. *American Psychologist, 47,* 1102–1114.
61. Rosenstock, I. M. (1966). Why people use health services. *Milbank Memorial Fund Quarterly, 44,* 94–94.
62. Snyder, M. P., Story, M., & Trenkner, L. L. (1992). Reducing fat and sodium in school lunch programs: The LUNCHPOWER! Intervention Study. *Journal of the American Dietetic Association, 92,* 1087–1091.
63. Sorensen, L. B., et al. (2003). Effect of sensory perception of foods on appetite and food intake: A review of studies on humans. *International Journal of Obesity and Related Metabolic Disorders, 27,* 1152–1166.
64. Steiner, J. E. (1977). Facial expressions of the neonate infant indicating the hedonics of food-related chemical stimuli. In J. Weiffenbach (Ed.), *Taste and Development: The Genesis of Sweet Preference* (DHEW Publication No. NIH 77-1068). Washington, DC: US Government Printing Office.
65. Stevens, V. J., et al. (2002). Randomized trial of a brief dietary intervention to decrease consumption of fat and increase consumption of fruits and vegetables. *American Journal of Health Promotion, 16*(3), 129–134.
66. Stubbs, R. J., et al. (1996). Breakfasts high in protein, fat or carbohydrate: Effect on within-day appetite and energy balance. *European Journal of Clinical Nutrition, 50,* 409–417.
67. Stubenitsky, K., & Mela, D. J. (2000). UK consumer perceptions of starchy foods. *British Journal of Nutrition, 83,* 277–285.
68. Wang, L., Martínez Steele, E., Du, M., Pomeranz, J. L., O'Connor, L. E., Herrick, K. A., Luo, H., Zhang, X., Mozaffarian, D., & Zhang, F. F. (2021). Trends in consumption of ultraprocessed foods among US youths aged 2–19 years, 1999–2018. *JAMA, 326,* 519–530.
69. Wardle, J., et al. (2000). Stress, dietary restraint and food intake. *Journal of Psychosomatic Research, 48,* 195–202.
70. Worsley, A., & Crawford, D. (1985). Awareness and compliance with the Australian dietary guidelines: A descriptive study of Melbourne residents. *Nutrition Research, 5,* 1291–1308.

INDEX

G

H

N

O

FSC
www.fsc.org
MIX
Papier | Fördert
gute Waldnutzung
FSC® C083411

Kolibri 360 GmbH
Ferdinand-Jühlke-Straße 7
99095 Erfurt, Deutschland
produktsicherheit@kolibri360.de